# Hands-On Introduction to
# LabVIEW™
## for Scientists and Engineers

# Hands-On Introduction to
# LabVIEW™
# for Scientists and Engineers

John Essick

New York   Oxford
OXFORD UNIVERSITY PRESS
2009

Oxford University Press, Inc., publishes works that further
Oxford University's objective of excellence
in research, scholarship, and education.

Oxford   New York
Auckland   Cape Town   Dar es Salaam   Hong Kong   Karachi
Kuala Lumpur   Madrid   Melbourne   Mexico City   Nairobi
New Delhi   Shanghai   Taipei   Toronto

With offices in
Argentina   Austria   Brazil   Chile   Czech Republic   France   Greece
Guatemala   Hungary   Italy   Japan   Poland   Portugal   Singapore
South Korea   Switzerland   Thailand   Turkey   Ukraine   Vietnam

Published by Oxford University Press, Inc.
198 Madison Avenue, New York, New York 10016

www.oup.com

Oxford is a registered trademark of Oxford University Press.

LabVIEW™ is a registered trademark of National Instruments. This book is an independent
publication. National Instruments Corporation is not affiliated with the publisher or the author,
and does not authorize, sponsor, endorse or otherwise approve this book. The information in
this book was written by John Essick and not by National Instruments.

**Library of Congress Cataloging-in-Publication Data**

Essick, John.
   Hands-on introduction to LabVIEW for scientists and engineers / John Essick.
     p. cm.
   Includes index.
   ISBN 978-0-19-537395-0 (pbk. : alk. paper)   1. LabVIEW.   2. Science—
Experiments—Data processing.   3. Scientific apparatus and instruments—
Computer simulation.   4. Computer graphics.   I. Title.
   Q185.E69   2009
   006.6—dc20                                                                      2008029986

Printed in the United States of America
on acid-free paper

# Dedication

*To my wife Katie*

# Contents ■ ■ ■ ■ ■ ■ ■ ■ ■ ■ ■ ■ ■ ■ ■ ■ ■ ■ ■ ■ ■

# Preface

*Hands-On Introduction to LabVIEW for Scientists and Engineers* provides a learn-by-doing approach to acquiring the computer-based skills used daily in experimental work. As a course textbook or as a self-study supplement, the text explores practical programming solutions for carrying out interesting and relevant projects. Readers, who are assumed to have no prior computer programming or LabVIEW background, begin writing meaningful programs in the first few pages.

Instructors adopting the book as a classroom text can easily choose the desired depth of coverage for their courses. The first four chapters, which form the foundation appropriate for all readers, focus on the fundamentals of LabVIEW programming as well as the basics of computer-based experimentation using a National Instruments data acquisition (DAQ) device. These opening chapters can be used as the basis of a three-week introduction to LabVIEW-based data acquisition. Subsequent chapters have been written as independently as possible so that an instructor can fill out a course as desired. A full-featured course that uses most of the text's chapters will bring students to an intermediate skill level in computer-based data acquisition and analysis.

The progression of topics in *Hands-On Introduction to LabVIEW* is as follows:

*Chapters 1–3:* Fundamentals of the LabVIEW Graphical Programming Language. Central features of LabVIEW, including its control loop structures, graphing modes, mathematical functions, and text-based MathScript commands, are learned in the course of writing digitized waveform simulation programs.

*Chapter 4:* Basic Data Acquisition: Concepts of digitized data such as resolution, sampling frequency, and aliasing are covered. Then, using LabVIEW's high-level Express VIs, programs are written that execute analog-to-digital, digital-to-analog, and digital input/output tasks on a National Instruments DAQ device. Computer-based instruments constructed include a DC voltmeter, digital oscilloscope, DC voltage source, waveform generator, and blinking LED array.

*Chapter 5–8:* More LabVIEW Programming Fundamentals. Implementation of data file input/output, local memory, and conditional branching in LabVIEW is investigated through the writing of several useful programs (e.g., spreadsheet data storage, numerical integration, and differentiation). Additionally, LabVIEW's control flow approach to computer programming is studied.

*Chapter 9–10:* Data Analysis. Proper use of LabVIEW's curve fitting and Fast Fourier Transform functions is investigated. Using Express VIs to control a DAQ device, two computer-based instruments—a digital thermometer and a spectrum analyzer—are constructed.

*Chapter 11:* Intermediate-Level Data Acquisition. Programs are written to carry out analog-to-digital, digital-to-analog, and digital counter tasks on a DAQ device using the conventions of DAQmx. This lower-level approach (in comparison to the high-level Express VIs) allows utilization of the full available range of DAQ device features. A DC voltmeter, DC voltage source, waveform generator, and frequency meter are constructed as well as a sophisticated digital oscilloscope based on the state machine architecture.

*Chapter 12:* Temperature-Control Project. A wide range of the LabVIEW skills acquired throughout the book is used to construct a Proportional-Integral-Derivative (PID) temperature control system. Appendix I gives a design for the hardware required for this project.

*Chapter 13:* Control of Stand-Alone Instruments. Using LabVIEW's VISA communication driver, control of a stand-alone instrument over the General Purpose Interface Bus (GPIB) is studied. An Agilent 34401A Multimeter is used to demonstrate the central concepts of GPIB communication.

Key features of *Hands-On Introduction to LabVIEW* include its emphasis on real-world problem solving, its early introduction and routine use of data acquisition hardware, its "Do It Yourself" projects at the end of each chapter, and its healthy offering of back-of-the-chapter homework problems.

**Real-World Problem Solving:** Chapter topics and exercises provide examples of how commonly encountered problems are solved by scientists and engineers in the lab. LabVIEW features, along with relevant mathematical background, are introduced in the course of solving these problems. The "best practice" strategies presented equip readers to optimize their use of LabVIEW.

**Data Acquisition Usage Throughout:** LabVIEW's Express VIs allow exercises involving DAQ hardware to appear early and then routinely in *Hands-On Introduction to LabVIEW*. Express VIs package common measurement tasks into a single graphical icon and so allow the user to write a program with minimal effort. Of particular note, following the book's first three software-only chapters that teach the fundamentals of the LabVIEW programming language, data acquisition using a DAQ device is covered in Chapter 4. For a professor who wishes to devote only three weeks (or so) to instruction in computer-based data acquisition, Chapters 1 through 4 will provide the needed instructional materials. For those planning a more comprehensive study of LabVIEW, the Express VIs allow construction of a computer-based digital thermometer and spectrum analyzer in Chapters 9 and 10, respectively. In Chapter 11, the control of a DAQ device via the more advanced programming DAQmx icons is covered. In contrast to the Express VIs, the DAQmx icons enable a user to utilize the full available range of the DAQ-device features. In Chapter 12, readers use a DAQ device to precisely control the temperature of an aluminum block, while in Chapter 13 data is acquired remotely from a stand-alone instrument.

**Do It Yourself Projects:** To allow a reader to gauge his or her understanding of the presented material, each chapter of *Hands-On Introduction to LabVIEW* concludes with a "Do It Yourself" project. Each of these projects poses an interesting problem and (loosely) directs the reader in applying the chapter's material to find a solution. In some chapters, this project involves writing a program that functions as a stopwatch

(Chapter 1) or determines a person's reaction time (Chapter 8); in other chapters the reader constructs a computer-based instrument including a digital thermometer (Chapter 9), a spectrum analyzer (Chapter 10), and a frequency meter (Chapter 11).

**Back-of-the-Chapter Homework Problems:** A selection of homework-style problems is included at the end of each chapter so that interested readers can further develop their LabVIEW-based skills. In some of these problems, readers test their understanding by applying the chapter topics to new applications (e.g., Bode magnitude plot); in others, readers use programs written within the chapter to explore important experimental issues (e.g., frequency resolution of a Fast Fourier transform). Finally, a number of problems introduce readers to features of LabVIEW relevant to, but not included, in the chapter's text (e.g., data storage in binary format, use of property nodes).

*Hands-On Introduction to LabVIEW* is compatible with both the Full Development System and the Student Edition of LabVIEW. An instructor might consider having students purchase personal copies of the low-cost Student Edition software and then having them perform non–hardware-related chapter sections and/or back-of-the-chapter problems as homework on their own computers.

To aid readers in creating their LabVIEW programs, the following conventions are used throughout the book: **Bold** text designates the features such as graphical icons, palettes, pull-down menus, and menu selections that are to be manipulated in the course of constructing a program. The descriptive names that label controls, indicators, custom-made icons, programs, disk files, and directories (or folders) are given the **straight** font. *Italic* text highlights character strings that the programmer must enter using the keyboard and also signals the first-time use of important terms and concepts.

Any suggestions or corrections are gladly welcomed and can be sent to John Essick, Reed College, 3203 SE Woodstock Blvd., Portland, OR 97202 USA or jessick@reed.edu. Updates and ancillary materials for *Hands-On Introduction to LabVIEW* are available at http://academic.reed.edu/physics/faculty/essick.html.

*Hands-On Introduction to LabVIEW* is built on work originally published as *Advanced LabVIEW Labs*. For their advice and assistance in reorganizing, enhancing, and updating *Advanced LabVIEW Labs* into this final form, I would like to thank John Challice, Adriana Hurtado, and Danielle Christensen of Oxford University Press, Erik Luther and Jim Cahow of National Instruments, and reviewers Marty Johnston, University of St. Thomas; Perry Tompkins, Samford University; Ian Robinson, University of Illinois at Urbana-Champaign; William DeGraffenreid, California State University, Sacramento; David Pellett, University of California, Davis; Hakeem Oluseyi, Florida Institute of Technology; and David Loker, Penn State Erie. I would also like to thank Andrew Dawes, Milton From, and Lucas Illing for providing helpful feedback based on using drafts of this text in their courses. I gratefully acknowledge the support of Reed College's sabbatical program, which allowed me the time to complete this work. Finally, to my family: thank you for being lively and loving while I worked on this project.

John Essick
Portland, Oregon

# Chapter 1 ■ ■ ■ ■ ■ ■ ■ ■ ■ ■ ■ ■ ■ ■ ■ ■

# The While Loop and Waveform Chart

Welcome to the world of *LabVIEW*, an innovative graphical programming system designed to facilitate computer-controlled data acquisition and analysis. In this world, you–the LabVIEW user–will be operating in a programming environment that is different from that offered by most other programming systems (for example, the C and BASIC computer languages). Rather than creating programs by writing lines of text-based statements, in LabVIEW you will code programming ideas by selecting and then properly patterning a collection of graphical icons.

A LabVIEW program consists of two windows, the *front panel* and the *block diagram*. Once you have completed a program, the front panel appears as the face of a laboratory instrument with your own design of knobs, switches, meters, graphs, and/or strip charts. The front panel is the program's user-interface, that is, it facilitates the interaction of supplying inputs to and observing outputs from the program as it runs.

The block diagram is the actual LabVIEW programming code. Here reside the graphical images that you have appropriately selected from LabVIEW's well-stocked libraries of icons. Each icon represents a block of underlying executable code that performs a particular useful function. Your programming task is to make the proper connections between these icons using a process called *wiring*, so that data flows amongst the graphical images to accomplish a desired purpose. Because the icon libraries are designed specifically with the needs of scientists and engineers in mind, LabVIEW enables you–the modern-day experimentalist–to write programs that perform all of the laboratory tasks required for your state-of-the-art research, including instrument control, data acquisition, data analysis, data presentation, and data storage.

To begin developing your skill in LabVIEW graphical programming, in this first chapter I will guide you through the steps of creating a LabVIEW program. Together, we will write a program that systematically increments the argument of the sine function and plots the resulting sine wave as time progresses. In the process of writing this program, you will learn about the *While Loop*, one of LabVIEW's four fundamental program control structures, as well as one of LabVIEW's three commonly used graphing modes, the *Waveform Chart*.

## 1.1 SINE WAVE PLOT USING A WHILE LOOP AND WAVEFORM CHART

Find the LabVIEW program (which may have a desktop icon or may be in a folder named National Instruments, LabVIEW, or LabVIEW Student Edition, depending on your particular computing system) and launch it by double clicking on its icon or name.

After a few moments, a **Getting Started** window will appear. Place your mouse cursor over the **Blank VI** selection, and then click. An untitled front panel will appear in the foreground of your screen with a slightly offset block diagram in the background. In addition, a menu bar for an array of pull-down menus that contain LabVIEW program editing items will appear at the top of the window.

There are three ways to toggle the block diagram between the background and foreground:

- Select **Show Block Diagram** (when it is in the background) or **Show Front Panel** (when the block diagram is in the foreground) from the **Windows** pull-down menu.
- Click your mouse cursor on a visible region of the window (block diagram or front panel) that is currently in the background.
- Use the keyboard shortcut by typing <*Ctrl+E*>.

Practice toggling the block diagram between the foreground and background using each of the three available methods.

## 1.2 BLOCK DIAGRAM EDITING

To begin programming, place the blank block diagram in the foreground. The objects to be placed on this block diagram as you write your LabVIEW program are found in a repository called the *Functions Palette*. If a Functions Palette is not already visible, activate one by selecting **Functions Palette** from the **View** menu.

A "floating" Functions Palette will appear on your screen, which can be placed in a convenient spot by "clicking and dragging" on its title bar.

The Functions Palette is an organized collection of *categories* (such as *Programming, Express*, and *Mathematics*), where each category contains a palette of related programming objects. The Functions Palette may be configured in a multitude of ways, and the particular configuration (called a *view*) that you see will depend on choices made by past users (if any) of your LabVIEW system.

For our work, configure the view of the Functions Palette as follows:

- Click on the **View** button ⬚ **View▼** near the upper right corner of the Functions Palette. In the menu that appears, select **Change Visible Categories**...In the dialog window that follows, click on **Select All**, then press the **OK** button.
- Click on the **View** button again. This time, in the menu, click on **Options**...In the dialog window that appears, select **Controls/Functions Palettes**, and then in the **Format** box, choose **Category(Standard)**. Finally, press the **OK** button.

The Functions Palette's view is now configured. A particular category's palette can be toggled between visible ("open") and hidden ("closed") by clicking on its name. Open the **Programming** category and close all others, so that your Functions

Palette appears as follows (the categories listed on your Functions Palette may differ slightly from this illustration if your system has some optional LabVIEW add-ons).

Introduce yourself to the contents of the Programming palette. By passing the mouse cursor over each button in this palette, the name of the button will appear in a small box called a *tip strip*. An example of a tip strip is shown below.

Now place the cursor over the **Structures** button, and click. The **Structures** sub-palette will then appear as shown below. To close this subpalette and return to the Programming palette, simply click on **Programming**.

Move the cursor over the objects in the Structures subpalette and watch their names appear in a tip strip. Select the **While Loop** structure from the subpalette by placing the cursor over this object (which looks like a gray arrow bent into the shape of a square), then clicking. Once you have made this selection, the cursor will appear as a miniature of the While Loop structure when it is placed within the block diagram window. To place a While Loop within the diagram, click where you want the upper left corner of the loop to be, then, while holding down the mouse button, drag the cursor to define the size of your loop.

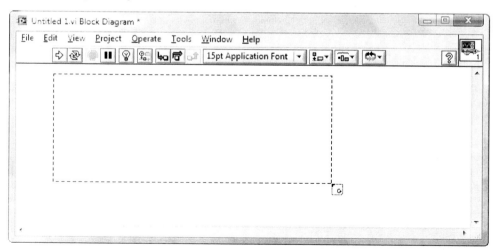

When you release the mouse button, the While Loop will appear as shown below.

If you are dissatisfied with the dimensions of your While Loop, this can be remedied by using the *Positioning Tool* ⭤, whose job it is to select, move, and resize objects. The ⭤ is one of the several available LabVIEW editing tools displayed in the *Tools Palette*, which may or may not be visible at the moment.

Activate the Tools Palette by selecting **Tools Palette** in the **View** pull-down menu.

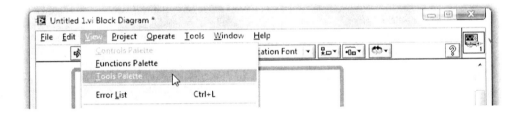

The "floating" Tools Palette will appear as shown next and, just like the Functions Palette, can be relocated to a convenient location by "clicking and dragging" its title bar.

The Positioning Tool may already be selected on your Tools Palette, in the manner shown above. If not, use the mouse cursor to select it now. A short description of each of the ten tools can be obtained from the tip strip that appears when placing the cursor over each tool's button.

The Positioning Tool can be used to resize your While Loop by the following procedure. Place the ⌖ at one of the loop's corners. At the corner, the ⌖ will transform into a *resizing handle* ⌐. Click and drag this cursor to redefine the dimensions of your While Loop. When you release the mouse button, the While Loop of desired size will appear.

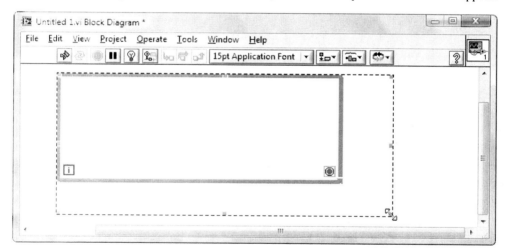

The While Loop structure is used to control repetitive operations. By default, it will repeatedly execute the subprogram written within its borders (called the *subdiagram*) until a specified Boolean value is no longer FALSE. Thus this structure is equivalent to the following code:

Do
   Execute subprogram (which sets *condition*)
While *condition* is FALSE

Within the While Loop, you will find the iteration terminal  as well as the conditional terminal ⊚, which is set to its **Stop if True** state by default. At the end of each loop iteration, LabVIEW checks the value of ⊚. If it is FALSE, the value of ▣ is incremented by one, and the loop begins another execution; if TRUE, the loop ceases execution. The initial value of ▣ (during the first iteration of the While Loop) is zero. Thus, in a While Loop where ⊚ is continuously FALSE until becoming TRUE during the tenth iteration, the loop will execute exactly ten times and the final value of ▣ will be nine. In this chapter, we will cause a While Loop to execute numerous iterations by connecting ⊚ to a Boolean control terminal that is usually FALSE.

Now, let's write a program that will generate and plot a sine wave. With a While Loop already on your block diagram, select the **Sine** icon from the Functions Palette. To find this icon, first open the Mathematics palette by clicking on the **Mathematics** category. Then, select the **Elementary & Special Functions** subpalette, and then finally the **Trigonometric Functions** sub-subpalette. From now on, such a sequence of choices will be indicated as follows: **Functions>>Mathematics>> Elementary & Special Functions >>Trigonometric Functions**.

Once you've selected the **Sine** icon from the Functions Palette by clicking the mouse, place the mouse cursor at the location within your block diagram at which you wish the icon to reside.

Then click the mouse and–Whoomp!–there the icon is.

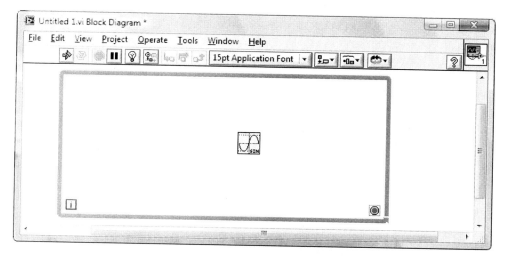

If you subsequently decide that you'd like to move this icon to some other location, you can do so through use of the Positioning Tool. With the ⏳, click on the **Sine** icon. When thus selected, the icon will become highlighted by a moving, dashed border called a *marquee*. With the mouse button depressed, drag the highlighted icon to the newly desired location within your While Loop. When properly placed, release the mouse button and move the cursor to an empty spot on the block diagram. Then click the mouse to deselect the icon.

Once an object is highlighted with a marquee, there are a couple of other handy movement techniques. If you hold down the <*Shift*> key and then drag the object, LabVIEW will only allow purely horizontal or vertical motion. Also, you can move the selected object in small, precise increments by pressing the keyboard's <*Arrow*> keys, rather than dragging with the mouse. Try moving the **Sine** icon using both of these techniques.

The function of the **Sine** icon is described in the following *Help Window:*

In a Help Window, the icon's input connections are shown on the left and output connections on the right. Thus we see that the **Sine** icon accepts an argument **x** in radians and returns values for **sin(x)** at its output.

You may see this Help Window for yourself by selecting **Show Context Help** from the **Help** pull-down menu. In our shorthand convention, this menu selection is **Help>>Show Context Help**.

When the Help Window appears, place it in a convenient position using the ⬉. Then place the ⬉ over the **Sine** icon to view its description.

When you have no further need for the Help Window, it can be toggled off in the **Help** menu. As an alternative to mouse control, try toggling the Help Window on and off using the following keyboard shortcut: <*Ctrl+H*>.

Now let's write a program that generates an array of points that follows the sine function. First, we need to configure the While Loop so that it will repetitively perform the operation defined within its borders. One (crude) method of accomplishing this goal is the following: Select a **False Constant** from **Functions>>Programming>> Boolean**.

Place the **False Constant** ⬚ near to the While Loop's conditional terminal ◉. If you place ⬚ too close to ◉, LabVIEW's *automatic wiring* feature will automatically connect these two icons with a green wire. In the future, you may want to take advantage of this effort-saving feature. At present, however, we are trying to gain experience in wiring two icons together manually, so if automatic wiring occurs, select **Undo Create** in the **Edit** pull-down menu. LabVIEW's **Undo** function will revert your block diagram to its state prior to the most recent editing action (i.e., the addition of the False Constant). You can then place a new ⬚ on your block diagram a bit farther from ⬤ so that autowiring does not occur.

As an aside, remember the **Edit>>Undo** editing trick, whose keyboard shortcut is $<Ctrl+Z>$. It's the easy way to erase editing mistakes when they occur.

We will now connect the **False Constant** to ⬤ by using the *Wiring Tool* ◆. Go to the Tools Palette and select the Wiring Tool by clicking on it.

Position the ◆ over the **False Constant** icon, so that ⬚ begins to blink. Then click the mouse. You have now tacked down one end of a wire to ⬚.

Smoothly move the Wiring Tool to the right until it is over the conditional terminal and the  is blinking. Click the mouse.

If all goes well, you will see a dotted green wire connecting the ⬛ and ⬤ icons. Such a green wire is LabVIEW's method of indicating that Boolean data is to be passed between two programming objects, in this particular case, from the False Constant ⬛ to the conditional terminal ⬤.

Here are two features of wiring, which are demonstrated in the next illustration:

First, while using the ✎, if you make a mistake and try to form an improper connection between two objects, the resulting *broken wire* will appear as a black dashed line with a ✖ through it. Such a mistake can be erased by selecting **Edit >>Remove Broken Wires** or, more simply, using the keyboard shortcut: <*Ctrl+B*>. Alternately, you can highlight the broken wire using the Positioning Tool, and then erase it by pressing the <*Delete*> key on your keyboard.

Second, while using the ✎, if at some location you wish to make a right-angle bend in the wire, click the mouse when you arrive at that location, then move the ✎ off in the newly desired (perpendicular) direction.

Our accomplishment so far is the creation of a While Loop that iterates continually. To see why this is so, remember that the conditional terminal is set to its default state **Stop if True**. Thus, during each and every loop iteration, a Boolean FALSE value will flow through the wire from ⬚🅵 to ⊙, setting the conditional terminal to FALSE. When LabVIEW checks the value of ⊙ at the end of the iteration, the While Loop will be instructed to re-execute (rather than stop).

We will now use the While Loop's iteration terminal 🅸 as a source of ever-increasing argument **x** for our sine function. Use the ⬚ to position 🅸 in the neighborhood of the **Sine** icon, then wire the 🅸 to the Sine's argument **x** input.

Here is a method of producing proper wiring. First, activate the Help Window by the keyboard shortcut <*Ctrl+H*>, then position the Help Window in a convenient, out-of-the-way place. Now, select the Wiring Tool from the Tools Palette, position the ✎

over 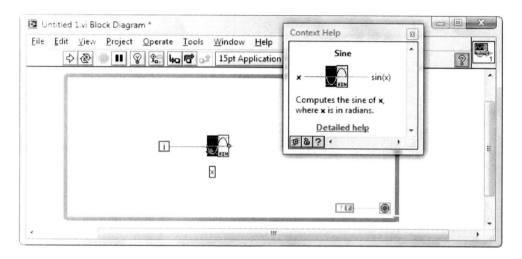 so that this icon blinks, and click to tack down one end of a wire to the. Move the ❖ smoothly over to the terminal for the Sine's argument **x** input, using the Help Window as your guide. When the ❖ is positioned correctly, the terminal for the **x** input will begin to blink both on the block diagram and in the Help Window. To erase any further doubt regarding the identity of the terminal to which you are wiring, the terminal's name is supplied by a tip strip. Click the mouse to complete the wiring. As an additional wiring aid, *whiskers* appear from each input and output as the Wiring Tool closely approaches the icon. These whiskers are especially helpful when wiring to icons with a multitude of inputs and outputs.

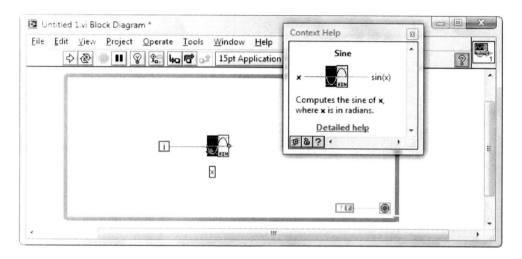

If done correctly, a colorful wire will connect to the **x** input, indicating that numerical data is to be passed between these two objects. Remember, if you make a mistake, the illegal wiring will result in a black dashed line that can be most easily erased by the **Remove Broken Wires** command's keyboard shortcut: *<Ctrl+B>*.

In LabVIEW, the color of a wire carrying numerical data indicates the manner in which the number is represented. Blue wires and icons denote integers, which come in one-, two-, four-, and eight-byte varieties, while orange indicates single- (four-byte), and double-precision (eight-byte) floating-point numbers. We will soon discover how to control the exact formatting of a particular numerical value. For now, simply note that and wires emanating from it are blue, indicating integer values. However, a red *coercion dot* appears where this blue wire connects to the Sine icon input, denoting that the Sine icon is automatically converting this integer input into the floating-point format it requires for its argument **x** input.

## 1.3 LABVIEW HELP

To obtain more detailed information about the inner workings of a particular icon, the online reference resource *LabVIEW Help* can be accessed by clicking on the **Detailed Help** text or the Question Mark [?] at the bottom of the icon's Help Window. Alternately, you can select **Help>>Search the LabVIEW Help...**, and then search for the Help Window describing your icon of interest.

A LabVIEW Help window is shown next.

The window informs us that the default data type for the **x** input (called its *representation*) is 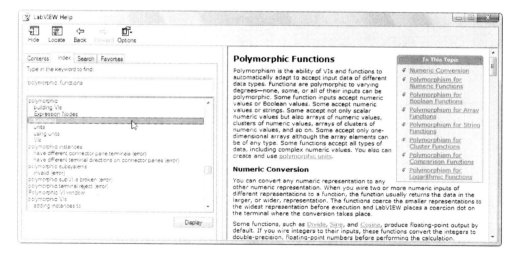, that is, a double-precision floating-point number. However, we also are told that Sine is a *polymorphic function*. The definition of polymorphic can be found by clicking on the **Index** tab and typing the keyword **polymorphic**, and then selecting the subtopic **functions**.

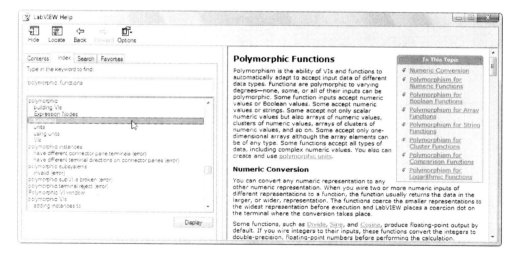

Reading the relevant sections of this window, we find that a polymorphic function can adapt (from its default settings) to accept a variety of data types at its input. In the case of a numeric trigonometric function like **Sine** (whose default input representation is ), the output of the icon will always have the same representation as the input. So, for example, a single-precision floating-point input will yield a single-precision floating-point output. The only exception to this rule is for an integer input, in which case the output is a double-precision floating-point number.

## 1.4 FRONT PANEL EDITING

Now, let's instruct LabVIEW to plot the sine function as each While Loop iteration generates a new value. A graph falls under the generic category of user-interface, which is the domain of the front panel. Switch to the front panel by using the keyboard shortcut <*Ctrl+E*>.

A user typically wishes to interact with a program by supplying inputs and observing outputs. In LabVIEW, these operations are facilitated by a wide choice of knobs, switches, meters, and graphs found in the *Controls Palette*. If it is not already visible, activate the Controls Palette by selecting **Controls Palette** in the **View** pull-down menu. Once activated, the Controls Palette will appear whenever the front panel is brought to the foreground. It will disappear and be replaced by the Functions Palette when the block diagram is subsequently toggled into the foreground. If only a few categories are visible in your Controls Palette, click the **View** button near its upper right corner, then select **Change Visible Categories...>>Select All**.

In **Controls>>Modern>>Graph**, select a **Waveform Chart**. The Waveform Chart, one of three commonly used graphical modes in LabVIEW, behaves like a laboratory strip chart, producing a real-time plot as each new data point is generated. In contrast, the **Waveform Graph** and **XY Graph** (which we will study later) display a full array of data that was produced at an earlier time.

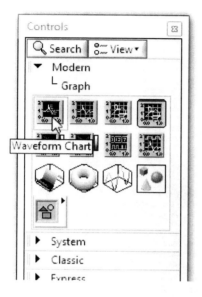

After you place the **Waveform Chart** on the front panel with a single mouse click, if you don't click the mouse further, a highlighted region will appear above the upper left corner of the plot containing the default text **Waveform Chart**.

This highlighted region is the Chart's *Label*. You can use the keyboard to give the plot a descriptive name, but let's just keep the default **Waveform Chart** for this program.

Secure this label by either clicking on the Enter button ☑ at the upper left end of the front panel, pressing *<Enter>* on the Numeric Keypad of your keyboard, or simply clicking the mouse cursor on an empty region of the front panel. In the future, if you accidentally click the mouse so that the label loses its highlighting before you have a chance to enter a name, the label region can be re-highlighted using the *Labeling Tool* A found in the Tools Palette.

In addition to the chart region and the Label, the Waveform Chart includes the *Plot Legend,*  Plot 0 ⟋⟍ . As we will discover shortly, the Plot Legend allows control over the plot style. Through it, one can choose such plot characteristics as whether data will be plotted as points or as interpolated lines and the shape, if any, of the data points.

Let's explore how to reposition and resize the Waveform Chart. First, using the ⬈, highlight the entire Waveform Chart (including the chart region, Label, and Plot Legend) with a marquee by placing the ⬈ over the chart region and clicking. With the mouse button depressed, drag the highlighted object to the newly desired position. Once properly placed, release the mouse button and move the cursor to an empty spot on the block diagram. Then click the mouse to deselect the object. In addition, the Label and Plot Legend can each be moved independently. To demonstrate this feature, highlight just the Plot Legend with a marquee by placing the ⬈ directly over this object and clicking. Then you will be able to drag this single object to a convenient location. Finally, to resize the chart region, position the $\mathbb{k}$ at one of the Chart's corners. At the corner, the $\mathbb{k}$ will transform into a resizing handle $\searrow$. Click and drag this cursor to redefine the dimensions of your chart region. Experiment with this resizing for a few moments. You will find that you can resize both the actual chart region as well as its background frame.

A more important adjustment is the manner in which the axes should be scaled. In a moment, we will plotting sine-wave values on the Waveform Chart's $y$-axis, so this axis must be prepared to chart data in the range of $-1.0$ to $+1.0$. Note that the default setting for the Waveform Chart's $y$-axis data range is $-10.0$ to $+10.0$.

It is the job of the *Operating Tool* $\sqrt[m]{}$ to change values appearing on both the front panel and block diagram. Select the $\sqrt[m]{}$ on the Tools Palette. Then, change the $y$-axis data range from its default setting through the following procedure: First, use the $\sqrt[m]{}$ to highlight the upper limit of the $y$-axis data range. Once highlighted, enter the newly desired value, which in our case is *1.0*, and then click on ☑ (or click the mouse on an open area of the front panel). Then in a similar manner, define the lower limit of the $y$-axis data range to be $-1.0$. LabVIEW will automatically redefine the intermediate $y$-axis labeling.

To save the redefined $y$-axis labeling scheme, simply select **Edit>>Make Current Values Default**.

## 1.5 POP-UP MENU

You are about to enter a hidden world and ascend to a higher level as a LabVIEW programmer. You might wish to pause and reflect upon your life prior to the enlightenment that you are about to attain. Here is the deep secret: Almost all LabVIEW objects have their own associated *pop-up menu* that is hidden from the uninitiated. By gaining access to a pop-up menu, you, the programmer, are empowered to control the functioning of the associated object. How does one gain access to a pop-up menu? By performing the simple operation of "*popping up*" on the object. To pop up on an object, place the mouse cursor directly over it, then click the right mouse button ("right-click"). You can also pop up on an object by using the *Pop-Up Tool* ⊟ found in the Tools Palette.

Pop up on the chart region of the **Waveform Chart**. As a first example of controlling the features of this object using its pop-up menu, toggle the Label on and off by selecting **Visible Items>>Label**. You will find that when a feature such as Label is activated ("on"), a check mark appears by its name in the pop-up menu.

Now, let's use the Waveform Chart's pop-up menu to assist us in appropriately scaling its *y*-axis. A few moments ago, we manually chose proper scaling for this axis using the Operating Tool. A simpler method to accomplish this same goal is via activation of the Waveform Chart's autoscaling feature: Pop up on the chart region of the Waveform Chart and inspect the **Y Scale>>Autoscale Y** option. You will find it checked, meaning it is already activated. By default, autoscaling on the Waveform Chart's *y*-axis (but not its *x*-axis) is activated.

Finally, let's use the pop-up menu to activate one of the Waveform Chart's as-yet unseen features. Pop up on the chart region and select **Visible>>Scale Legend**. The *Scale Legend*, which allows access to several useful functions that determine the scaling and labeling of the *x*- and *y*-axis, will then appear below the Waveform Chart.

Here, we see that, by default, the *x*-axis and *y*-axis are labeled **Time** and **Amplitude**, respectively. We, of course, will be plotting *sin(x)* vs. *x* so these default labels are inappropriate choices for our plot. Use the 🖑 to highlight **Time**, replace it with **x (radians)**, and press <*Enter*>. Similarly, replace **Amplitude** with **sin(x)**.

Additionally, the Scale Legend provides a simple method for activating autoscaling of axes. Simply use the 🖑 to click on the **Scale Lock** button of the desired axis. When the Scale Lock is open [icon], autoscaling is OFF. When it is closed [icon], autoscaling is continually ON. If you just want to autoscale, say, the *x*-axis once (i.e., not continuously), click on the *Autoscale* [icon] button. The small green indicator in the Autoscale button lights when autoscaling is activated.

## 1.6 FINISHING THE PROGRAM

With one last editing step, you'll be ready to run your first LabVIEW program! Use the keyboard shortcut *<Ctrl+E>* to return to your block diagram.

There you will find the Waveform Chart's *icon terminal* [icon]. This icon terminal is the Waveform Chart's block-diagram portal; that is, it accepts the block-diagram data to be plotted on the front-panel Chart. The icon terminal contains a small picture of plotted data (indicating it is associated with the Waveform Chart) and has an orange border with text **DBL** denoting that the data input to it should be double-precision floating-point numbers. It is further identified by the **Waveform Chart** label that you entered on the front panel. By popping up on the icon terminal and selecting **View As Icon**, the terminal will morph into its *data-type terminal* [icon] guise, whose smaller size can be helpful if you need to conserve block-diagram real estate. To toggle back to the icon terminal form, simply select **View As Icon** in the pop-up menu. Throughout this book, icon terminals will always be used in block-diagram illustrations. However, feel free to use data-type terminals on your diagrams, if you wish.

Using the [icon], place the Chart's icon terminal near the **sin(x)** output of the **Sine** icon, and then use the [icon] to connect the **sin(x)** output to [icon]. Remember that the Help Window can be activated with *<Ctrl+H>* to aid in this wiring operation.

### 1.7 PROGRAM EXECUTION

Now return to the front panel. You are ready to run your program, with the help of the *toolbar* shown below.

The leftmost button on the toolbar is the **Run** button . To start your program, simply click on the . As your program executes, you will observe a rather jagged-looking sine wave produced on your Waveform Chart in strip chart fashion.

After watching the sine wave move across the Waveform Chart for a while, you may start to wonder about a few things. For example, the Waveform Chart's block-diagram icon is supplied only with *y*-axis (sine-wave) values, so how is the Waveform Chart producing the associated *x*-axis values? What determines the speed at which the sine wave appears to move? And, now that the program is running, how do we turn it off?

Let's answer the first question first. Concisely stated, the *x*-axis denotes the *count indices* of the plotted data. That is, the Waveform Chart keeps track of the number of data values it has received and associates a count index with each datum. For the $i^{th}$ data value, the plot displays an $(x, y)$ point, where $x = i$ and $y = $ the actual data value. In its default setting, the Waveform Chart's $x$-axis values are the indices of the last 101 $y$-axis (in our case, sine-wave) values supplied to its block-diagram icon.

In regard to the second question, the speed at which the sine wave moves across the Waveform Chart region is determined by the calculational time delay necessary to produce each new data point. So this speed simply reflects the time it takes for each iteration of your block diagram's While Loop. Because we are allowing this program to run freely, this time per iteration is determined by the speed of your computer's processor, which (most likely) is very fast. On my computer, the $x$-axis values are on the order of 30 million after ten seconds of runtime. Thus, the time per iteration is approximately $10s/3 \times 10^7 \approx 3 \times 10^{-7}$ s.

Finally, how can the program be turned off? On your block diagram, a False Boolean value is constantly being read by the While Loop's conditional terminal at the end of each loop iteration, so your program will run forever. Since there is no way provided within the program to halt execution, your only recourse at the moment is to click on the **Abort Execution** button in the toolbar. Halt your program by clicking on the **Abort Execution** button.

Now that you have used the **Abort Execution** button, let me mention that it is bad practice to do so. This button, at the moment it is activated, causes your computer to immediately cease operation of your program. For the present program, such an action is probably no big concern. However, in more sophisticated programs, pressing **Abort Execution** at the wrong time could halt execution while reading data from a file or during communication with data acquisition devices connected to your computer. These situations lead to data corruption and other undesirable consequences. Thus it is always best to code a built-in stopping mechanism into your program.

## 1.8 PROGRAM IMPROVEMENTS

Based on the preceding observations, let's improve your program in the following two ways: (1) provide a front-panel button that, when pressed, allows the program to complete its current While Loop iteration, and then halts the program, and (2) provide a control over the speed at which at the While Loop iterates.

**Front Panel Switch.** LabVIEW provides a multitude of front-panel switches that can be used to bring your programs to a graceful stop. Halt your program so that you may edit its front panel. In **Controls>>Modern>>Boolean**, select a **Stop Button**.

Using the , put this button in a convenient location on your front panel. The default Boolean value for the Stop Button is FALSE.

There are six modes in which a LabVIEW Boolean switch can behave in response to being pressed. These modes are listed in the switch's pop-up menu under the **Mechanical Action** option. Pop up on the **Stop Button** and select **Mechanical Action.** You will find that this switch is set to **Latch When Released** by default. In this mode, the user places the Operating Tool over the **Stop Button** switch and depresses the mouse. Then, at the later time when the user releases the mouse, the switch changes from its default value to the opposite Boolean value. The switch retains ("latches") this new value until the program reads it once, at which time the switch returns to its default setting. You can find detailed explanations of each mode by selecting **Help>>Search the LabVIEW Help** and then searching for **Mechanical Action**.

Now switch to the block diagram. There, you will find the icon terminal that feeds the value of the front panel's **Stop Button** to the block diagram. Our plan is to connect to the While Loop's conditional terminal. To accomplish this, select the wire connecting the **False Constant** to the conditional terminal by clicking on it with the . Once it becomes highlighted with a marquee, erase this wire by pressing <*Delete*> on your keyboard.

In a similar way, delete the **False Constant** [icon]. Now drag the Stop Button's icon terminal [icon] near to [icon] and wire these two objects together. Since [icon] is FALSE by default, the While Loop will continually iterate when the program is started. At some later time, almost assuredly during some intermediate portion of the While Loop cycle, the Stop Button will be pressed and then released, thus changing to the TRUE state. Since the While Loop only checks the value of [icon] at the end of an iteration, the Loop will fully execute its calculations during that final iteration, before ceasing operation. Once the Stop Button's TRUE value is read, it will return to its default value of FALSE so the program is ready for the next time it is run.

**Controlling Iteration Rate.** While we are on the block diagram, let's also provide a method for controlling the rate at which the While Loop iterates. Through an icon called **Wait (ms)**, LabVIEW provides a method of delaying subsequent program operations for a specified time period. In **Wait (ms)**'s Help Window shown next, we see that, by specifying a numerical constant at the icon's input, the program will wait for that specified number of milliseconds.

In **Functions>>Programming>>Timing**, select the **Wait (ms)** icon and position it within the While Loop. In the Help Window above, the blue wire emanating from the **milliseconds to wait** input indicates that this terminal should be wired to an integer value. Select a **Numeric Constant** from **Functions>>Programming>>Numeric**. When the **Numeric Constant** is first selected, its interior is highlighted and ready for you to enter a number. If you click the mouse, this highlighting will disappear. To restore the highlighting, use the 🖑. Type the integer *100* into the **Numeric Constant**, press <*Enter*> (or click on a blank region of the block diagram), then wire this icon to the input of **Wait (ms)**.

The block diagram above causes one to ponder an obvious question. During each loop iteration, does the wait precede or succeed the sine-wave plot? The surprising answer is that LabVIEW effectively processes these two operations in parallel. That is, the wait and the sine-wave plot begin at the same time and the loop iteration concludes when both of these operations are complete. Thus, the time for an iteration is determined by whichever of the two is the slowest. Earlier we found that when only the sine-wave plotting was within the While Loop, the loop iterated much faster than ten times a second. Thus, by including a delay of 100 milliseconds, this wait operation should provide the "rate-limiting step" that determines the time per iteration.

Now return to the front panel and run your new, improved program by pressing ⇨. Because of the **Wait (ms)** icon, you will find that the sine wave is now generated in an easy-to-observe manner. Use the front panel's **Stop Button** to stop your program gracefully. After halting the program, if you would like to clear your plot, pop up on the Waveform Chart and select **Data Operations>>Clear Chart** (this pop-up menu is altered while the program runs, but the "runtime" menu also allows the chart be cleared).

**Improving Sine-Wave Resolution.** To see why our sine wave has a jagged appearance, pop up on the black region of the Waveform Chart's **Plot Legend**

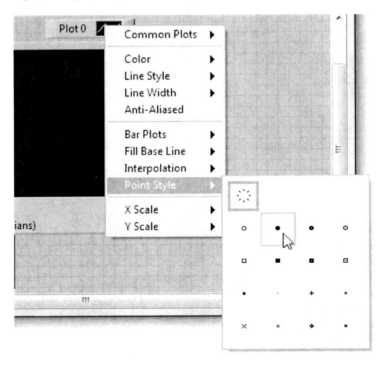

. Under **Interpolation** we discover that, by default, the Waveform Chart presents data in a "connect-the-dots" fashion. That is, a straight line is drawn from each data point to its neighbor, and the entire collection of these lines represents the waveform. Also, the default **Point Style** is **None**, which means that no symbol is placed at each data point. However, we can make the actual data points visible by selecting, for example, a large **Solid Dot** for the **Point Style**. While you're at it, you might like to spend a few minutes exploring the effect of selecting the various **Point Style**, **Line Style**, **Interpolation**, and **Color** options available.

With the data points now visible, we see that each cycle of the sine wave is only represented by a few sampled locations. It is this sparse sampling that leads to the jagged waveform appearance when lines are interpolated between adjacent data points.

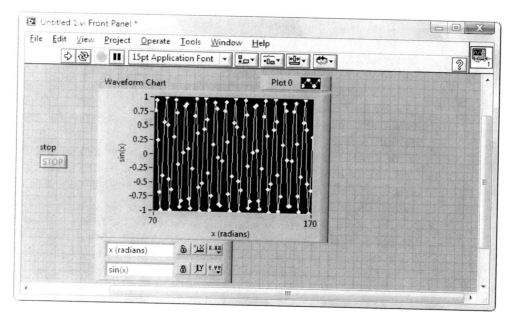

The reason for this paucity of data per cycle is, of course, due to our programming decision to take the sine function argument **x** to be the value of the While Loop's iteration terminal . Because  increments in steps of one and the sine function completes each new cycle every time **x** increases by $2\pi \approx 6$, only about six locations are sampled in each sine-wave cycle.

To sample the sine function with higher resolution, let's take **x** to be one-fifth of , rather than simply . Then it will take five times more iterations of the While Loop for **x** to increase by $2\pi$, resulting in five times more sampled points per cycle. To accomplish this feat, select the **Divide** icon from **Functions>>Programming>>Numeric** and place it on your block diagram.

Then, also from **Functions>>Programming>>Numeric**, obtain a **Numeric Constant**. When you first position the **Numeric Constant** on your diagram (before the next mouse click), it is a blue-bordered rectangular box with its interior highlighted, ready to receive the numerical value of your choosing (remember that if the mouse is clicked accidentally, the interior of the **Numeric Constant** can be highlighted again by using the ). Enter the number *5.0*. The Numeric Constant's border will change from blue to orange, indicating that it contains a floating-point (as opposed to integer) number.

If instead you had programmed the **Numeric Constant** with the number *5* (rather than *5.0*), the icon would remain blue, indicating that it contains an integer. On your diagram, even though you programmed the **Numeric Constant** with *5.0*, it may appear as an orange 5 rather than an orange 5.0 because an option called **Hide tailing zeros** is activated. To deactivate this option, pop up on the **Numeric Constant** and select **Display Format...** In the dialog box that appears, choose **Default editing mode**, then make **Digits** and **Precision Type** equal to **1** and **Digits of precision**, respectively; uncheck the **Hide trailing zeros** box; and finally press **OK**. In this book, I will always take the trouble to uncheck **Hide trailing zeros** so that it is apparent when a **Numeric Constant** contains a floating-point number. On your block diagrams, you may or may not wish to go to this trouble. The icon's blue or orange border will tell you whether it is programmed with an integer or floating-point number, respectively.

Position the **Divide** and 5.0 icons conveniently and wire them so that one-fifth of i is input to the Sine's **x** input. Note the red coercion dot at the iteration terminal's connection to the **Divide**. This dot denotes that the integer from i is converted to a floating-point number. Thus the Divide icon performs a floating-point divide operation.

Return to the front panel. There is one thing we still need to fix—the *x*-axis calibration. If we make no further changes, the quantity plotted on the *x*-axis will be the count index associated with each sine-wave value on the *y*-axis. Remembering that we have modified our program so that it takes the sine function argument for a particular data point to be the count index value *i* divided by 5, we see that the count index is no longer equal to the radian argument **x,** and so the wrong quantity is being plotted on the *x*-axis. However, this axis can be calibrated as **x** in radians by simply multiplying each count index by 0.2. Such calibration by a multiplicative constant is a common need when using the Waveform Chart, and thus LabVIEW provides a mechanism for implementing it. Pop up on the **Waveform Chart** and select **Properties**. In the **Chart Properties** dialog window that appears, click on the **Scales** tab. Then under the **x (radians) (X-Axis)** menu, set **Multiplier** equal to *0.2*, then press the **OK** button.

Run this final program and enjoy the beautifully realized sine-wave being generated. While your program is running, imagine writing the mountains of code necessary to replicate this real-time sine-wave plot, complete with the attractive user-interface, using a text-based language such as C. I think you'll agree, the graphical-based LabVIEW programming language is simple, yet very powerful.

## 1.9 DATE-TYPE REPRESENTATIONS

There's one last perplexing detail to resolve before our block diagram is complete. You may have noticed that a coercion dot appears at the input of the **Wait (ms)** icon, indicating a numeric-format mismatch, although we have seemingly wired the requisite integer-formatted number to this input. This puzzle can be solved by consulting the LabVIEW Help, which can be accessed by clicking on the ? in the **Wait (ms)** Help Window.

The LabVIEW Help window for **Wait (ms)** is shown below.

Here, we find that the **milliseconds to wait** input is configured to accept an unsigned four-byte (or 32-bit) integer, a numeric format whose shorthand name is **U32**. This type of integer is always positive and can range in value from 0 to $(2^{32}-1) = 4{,}294{,}967{,}295$. The unsigned integer-type is in contrast to the signed integer, which sacrifices one of its bits for use as a plus or a minus sign. For example, the four-byte signed integer, a format called **I32**, can range from $-2^{31}$ to $+ (2^{31}-1)$.

To discover the format of the integer on your block diagram, popup on 100 , then select **Representation** from the pop-up menu. You will find that this integer is of type **I32**, the default integer format for a **Numeric Constant**.

Change the data type of this number to an unsigned integer by selecting **U32** from the **Representation** palette. You will then find that the coercion dot disappears. Another mystery solved!

## 1.10 AUTOMATIC CREATION FEATURE

Now that you understand some of the subtleties involved in wiring up constants to icon terminals, here's a time-saving shortcut I think you'll appreciate. First, delete the Numeric Constant 100 and its wire from your block diagram. Next, place the cursor over the **milliseconds to wait** input of **Wait (ms)**, then pop up and select **Create>>Constant** from the menu that appears.

Like magic, an already-wired **Numeric Constant** of the correct data type (in this case, **U32**) will appear, with its interior highlighted.

All you need to do is type in the desired integer value of *100* from the keyboard, press *<Enter>*, and you're finished.

Automatic creation of "terminal-appropriate" objects is a universal feature of LabVIEW icons. Use of these time-saving "creation" options, available in the pop-up menus at the input and output terminals of all block diagram icons, will speed up your program development time.

## 1.11 PROGRAM STORAGE

LabVIEW programs simulate the function of laboratory instruments. As such, these programs are given the name *virtual instruments* (VIs). Your While Loop–based VI behaves as a strip chart plotting a sine wave, so let's save this VI under the descriptive name **Sine Wave Chart (While Loop).**

In the **File** pull-down menu, select **Save**, which will activate the **Name the VI** dialog box shown next. Since you will be writing and saving many VIs in your study of LabVIEW, let's first create a folder in which to store this work. Use the **Save in:** box to navigate to a desired storage place on your computing system (e.g., **Desktop, Documents** folder, flash drive), and then click on the **Create New Folder** button.

In the highlighted box that appears next to the folder icon, title the new folder **YourName**, and then click **Open** to open this folder.

The VI you have written is associated with the first chapter of this book. Thus, within the **YourName** folder, create a subfolder named **Chapter 1** and save the VI there as follows: With **YourName** open, click on the **Create New Folder** button. In the highlighted box that appears next to the folder icon, name the new folder **Chapter 1**, and click **Open** to open this folder. Finally in the **File name:** box, type **Sine Wave Chart (While Loop)**, and then press **OK**. Your VI will be saved in the **Chapter 1** folder, which is within the **YourName** folder (a location we will denote as **YourName\Chapter 1**) with the extension **.vi** appended to its name.

## DO IT YOURSELF

Write a VI called **Stopwatch** that functions as a stopwatch with a precision of 0.01 second. Design **Stopwatch** so that it continuously displays the elapsed time from when its **Run** button is pressed until its **Stop Button** is pressed. The front panel of **Stopwatch** should appear as shown below. Here, the elapsed time is displayed on a **Numeric Indicator** (found in **Controls>>Modern>>Numeric**) labeled **Elapsed Time (second)**. Choose the Stop Button's **Mechanical Action** appropriately.

## Problems

1. Write a VI called **Greater Than Ten** that lights a **Round LED** indicator (found in **Controls>>Modern>>Boolean**) on its front panel when the iteration terminal 🔢 of a While Loop on its block diagram has a value of greater than 10. You may find the following icons useful: **Select** and **Greater?** in **Functions>>Programming>> Comparison**.

2. Write a VI called **Single Sine Cycle** that stops automatically after the While Loop on its block diagram has iterated exactly 1000 times. As the While Loop performs these 1000 iterations, make a Waveform Chart plot one cycle of a sine wave (where the final data point is at the equivalent point on the sine wave as the initial point). You may find the following icons useful: **Equal?** in **Functions>>Programming>> Comparison** and **Pi Multiplied By 2** in **Functions>>Programming>> Numeric>>Math & Science Constants**.

3. On the block diagram of **Sine Wave Chart (While Loop)**, pop up on the While Loop's conditional terminal and select its **Continue if True** mode. Then, modify the program as needed so that it executes in the original manner (i.e., charts the sine wave until the user clicks the **Stop Button**).

4. Write a VI called **Metronome**, whose front panel includes a **Numeric Control** (found in **Controls>>Modern>>Numeric**) labeled **Beats Per Minute,** as shown below.

When run, construct **Metronome** so that it produces *N* beeps every minute until the **Stop Button** is clicked, where *N* is the value entered in the **Beats Per Minute** control and a beep is a 264-Hz sound wave of 100-ms duration. To create a beep, use **Beep.vi**, which is found in **Functions>>Programming>>Graphics and Sound**, with its **use system alert?** input set to FALSE.

5.  The odd integers are given by $2i + 1$, where $i=0,1, 2,\ldots$ Write a program called **Odd Integer Search** that answers the following question: What is the smallest odd integer that is divisible by 3 and also, when cubed, yields a value greater than 4000? Display the answer to this question in a front-panel **Numeric Indicator**. You may find icons in the following palettes useful: **Functions>>Programming>>Numeric** (e.g., **Quotient & Remainder**), **Functions>>Programming>>Comparison, Functions>> Programming>>Boolean,** and **Functions>>Mathematics>>Elementary & Special Functions>>Exponential Functions**.

6.  Construct a VI named **EvenOdd** whose front panel has two **Round LED** indicators and a **Stop Button** (all found in **Controls>>Modern>>Boolean**). Label one indicator **Even** and the other **Odd,** as shown below. Place a While Loop on the block diagram, which iterates once every 0.5 seconds, and construct a subdiagram within it so that the **Even** indicator is lit (and **Odd** is unlit) during iterations for which the value of the iteration terminal $\boxed{i}$ is even, while the **Odd** indicator is lit (and **Even** is unlit) during iterations for which the value of $\boxed{i}$ is odd. You may find the following icons useful: **Select** and **Equal To 0?** in **Functions>>Programming>>Comparison** as well as **Quotient & Remainder** in **Functions>>Programming>>Numeric**.

7.  Explore the impact of While Loop execution on CPU usage. Open the **Windows Task Manager** by right-clicking an empty area on the taskbar at the bottom of your monitor and then selecting **Task Manager**, or by pressing <*Ctrl+ Shift+ Esc*>. You will find that the **Task Manager** displays CPU usage (expressed as a percentage of your computer's maximum value) at the bottom of its window.

    (a)  Open **Sine Wave Chart (While Loop)**, and run this VI with **Wait (ms)** programmed to produce one While Loop iteration every 100 ms. What is the approximate increase in CPU usage that results when the program is running?

(b) Next, program **Wait (ms)** to produce one While Loop iteration every 1 ms, then run the VI. What is the approximate increase in CPU usage that results when the program is now running?

(c) Finally, delete **Wait (ms)** from the VI's block diagram so that the While Loop will iterate as fast as possible on your computing system (e.g., millions of iterations per second), then run the program. What is the approximate increase in CPU usage that results when the program is now run? (Note that such "memory hogging" should be avoided unless your program is executing a high-speed task, that is, not simply updating a plot, as in this case.)

8. Write a program called **Iterations Until Integer Equals Five**, which randomly generates an integer in the range from 1 to 10, and iterates this random process until the integer equals 5. The number of iterations required until the randomly generated integer equals 5 is then displayed on the front panel in a **Numeric Indicator** (found in **Functions>> Programming>>Numeric**), which is labeled **Required Iterations**, as shown below.

To create a random integer in the range from 1 to 10, use **Random Number (0–1)**, to generate a random floating-point number in the range from 0 up to (but not including) 1, multiply this number by 10, and then use **Round Toward + Infinity** to round the floating-point number to the next highest integer (note **Round Toward + Infinity** rounds $x.000$ to $x$). All of these icons are found in **Functions>>Programming>> Numeric**. Think carefully about how to determine the value of **Required Iterations**. Also, you will need to use the **Equal?** icon, which is found in **Functions>> Programming>>Comparison**.

Knowing that the integers are created randomly (i.e., equal probability for producing integer 1 through 10 each iteration), what do you expect the value of **Required Iterations** to be on average? Run **Iterations Until Integer Equals Five** 20 times and record the value of **Required Iterations** resulting from each run. Is the average of these 20 values close to the value that you expected?

9. A characteristic of the Waveform Chart can be controlled within a program using a **Property Node**.

(a) In this chapter, you set a Waveform Chart's **X-Axis Multiplier** equal to *0.2* using the **Properties** dialog window. Alternately, this property of the Chart can be set on the block diagram via a Property Node as follows: With **Sine Wave Chart (While**

**Loop)** open, pop up on the Chart's icon terminal and select **Create>>Property Node>>X Scale>>Offset and Multiplier>>Multiplier**, then place the resulting Property Node on the block diagram. This Property Node is created as an indicator, hence the small outward-directed black arrow at its right side. Change this icon to a control by popping up on its lower section and selecting **Change To Write**. The black arrow will now be inward-directed on its left side. Using **Create>>Constant**, wire a value of *0.2* to the **XScale.Multiplier** property control as shown below.

Waveform Chart

Run **Sine Wave Chart (While Loop)** and assure yourself that the Property Node is indeed setting the **X-Axis Multiplier** to 0.2.

(b)    Program **Sine Wave Chart (While Loop)** so that its Waveform Chart is cleared at the start of each run. The appropriate Property Node is made by selecting **Create>>Property Node>>History Data**. After changing this Property Node to a control, **Create>>Constant** will produce the needed input called an **Empty Array** as shown below.

Waveform Chart

In order to clear the Chart at the start of each program execution (i.e., immediately after the Run button is pressed), should this Property Node be placed inside or outside of the While Loop? Run **Sine Wave Chart (While Loop)** several successive times and demonstrate that the Property Node performs as intended.

(c)    Change the background color **Sine Wave Chart (While Loop)**'s Waveform Chart during runtime as follows: Place a Property Node within the While Loop using **Create>>Property Node>>Plot Area>>Colors>>BG Color**, then change this Property Node to a control. On the front panel, place a **Framed Color Box**, found in **Controls>>Modern>>Numeric**, then on the block diagram wire its terminal to the Property Node. Run **Sine Wave Chart (While Loop)** and demonstrate that the Chart's background color can be controlled using the Operating Tool.

# Chapter 2

# The For Loop and Waveform Graph

For the remainder of the book, only the code to be written on block diagrams will be illustrated, rather than the entire block-diagram windows. Front-panel windows, however, will be shown in their entirety.

The LabVIEW programming language provides two possible loop structures to control repetitive operations in a program. In the previous chapter, we explored the While Loop, which (by default) executes the subdiagram within its borders until the Boolean value wired to its conditional terminal is TRUE. That is, the While Loop executes until a specified condition is no longer FALSE. Now we will focus our attention on the *For Loop,* LabVIEW's other available loop structure. In contrast to the While Loop, which is controlled by the value of a specified condition, the For Loop simply executes the subdiagram within its borders a specified number of times. This loop structure is found in **Functions>> Programming>>Structures** and appears on your block diagram as shown below.

In this graphical structure, your "to-be-iterated" subprogram is written in the presently blank region within the borders. Once this subdiagram is written, the value of the *count terminal* ⊡ determines the total number of times $N$ that the For Loop will iterate. The value of count terminal is set by wiring a **Numeric Constant** (located outside the loop) to the ⊡. The For Loop's other internal icon is familiar from your work in the previous chapter. As in the While Loop, the iteration terminal ⊡ contains the current number of completed loop iterations. The value of ⊡ is 0 during the first loop iteration, 1 during the second, ... and $N-1$ during the last iteration. Thus the For Loop is equivalent to the following text-based code:

For i=0 to $N-1$
Execute Subprogram

In this chapter, you will write a sine-wave plotting program based on a For Loop. In the process of writing this VI, you will explore LabVIEW's method of storing numerical data in the form of an array and learn to operate a *Waveform Graph,* the second in LabVIEW's triad of commonly used graphing options. Additionally, you will further hone your LabVIEW editing skills.

## 2.1 SINE WAVE PLOT USING A FOR LOOP AND WAVEFORM GRAPH

Create a fresh VI by selecting **Blank VI** in the **Getting Started** window or, if a program is already open, **File>>New VI**.

Here's a shortcut for activating the Controls Palette, if it isn't already visible: Pop up (right-click the mouse button) on the blank front panel and a Controls Palette will appear. This palette is temporary, in that it will disappear as soon as you release the mouse button. However, you may secure the palette by placing the mouse cursor over the thumbtack located in the upper left corner of its window, then releasing the mouse button. This "tacked-down" Controls Palette can then be placed in a convenient location by "clicking and dragging" on its title bar. By the way, a similar "thumbtacking" procedure can be used to make any frequently used subpalette of the Controls or Functions Palette continually visible.

Place a **Waveform Graph** on the front panel from **Controls>>Modern>>Graph**. Secure its default label **Waveform Graph** by pressing *<Enter>* and use the ⬀ to position the entire graph nicely on the front panel. Pop up on the **Waveform Graph** and select **Visible Items>>Scale Legend**. You will be creating a program to plot *sin(x)* vs. *x,* so (with the help of the ⬚) use the Scale Legend to label the *x-* and *y*-axis as **x (radians)** and **sin(x)**, respectively.

By default, autoscaling is activated on both the *x-* and *y*-axis for a Waveform Graph. You can verify this fact yourself by noting that the **Scale Lock** buttons for both axes are closed 🔒. Keep autoscaling activated on both axes by changing nothing, then toggle off (i.e., hide) the Scale Legend by popping up on the Graph and selecting **Visible Items>>Scale Legend**.

If, as in this present case, labeling the axes is the only required task, rather than using the Scale Legend you can simply highlight each axis-label using the Labeling

Tool  from the Tools Palette, and type in the desired text directly. Try this time-saving tip.

## 2.2 WAVEFORM GRAPH

What is the Waveform Graph and how does it differ from the Waveform Chart? In our previous work, we saw that a Waveform Chart acts as a real-time data plotter. As each new data point is generated on the block diagram, this single numerical value is passed to the Waveform Chart's terminal and then immediately displayed in the front-panel plot. The new data point is appended to the already existent data plot, so you can view the current value in the context of the previous values.

In contrast to the Chart's interactive method of plotting data, the Waveform Graph (as well as the yet unstudied XY Graph) accepts an entire block of data that has been generated previously. The Waveform Graph's terminal accepts this data block as a one-dimensional (1D) array of $N$ numerical elements with the order of the elements indexed by integers in the range of 0 to $N-1$. For example, a 10-element array would have the following form.

| Index | 0 | 1 | 2 | 3 | 4 | 5 | 6 | 7 | 8 | 9 |
|---|---|---|---|---|---|---|---|---|---|---|
| Array | 1.20 | 1.30 | 1.40 | 1.50 | 1.60 | 1.70 | 1.80 | 1.90 | 2.00 | 2.10 |

The Waveform Graph then assumes that the numerical value of each element is the y-value of a data point to be plotted. The x-value of each plotted point is taken to be the index of that point within the 1D array. That is, the x-value of the array's $i^{th}$ element is $i$ and the y-value is the numerical value of the $i^{th}$ element. Thus the Waveform Graph implicitly assumes that data points are evenly spaced along the x-axis, a situation that is often realized in practice. For example, the time-varying nature of an analog signal is commonly represented by a set of discrete data points that were digitized ("sampled") during a sequence of equally spaced time intervals. You will find later that the XY Graph allows one to plot the more general case of an array of nonequally spaced data points (as well as multivalued functions such as circular shapes).

Switch to the block diagram. There you will find the Waveform Graph's terminal, in its icon terminal 🔲 guise, with its label located nearby. Note that the icon terminal contains a distinctive small picture of plotted data (indicating it is associated with the Waveform Graph) and has an orange border with text **DBL** denoting that the data input to it should be double-precision floating-point numbers. If you pop up on this icon terminal and select **View As Icon**, the terminal will change to its data-type terminal ▸DBL⟩ form, where we note that the DBL data-type specifier is enclosed within a pair of square brackets. The square brackets indicate that the Waveform Graph terminal should be wired to accept a numerical array, rather than just a single number, as in the case of the Waveform Chart discussed above. To toggle back to the icon terminal manifestation, pop up and select **View As Icon**.

## 2.3 OWNED AND FREE LABELS

Let me digress for a moment and discuss LabVIEW labels. The Waveform Graph's label presently on your block diagram (which contains the text **Waveform Graph**) is what is known as an *owned label*. An owned label belongs to a particular object (in this case, the Waveform Graph's terminal) and moves with that object. To demonstrate this fact, use the ⬉ to highlight the terminal, and then drag it to a new location on the block diagram. Note that the label accompanies the icon to its new location. Deselect the icon by clicking on an empty spot of the block diagram, then click on top of the label itself. A marquee will then appear solely around the label, and you will be able to drag just the label to some newly desired location relative to its owning-icon. An owned label can be made visible or invisible using the **Visible>>Label** option in the object's pop-up menu. Experiment a bit on the block diagram until you understand the functioning of an owned label.

An alternate form of LabVIEW annotation is called the *free label*. Free labels are not attached to any object and can be created, moved, or disposed of independently. They can be used as explanatory comments on your front panels and block diagrams.

To create a free label, select the Labeling Tool **A** from the Tools Palette and click on the desired location for the label. A small, bordered box appears, ready to accept text input. After you type the text message, press *<Enter>*, then the free label is complete. Try producing a free label on your block diagram. After you succeed in creating a free label, click on it with the ⬉ and delete it.

## 2.4 CREATION OF SINE WAVE USING FOR LOOP

Now select a **For Loop** from **Functions>>Programming>>Structures** and add it to your block diagram (the Functions Palette can be activated by popping up on a blank region of the block diagram). If you are dissatisfied with your initial choice of dimensions, you can place the ▶ at corner of the For Loop where it will morph into the resizing handle and allow you to reshape the loop borders.

Our programming strategy is this: Use the For Loop to create a one-dimensional array of data points that represent a few cycles of the sine function, and then pass this array to the Waveform Graph's icon terminal for plotting on the front panel. Since we desire data to be passed to the Graph's terminal only after the For Loop has completed its final iteration, we must place the terminal in a region outside the Loop's boundary. Use the ▶ to build the following block diagram.

Waveform Graph

Now let's write the code for creating the sine-wave values within the For Loop. To achieve this goal, obtain a **Sine** icon from **Functions>>Mathematics>>Elementary & Special Functions>>Trigonometric Functions** and place it inside the For Loop.

Waveform Graph

Now use **Divide**, from **Functions>>Programming>>Numeric**, to write the sine-wave creation code shown below. After placing **Divide** on your diagram, pop up on its bottom input and select **Create>>Constant** to automatically create the **Numeric Constant** 5.0 (if the constant appears as an orange 5, you can change it to 5.0 using **Display Format** in the pop-up menu, if desired). Next, wire i to the top input of **Divide**, then wire Divide's output to the **x** input of Sine. If you reverse the order in which 5.0 and i are wired to **Divide**, the **Numeric Constant** will be autocreated as an integer rather than a floating-point number (which can be corrected by selecting **Representation>>DBL** in the Numeric Constant's pop-up menu).

Ready for another editing shortcut? When coding a block diagram, you will find that the most commonly used tools are ⬆ and ✶. Rather than having to go to the Tools Palette each time to switch back and forth between these tools, simply press the <*Spacebar*> on your keyboard. With each depression of the <*Spacebar*> key, the mouse cursor will toggle back and forth between the Positioning and Wiring Tool. If you require some other feature in the Tools Palette, press your keyboard's <*Tab*> key. As you press <*Tab*> repeatedly, the mouse cursor will cycle from 🖑 to ⬆ to **A** to ✶ in the Tools Palette. Finally, by clicking on the **Automatic Tool Selection** button

✖ ▬ near the top of the Tools Palette, its green LED indicator will illuminate, denoting that the *automatic tool selection* option is activated. Then, when you move the cursor to a position on the block diagram (or front panel), LabVIEW will automatically select the tool that is most appropriate for that location. When this option is activated (indicated by the illuminated green LED), it can be toggled off by clicking on the **Automatic Tool Selection** button.

Now we need to tell the For Loop the number of iterations $N$ we would like it to perform. We know from our experience of writing similar block-diagram code in the last chapter that about 30 loop iterations will be required to generate one cycle of the sine function on this block diagram. Let's set $N=100$, so that we generate (a little more than) three sine-wave cycles. To accomplish this setting, wire a **Numeric Constant** (defined to be *100*) to the . You already know two methods for acquiring the required **Numeric Constant** icon–the "hard" way by obtaining it directly from **Functions>>Programming>>Numeric** and the "easy" way by simply popping up on the  and selecting **Create>>Constant**. Be good to yourself–use the "easy" way.

## 2.5 CLONING BLOCK-DIAGRAM ICONS

Here's one final editing option that you can add to your bag of tricks. It's called *cloning* and can be used in your present situation to produce the required Numeric Constant (to be wired to the ) because an equivalent icon already exists on the block diagram. To clone the existing **Numeric Constant**  on your block diagram, place the Positioning Tool over this icon.

Waveform Graph

Now click the mouse button while depressing *<Ctrl>* on your keyboard. Then, by moving the mouse while still holding down the button, you will drag a copy of the icon with you, while the original stays in place. Once you arrive at the newly desired location, release the mouse button and the cloned icon will reside there (if you place  within the autowiring range of the , a wire will automatically connect the two).

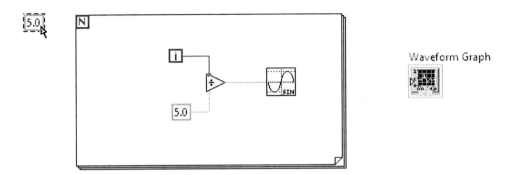

Use the Operating Tool to change the value of this new **Numeric Constant** to *100* and then wire it to the 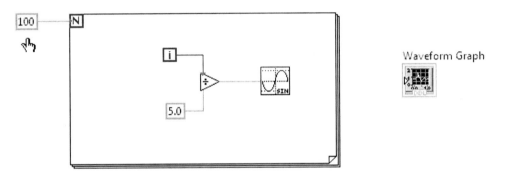 .

The For Loop's count terminal is configured to accept the numeric format **I32** (signed four-byte integer, also called *long integer*). The red coercion dot, indicating a data-type mismatch, can then be erased by popping up on the **Numeric Constant** and selecting **Representation>> I32**.

## 2.6 AUTO-INDEXING FEATURE

Finally, we need to store the sequence of sine-function values produced by the For Loop in a 100-element array so that we can pass this array to the Waveform Graph's terminal for plotting. Does it sound like there's some hard programming ahead? Surprisingly, there's not! Array storage of the sequence of values generated by the multiple iterations of a loop structure is such a commonly required operation that LabVIEW provides it as a built-in optional feature of both the For Loop and the While Loop. Perhaps it's easiest if we first implement this feature and then decipher what we've done.

Using �帝, simply wire the **sin(x)** output of the **Sine** icon to the Waveform Graph's terminal. You're finished!

Here, we've accomplished our goal by implementing the *auto-indexing* capability of the For Loop. With this feature, as the For Loop creates a new value with each loop iteration, this new value is indexed and appended to an array at the loop boundary. After the loop completes its final iteration, the array is passed out of the loop to the Graph's terminal. Notice that, in the region within the For Loop, the wire emanating from the Sine output is thin, denoting the fact that this icon produces just a single numerical value with each loop iteration. However, when this wire passes through the black-bordered square ☐ at the loop's boundary (called a *tunnel*), the wire becomes much thicker. This thick wire is LabVIEW's way of denoting a one-dimensional array of data values.

Let's explore the auto-indexing feature a little further. First, note the pair of square brackets contained within the tunnel icon ☐. Such a bracket pair is LabVIEW code for an array. Thus these brackets are the graphical indicator that an array is being created at the tunnel as the For Loop iterates; that is, auto-indexing is activated.

Place the Positioning Tool over the tunnel at the For Loop's boundary.

Then, pop up on this tunnel to reveal its pop-up menu.

This menu offers you the ability of toggling the loop's auto-indexing feature on and off. As we surmised earlier, a For Loop's default setting is **Enable Indexing**, so the menu now offers the option of toggling this feature off with the selection **Disable Indexing**. Toggle the auto-indexing off by selecting **Disable Indexing**.

Now, rather than accumulating all the loop's iterated values into an array, the tunnel will only pass out a single value, the value of the sine function determined during the last loop iteration. To reflect this functional change for the tunnel, the bracket pair is no longer present within its icon, replaced instead by a solid color. As shown below, with the auto-indexing disabled, a broken wire now appears connected to the Waveform Graph's terminal because this icon expects to receive an entire array, not just a single numerical value.

To restore the proper programming, pop up on the tunnel and toggle the auto-indexing back on by selecting **Enable Indexing**.

## 2.7 RUNNING THE VI

Let's see if your program works. Return to the front panel. As mentioned previously, autoscaling is activated by default on both axes of a Waveform Graph (you can, of course, turn off this feature in the Graph's pop-up menu or in the Scale Legend). Before running the VI, it is a good idea to pause for a moment and anticipate how the axes will be autoscaled when the program is functioning properly. The code is designed to plot a sine wave, which is discretely sampled at 100 points. Thus, the $y$-axis values should range from $-1.0$ to $+1.0$. The $x$-axis value of each sine-wave value is taken to be its index within the one-dimensional array of data points. Because LabVIEW ascribes an index value of 0 to the first array element, the index of our last data is 99. Thus the $x$-axis values range from 0 to 99. Perhaps the scaling will be more pleasing if LabVIEW's autoscaling feature chooses 100 as the *x*-axis upper limit (which is the default value for the Waveform Graph).

OK, run your program. If all goes well, it will produce a plot as shown below, where the axes are scaled as we anticipated.

## 2.8 *x*-AXIS CALIBRATION OF THE WAVEFORM GRAPH

Pretty nice program, eh? There is one thing, however, we need to fix—the *x*-axis calibration. As is, the number-labels on this axis reflect the integer indexing derived from the array storage of your data, while the text-label indicates that this axis is intended to plot the radian argument of the sine function being graphed on the *y*-axis. Remembering that our program takes the sine function argument for a particular data point to be the array index value *i* divided by 5, we see that the *x*-axis can be calibrated in radians by simply multiplying each number-label by 0.2. Alternately, we can re-express this fact by saying that, by default, the Waveform Graph assumes the *x*-axis spacing between adjacent data points $\Delta x$ is 1. Calibration is then accomplished by defining $\Delta x$ to be 0.2.

Such calibration by a multiplicative constant is a common need when using the Waveform Graph, and therefore LabVIEW provides mechanisms for this procedure. In the previous chapter, we accomplished *x*-axis calibration of a Waveform Chart by defining a Multiplier in the **Chart Properties** dialog window. This mechanism can be implemented on the Waveform Graph also. Here we will introduce an alternative calibration mechanism, which appears explicitly on the block diagram. This calibration mechanism involves a new LabVIEW process: bundling together several objects into a *cluster*. The resulting cluster in not a mathematical object, but rather a LabVIEW convenience. By grouping a number of objects together into a single cluster, one is able to transport a large quantity of data across a block diagram using a single cluster wire. Thus clustering greatly simplifies the appearance and readability of your block diagrams.

Return to the block diagram, place the mouse cursor over the Waveform Graph's terminal, and activate its Help Window.

This Help Window, which is divided into a top and bottom portion, appears above. In the top portion, we find that a Waveform Graph with default data-point spacing $\Delta x = 1$ is created by simply connecting an array wire to the Waveform Graph terminal. This is the manner in which we've used the Waveform Graph thus far. We produced a single plot by feeding a one-dimensional data array to the Waveform Graph's terminal. Note that two (or more) waveforms can be displayed simultaneously on a Waveform Graph by connecting a two-dimensional (2D) array to the terminal. The bottom portion of the Help Window indicates that, when given an appropriate cluster, the Waveform Graph produces a plot on the front panel with its *x*-axis properly calibrated. The appropriate cluster is formed by bundling together the following sequence of objects: a value for the *x*-axis origin $x_0$, a value for the *x*-axis spacing $\Delta x$, and the array of numerical data to be used as the *y*-axis values of the waveform's points.

Here's how to include the appropriate cluster in your code. First, remove the array wire connected to the Waveform Graph terminal. Using the Positioning Tool, click on this wire to produce a marquee.

Then press your keyboard's <*Delete*> key to remove this wire.

Select **Bundle** from **Functions>>Programming>>Cluster, Class, & Variant**. The **Bundle** icon is used to collect several input objects into a cluster.

Then place the **Bundle** icon on your block diagram. You will find that its leftmost section consists of two inputs by default. We require three inputs, so you must resize the icon.

To resize the **Bundle** icon so that it has three inputs, place the ▶ at the icon's bottom-center (or top-center), so that the tool morphs into a resizing handle.

Then drag the handle until an additional input appears.

Secure this choice of inputs by releasing the mouse button. Use the Positioning Tool to position the Bundle and Waveform Graph terminal as shown.

Wire the For Loop's tunnel to the Bundle's bottom input. Once this connection is made, square brackets will appear within this input, the graphical specifier for an array data-type.

Now wire a **Numeric Constant** of *0.0* and *0.2* to the Bundle's top and middle inputs to define $x_0$ and $\Delta x$, respectively. Note that unbracketed **DBL**s appear within these inputs, indicating each consists of a solitary (i.e., non-array) double-precision floating-point number.

The **Bundle** icon receives these three inputs on its left side, bundles them together, and outputs the resulting cluster at its rightmost section. Wire this cluster output to the Waveform Graph's terminal.

You will find that as soon as this connection is made, the icon terminal's data-type specifier morphs from a floating-point numeric specifier into a cluster specifier. In addition, the wire connecting the Bundle output to the Waveform Graph's terminal has a pink banded appearance, denoting cluster data.

Now return to the front panel and run your final program. Note the *x*-axis is cali-brated in radians.

Using **File>>Save**, create a folder named **Chapter 2** within the **YourName** folder, then save this VI under the name **Sine Wave Graph (For Loop)** in the **YourName\Chapter 2** folder.

## 2.9 SINE WAVE PLOT USING A WHILE LOOP AND WAVEFORM GRAPH

Let's try accomplishing the above goal of plotting a sine wave on a Waveform Graph, but this time using a While Loop rather than a For Loop to generate the data array. Select **File>>New VI**, then place a **Waveform Graph** (from **Controls>>Modern>>Graph**) with its default label **Waveform Graph** on the fresh front panel. Autoscaling will be enabled on both axes by default. Using the **A** (or the Scale Legend), label the *x*- and *y*-axis as **x (radians)** and **sin(x)**, respectively.

Switch to the block diagram and write the following code.

Waveform Graph

The **Equal?** icon is found in **Functions>>Programming>>Comparison**. Its function is described by the Help Window shown next. Remember you can always access assistance from such Help Windows while wiring by toggling on **Help>>Show Context Help** or, more simply, through the keyboard shortcut <*Ctrl+H*>.

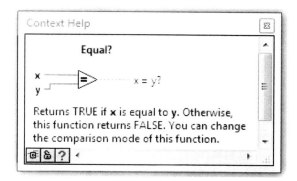

The While Loop on your block diagram will generate 100 values of the sine function before ceasing execution. Using your knowledge of the detailed operation of a While Loop, can you explain why this is so? In particular, explain why the proper value for the **Numeric Constant** in the comparison statement is *99* (and not *100*).

You now need to store the 100 sine-function values in a one-dimensional array and pass this block of data to the Waveform Graph's terminal for plotting on the front panel. Making use of the auto-indexing feature of the loop structure, we need simply to wire the **sin(x)** output of the Sine icon to the Waveform Graph terminal. Perform this wiring as shown here.

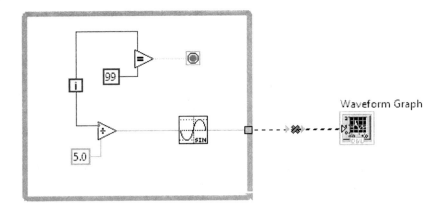

You will note that this seemingly simple connection produces a broken wire. To find out why, pop up on the tunnel at the While Loop's border.

In the tunnel's pop-up menu, you will be presented the option of enabling the loop's auto-indexing feature. Thus we see that, in contrast to a For Loop, the auto-indexing feature of a While Loop is toggled off by default. Turn the auto-indexing on by selecting **Enable Indexing**. Also include a **Bundle** icon to scale the *x*-axis in radians, as shown below.

Your program is now complete. Return to the front panel and run it. You might try using the keyboard shortcut for the **Run** command: $<Ctrl+R>$.

## 2.10 ARRAY INDICATORS AND THE PROBE

Are you unconvinced that there are actually 100 elements in the array you produced? You can easily check this fact by inspecting the array using an *Array Indicator.* An Array Indictor is constructed as follows. Select an **Array** shell from **Controls>>Modern>>Array, Matrix & Cluster**. As described in the next paragraph, the **Array** shell can be made to operate in either of two modes: to input an array from the front panel to the block diagram or to output an array from the block diagram to the front panel.

Place the **Array** shell on your front panel, and then enter the label **Array**. You will find that the shell consists of an *index display* on the left and a large blank region (called the *element display*) on the right. By filling the blank region with the appropriate object, one may choose whether the **Array** shell will function as an input or output device. In the present situation, we wish to output an array to the front panel, so we will occupy this blank region with a **Numeric Indicator**. If instead we wished to input an array to the block diagram, we would occupy this region with a **Numeric Control**.

To place a **Numeric Indicator** within the **Array** shell, do the following. Select a **Numeric Indicator** from **Controls>>Modern>>Numeric**.

Then place the cursor in the large blank region of the **Array** shell.

When you click the mouse button, the Array Indicator will be complete, as shown below.

Now go to the block diagram and place the Array Indicator's terminal in a convenient location. In this program, the While Loop creates a 1D array of data. Using the Wiring Tool, connect any point on thick wire representing this one-dimensional array to the Array Indicator's terminal.

Return to the front panel and run your program. After the VI completes execution, you can inspect the array generated via the Array Indicator (if you need to resize it, use the 🔺). Using the 👆, look at various elements in the array. To scan the array, you may increment or decrement the index number in steps of 1 by repeatedly clicking on the index display's up/down arrows. Alternately, to view a particular array element, you can use the 👆 to highlight the index display, then enter the desired element's index value from the keyboard, and press *<Enter>*, or click on ✔.

Through such exploration, you should find that your array consists of 100 elements, with indices running from 0 to 99.

LabVIEW's *Probe* provides another method for checking the contents of your array. This handy feature allows you to examine the contents of any wire on your block diagram. To inspect the elements of the sine-wave array using the Probe, place the mouse cursor on the wire carrying this data.

Then pop up on this wire and select **Probe**.

A numbered Probe Window will appear, along with an associated numbered rectangle marking the wire location being probed. Probe Windows can also be activated by clicking on the wire of interest with the *Probe Tool* ⊕ found in the Tools Palette.

Run your program, and then use the Probe to inspect the elements of the resulting array. When you're done with the Probe, close its window and it (as well as its associated numbered rectangle) will disappear. Probe Windows are an invaluable tool for debugging LabVIEW programs that execute but are producing questionable or unexpected results.

Finally, you might try producing an Array Indicator the "painless way." For example, pop up on the thick orange wire (or the tunnel) that carries the sine-wave array output from the While Loop, then select **Create>>Indicator**.

Now check the front panel. You will find an automatically produced Array Indicator there. Run your program and discover if it works as expected.

Using **File>>Save**, save this VI in **YourName\Chapter 2** (that is, the **Chapter 2** folder within the **YourName** folder) under the name **Sine Wave Graph (While Loop)**.

## DO IT YOURSELF

A flipped coin can land in one of two states—*heads* or *tails*. Write a VI called **Four Flipping Coins** that simulates simultaneously flipping four coins, and determines how many of these coins land in the heads state.

(a) First, build the front panel of **Four Flipping Coins** as shown here, where the state of each coin is represented by a **Round LED** Boolean indicator, which is found in **Controls>>Modern>>Boolean**. A lit (unlit) LED corresponds to heads (tails). Also include a **Numeric Indicator** labeled **Number of Heads** that gives the number of coins in the heads state.

On the block diagram, model the flip of each coin using a **Random Number (0–1)** icon, which is found in **Functions>>Programming >>Numeric.** When run, if this icon's **number (0 to 1)** output is greater than or equal to 0.5, assign the corresponding coin to the heads state; otherwise, that coin is in the tails state. Tally the results of all four coins to determine the number of coins in the heads state and output this integer to the **Number of Heads** front-panel indicator. Note that there are five possible outcomes—0, 1, 2, 3, and 4 heads. In writing the block diagram, you may find the following icons useful: **Greater or Equal?** in **Functions>>Programming>>Comparison, Boolean To (0,1)** in **Functions>>Programming>>Boolean,** and **Compound Arithmetic in Functions>>Programming >>Numeric.**

Run **Four Flipping Coins** to verify that it is working correctly.

(b) The present code on your block diagram simulates one *trial* of simultaneously flipping the four coins. Now simulate repeating such a trial over and over again. Place this code within a **For Loop** so that you can execute $N$ trials, where $N$ is given by the value in a front-panel **Numeric Control** labeled **Total Number of Trials.** Store the value of **Number of Heads** from each trial in an array using the For Loop's auto-indexing feature. Then after the $N$ trials are complete, plot a histogram on a **Waveform Graph** that displays the number of trials whose outcome was 0, 1, 2, 3, and 4 heads. To create a histogram, use the **Histogram.vi** icon found in **Functions>>Mathematics>>Probability & Statistics** with its **intervals** input wired to an integer *5* (since there are five possible outcomes for **Number of Heads**). The front panel of **Four Flipping Coins** will then appear as shown next.

Run **Four Flipping Coins** with several choices of **Total Number of Trials**. In the Waveform Graph's **Plot Legend**, you may wish to select a **Bar Plots** option in its pop-up menu. From statistical theory, it is expected that when the number of trials $N$ is large, the number of trials with an outcome of zero or four heads should equal $N\left(\dfrac{1}{2}\right)^4 = \dfrac{N}{16}$

and the number of trials with an outcome of two heads should be $N\dfrac{4!}{2!2!}\left(\dfrac{1}{2}\right)^4 = \dfrac{3}{8}N$

Above what (approximate) value of $N$ are your results consistent with this expectation?.

## Problems

1. I have agreed to hire you for a 30-day month. I pay you one penny on the first day, two pennies on the second day, and continue to double your daily pay on each subsequent day up (and including) to day 30. How many total dollars do you earn by the end of the month? Write a For Loop-based program called **Month's Total Pay** that determines the answer to this question. You may find **Add Array Elements** in **Functions>>Programming>>Numeric** useful as well as the **Functions>>Programming>>Mathematics>>Elementary & Special Functions>>Exponential Functions** subpalette.

2. A finite geometric series of $n$ terms is given by $y_n = 1+x+x^2+x^3+... \ x^{n-1}$, where $|x| < 1$.
   It is well-known that when $n = \infty$, $y_\infty = \dfrac{1}{1-x}$.

   (a) Write a For Loop–based program called **Finite Geometric Series** that, given $n$ and $x$ as inputs via front-panel **Numeric Controls**, determines $y_n$, calculates its percent deviation from $y_\infty$, and displays these two results in **Numeric Indicators**. You may find **Add Array Elements** in **Functions>>Programming>>Numeric** useful as well as the **Functions>>Programming>>Mathematics>>Elementary & Special Functions>>Exponential Functions** subpalette.

   (b) Use your VI to answer the following question: If $x=0.5$, and one wishes $y_n$ to deviate from $y_\infty = 2.000 \ldots$ by less than 0.1%, what is the minimum required value of $n$?

3. Write a program called **Chirp**, which outputs a quick succession of $n$ sound waves, each with a unique frequency $f$ and duration of 2 ms. Make the frequencies of the first through last sound wave obey the following pattern: 1000 Hz, 1010 Hz, 1020 Hz,..., 2990 Hz, 3000 Hz. To create each sound wave, use **Beep.vi**, which is found in **Functions>>Programming>>Graphics and Sound**, with its **use system alert?** input set to FALSE.

4. A unity-amplitude square wave $y(t)$ can be approximated by the following sum of three sine waves:

$$y(t) \approx \frac{4}{\pi}\left\{ \sin[x] + \frac{1}{3}\sin[3x] + \frac{1}{5}\sin[5x] \right\}$$

where $x$ is in radians. Write a VI named **Sum of Three Sines** that evaluates this sum at 300 equally spaced $x$-values in the range from $x = 0$ to $x = 6\pi$ (i.e., for three cycles of the first sine function in the sum), and then plots the resulting *y* versus *x* on a Waveform Graph, whose *x*-axis is calibrated in radians. You may find **Compound Arithmetic** in **Functions>>Programming>>Numeric** useful as well as the **Functions>>Programming>Numeric>>Math & Scientific Constants** sub-palette. When **Sum of Three Sines** is run, it should produce the front-panel plot shown next.

5. Write a VI called **Wait Test** that performs the following task *N* times in a row: Execute the **Wait (ms)** icon with **milliseconds to wait** equal to *1,* then store the resultant value of **millisecond timer value**. After this sequence of *N* operations, make your program plot the resultant *N*-element array of **millisecond timer value** values. Also, display this array in an Array Indicator called **Array** (this indicator can be resized using the ⬉). The front panel of **Wait Test** should appear as shown.

(a)   When **Wait Test** is run, a straight-line plot should result. What is the expected slope for this straight line?

(b)   Setting *N* equal to, say, 20, run the program several times under different operating conditions (e.g., several other programs besides LabVIEW running, holding down a key on the keyboard, moving the mouse, etc.). Does the resulting plot yield a straight line with the expected slope most of the time?

(c)   You should observe anomalies in the plot caused by your processor attending to other business than running **Wait Test**. Let's call such an event a *lapse*. Using **Array**, determine the typical length (in milliseconds) of a lapse. Is the frequency with which lapses occur dependent on prevailing operating conditions? If so, what conditions seem to enhance the occurrence of a lapse?

6.   It's moving day and you are dragging a heavy dresser at constant velocity across the horizontal floor of your new apartment as shown in Figure 2.1. The force you are applying to the dresser has magnitude *F* and it is directed at angle $\theta$ above the horizontal. The weight of the dresser is *W* and the coefficient of kinetic friction between it and the floor is $\mu$.

Newton's laws can be used to show that

$$F = \frac{\mu W}{\cos \theta + \mu \sin \theta}$$

or, expressing the force as a fraction of the dresser's weight, i.e.,

$$F' = F / W,$$

**Fig. 2.1    Problem 6.**

$$F' = \frac{\mu}{\cos\theta + \mu\sin\theta}.$$

(a)  Write a For-Loop based program called **Dragging Dresser** that, given $\mu$ from a **Numeric Control** labeled **Coefficient of Kinetic Friction**, calculates $F'$ from $\theta=0$ to $\theta=90°$ in $1°$ increments, and then displays the resulting array of $F'$-values on a Waveform Graph as well as in an Array Indicator labeled **Force.**

(b)  Run **Dragging Dresser** with $\mu=0.1$, $\mu=0.4$, and $\mu=1.0$ and determine the angle $\theta_{min}$ at which $F'$ is smallest in each case (i.e., $\theta_{min}$ is the angle at which the minimum magnitude force is required to drag the dresser at constant velocity for the given $\mu$).

7.  (a)  Write a program called **Mean of Random Array**, which generates an array of 100 random double-precision floating-point numbers in the range of 0 to 1, and then determines the mean of this array of numbers. Display the mean in a front-panel indicator called **Mean** with its **Digits of precision** set to 3. You will find the following two icons helpful: **Random Number (0–1)** in **Functions>>Programming>>Numeric** and **Mean.vi** in **Functions>> Mathematics>>Probability & Statistics.**

(b)  It is expected that the nominal value of **Mean** should be 0.500. Run **Mean of Random Array** $N$ times in a row, where $N=10$. Due to the random manner in which the array is generated each time, you will find most (say, 7 out of 10) of the **Mean** values fall within a range given by $0.500 \pm \Delta x$ What value do you observe for $\Delta x$? What then is the observed *percent fluctuation of the mean* in this data set, where percent fluctuation is defined as $\dfrac{\Delta x}{0.500} \times 100$ ?

(c)  Automate this process. Place a front-panel control called **Number of Times**, which will allow a user to enter a value for $N$. When run, the VI will determine the mean of a 100-element array of random numbers $N$ times, and save these mean values in an $N$-element array. The fluctuation of the mean can then be defined as the standard deviation $\sigma$ of this array of mean values. Use **Std Deviation and Variance.vi** (in **Functions>>Mathematics>>Probability & Statistics**) to determine $\sigma$ and display it in a front-panel indicator labeled **Fluctuation**. Save your work, and then run the VI with **Number of Times** set to *10*. Is the value of **Fluctuation** consistent with the value you determined for $\Delta x$ in part (b)?

8   Problem 8 in Chapter 1 explains how to code the following task: Randomly generate an integer in the range from 1 to 10, and iterate this random process until the integer equals 5. The number of iterations required until the randomly generated integer equals 5 is output as a quantity called **Required Iterations**.

Place the code that carries out the above-described task as a subdiagram within a For Loop so that, when executed, the For Loop will repeat the task *N* times, and store the resulting *N* values of **Required Iterations** in an array. Then, use **Mean.vi** (found in **Functions>> Mathematics>>Probability & Statistics**) to calculate the average of the *N* values of **Required Iterations** and display this average in a front-panel **Numeric Indicator** labeled **Average Number of Required Iterations**. Take $N = 1,000,000$. Save this VI under the name **Iterations Until Integer Equals Five (For Loop)**.

Run **Iterations Until Integer Equals Five (For Loop)**. Is the value you obtain for **Average Number of Required Iterations** equal to the value that you expected?

9.  The *waveform* data type provides an alternate method for producing a Waveform Graph with its *x*-axis calibrated properly, where it is assumed that the *x*-axis quantity is *time*. To investigate the use of this data type, write a program called **Sine Wave Graph (Waveform)**, which plots the function $y = \sin(2\pi t)$ from time $t = 0$ to $t = 0.99$ in increments of $\Delta t = 0.01$. The block diagram for this VI is shown below. Here, **Build Waveform** (found in **Functions>>Programming>>Waveform**), combines an initial time **t0** (called a *time stamp*), the sampling time increment **dt**, and the array of *y*-values. **Build Waveform** must be resized to expose its three inputs. The time stamp is created using **Get Date/ Time In Seconds** (found in **Functions>>Programming>>Timing**) and **Pi Multiplied by 2** is found in **Functions>>Programming>>Numeric>>Math & Scientific Constants**.

Run **Sine Wave Graph (Waveform)** and verify that it executes as expected.

# Chapter 3

# The Mathscript Node and XY Graph

In previous chapters, we have seen that the graphical nature of the LabVIEW programming language greatly simplifies the coding of desired computer operations. However, in the case of encoding mathematical relations, graphical programming is much more cumbersome than text-based languages. For this reason, LabVIEW offers the *Mathscript Node*, a resizable box that you can use to enter text-based mathematical formulas directly on the block diagram. To demonstrate the utility of the Mathscript Node, consider the relatively simple equation $y = 3x^2 + 2x + 1$. If you code this equation using LabVIEW arithmetic icons from **Functions>>Programming>>Numeric**, the subdiagram will appear as below.

Alternately, you can program the same equation using a Mathscript Node, which is found in **Functions>>Programming>>Structures**. The resulting subdiagram is as shown in the following illustration. As you can see, the Mathscript Node is much easier to write and to read.

An alternate (but less powerful) text-based structure called the *Formula Node* is also available in **Functions>>Programming>>Structures**. The Mathscript Node and Formula Node are similar in that both can be used to carry out common mathematical operations, including evaluating trigonometric and logarithmic functions as well as implementing Boolean logic, comparisons, and conditional branching. Here the parity between the two structures ends, however. Whereas the Formula Node is capable of performing only these relatively simple tasks, we will see that the Mathscript Node functions as a high-level computing language (called *Mathscript*) with wide-ranging functionality in digital signal processing and data analysis.

To illustrate the power of Mathscript-based programming, consider coding the problem we solved in the previous chapter: Evalute $\sin(x)$ at 100 evenly spaced $x$-values in the range from $x = 0$ to $x = 99/5 = 19.8$. Using graphical programming solely, we found that the solution to this problem is as follows.

The Mathscript language contains the text-based sine function $\sin(x)$, so we can simply transliterate the above code using the Mathscript Node as follows. The **To Double Precision Float** icon **DBL** (found in **Functions>>Programming>>Numeric>> Conversion**), which converts the value of [i] from **I32** to **DBL**, is necessary because Mathscript only recognizes double-precision floating-point numerics.

The above program will indeed execute as desired, but it fails to take advantage of the full power of the Mathscript language. Namely, in Mathscript, the command $x=start:step:stop$ generates a one-dimensional array of $x$-values, beginning at $x=start$, ending at $x=stop$, with the $x$-values evenly spaced by intervals $\Delta x=step$. Also, given $x$ as an array of $N$ elements, the command $y=\sin(x)$ will evaluate the sine function at each of the $N$ $x$-values and return the results in an $N$-element array called $y$. Thus, when using a Mathscript Node to solve our stated problem, a graphical For Loop is unneeded. The following diagram is sufficient.

If you would like to explore some of the available Mathscript functions, select **Help>>Search the LabVIEW Help**... When the LabVIEW Help window opens, click on the **Index** tab, and then search for *Mathscript*. Under the Mathscript heading, you will find many topics of possible interest, including *Classes of Functions*, which provides an organized listing of the (over 500) available functions, along with the proper syntax for implementing them.

In this chapter, you will rewrite the **Sine Wave Graph** VI using a Mathscript Node. You will then use this subdiagram to supply data to an *XY Graph*, the most general of LabVIEW's three commonly used graphing options. This new program will become the basis for **Waveform Simulator**, a VI you will use to simulate discretely sampled data in your future work with File Input/Output, Waveform Generation, and Fast Fourier Transform–based programs. Finally, you will learn how to create your own custom-made icon that can then be used as a *sub VI* in larger "calling" programs.

## 3.1 SINE WAVE PLOT USING A MATHSCRIPT NODE AND WAVEFORM GRAPH

To introduce ourselves to programming the Mathscript Node, let's first write the program just discussed, that is, a VI that evaluates $\sin(x)$ at 100 evenly spaced $x$-values in the range from $x=0$ to $x=19.8$. Open a new VI, and then place a **Waveform Graph** on its front panel. Using the $\mathbf{A}$ (or, alternately, the Waveform Graph's **Scale Legend**), label the $x$- and $y$-axis as **x (radians)** and **sin(x)**. By default, autoscaling will be activated on both of the axes.

Switch to the block diagram and place a **Mathscript Node** on it (found in **Functions>>Programming>>Structures**).

Next, place the ⬥ at a location on the right border of the **Mathscript Node**.

Then pop up on this border and select **Add Output** from the Mathscript Node's pop-up menu.

An indicator for this output will appear in the form of a small rectangular box perched on the Mathscript Node's border. The interior of this box will be highlighted, awaiting your character-based name for this equation variable. If this highlighting disappears (due to an inadvertent click of the mouse), use the 🖑 to restore access to the output box's interior. You can be as creative as you like in naming your output variable within the limitations of the following rules: A valid variable name must begin with a letter and, after that, be composed of only letters, digits, or underscores. Names are case-sensitive (that is, the capitalized version of a particular letter is distinct from its lowercase rendering) and, if desired, can be very long (up to a maximum of about 60 characters). Long names, of course, have the disadvantage of occupying considerable diagram space. Let's conserve valuable diagram real-estate by giving the newly created output variable an uncreative short name–how about $y$. Secure this choice by pressing *<Enter>* or clicking the mouse on an empty region of the block diagram.

Waveform Graph

Now, using the , click on the interior of the Mathscript Node and enter the following two lines of text-based code:

$$x = 0:1/5:19.8$$

$$y = \sin(x)$$

This code will create the two desired one-dimensional arrays $x$ and $y = \sin(x)$.

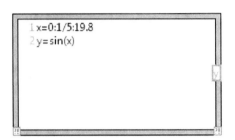

A variety of data types are possible for variables at the border of a Mathscript Node (e.g., real and complex scalars, real and complex arrays). Thus, we must alert our Mathscript Node to the data type of $y$, which is a one-dimensional array of double-precision floating-point numerics. Pop up on the output $y$, and select **Choose Data Type>>1D-Array>>DBL 1D**.

Finally, complete the block diagram as shown below. **Bundle** is found in **Functions>>Programming>>Cluster, Class, & Variant**. The **y array** indicator can be created with **Create>>Indicator** after popping up on the thick array wire. If you get a broken wire when wiring to the $y$ output, pop up on this object and make sure that the correct data type has been selected and also that it is indeed an output (that is, it wasn't accidentally created as an input).

Return to the front panel. Using **File>>Save**, create a folder named **Chapter 3** within the **YourName** folder, then save this VI under the name **Sine Wave Graph (Mathscript Node)** in **YourName\Chapter 3**.

Now run your program. It should produce a few beautiful cycles of the sine function, as shown. You may then view the actual values of the sine function using the **y array** indicator.

## 3.2 DEBUGGING WITH ERROR LIST

Let's digress for a moment and explore one of LabVIEW's handy program debugging features. Intentionally put an error in your block diagram by popping up on the Mathscript Node's *y* output and selecting **Choose Data Type>>Scalar>>DBL**.

Now scrutinize at the toolbar. There you will find that the **Run** button appears broken , which is LabVIEW's way of communicating that an error exist within your program. But this indicator isn't the only help available.

Click on the  (a tip strip alerts you that this action will produce a list of program errors). A dialog box will appear, listing all of the errors in your program. In the present case, two errors exist, both related to a data-type mismatch, namely, a scalar wire connected to an input formatted to accept an array.

In the **Details** box, a detailed description is given of the error that is highlighted in the center box. To highlight a different error, simply click on that error in the center box. Clicking on the **Show Error** button will take you to the error's location (via a marquee) on the block diagram, where you can then make the required repair. With all of this available online help, quite commonly a LabVIEW program can be debugged without having to flip through a user's manual.

Repair the error in your VI by popping up on the *y* output, and selecting **Choose Data Type>>1D-Array>>DBL 1D**. You will note that the **Run** button ⬜ becomes whole again, signaling that the program is now executable.

## 3.3 WAVEFORM SIMULATOR USING A MATHSCRIPT NODE AND XY GRAPH

For our studies ahead, we will need a data simulator VI, that is, a program that outputs digitized data of the type produced by a data acquisition device (or digital oscilloscope) in a real experiment. In a typical laboratory experiment, a time-varying analog (continuous) signal $x(t)$ is input to an instrument, which is programmed to measure ("sample") this signal at $N$ equally spaced times $t_i$. If $\Delta t$ is the time interval between one sample and the next, then $t_i = i\,\Delta t$, where $i = 0, 1, 2, \ldots, N-1$. Since there will be one data sample taken per time interval $\Delta t$, the *sampling frequency* is defined to be

$$f_s = \frac{1}{\Delta t} \qquad\qquad [1]$$

In this section, we will begin work on a data simulation VI based on a Mathscript Node. Here, we will assume that a physical system under investigation is producing a pure sine-wave signal with a time-varying *displacement* given by $x(t) = A\sin(2\pi f t)$, where $f$ and $A$ are the sine wave's frequency and amplitude, respectively. Further, we will assume that this analog signal is input to a digitizing instrument, which samples the signal at $N$ equally spaced times at a rate $f_s = 1/\Delta t$, and then outputs the result. Defining $x_i$ to be the sample taken at time $t_i$, the resulting data set will be given by the following two relations:

$$t_i = i\,\Delta t \qquad\qquad [2]$$

$$x_i = A\sin(2\pi f t_i) \qquad\qquad [3]$$

where $i = 0, 1, 2, \ldots, N-1$ and $\Delta t = 1/f_s$. Note that the first and last data sample is taken at time $t_0 = 0$ and $t_{N-1} = (N-1)\Delta t$, respectively. Our goal then is to write a software program that creates the data set described by Equations [2] and [3] and thus simulates the $(t, x)$ output of the above-mentioned digitizing equipment.

To display your VI's output, you will use an *XY Graph*. The XY Graph is a general-purpose graphing option that produces Cartesian-style plots. That is, each point on the plot is located by its $(x, y)$ value. The entire set of $(x, y)$ values is presented to this routine in the form of a cluster consisting of two 1D arrays of numerical values. The first and second array in the cluster correspond to the sequence of $x$ and $y$ values, respectively.

Open a new VI, and place an **XY Graph** (from **Controls>>Modern>>Graph**) on its front panel. Label the $x$- and $y$-axis as **Time** and **Displacement**, respectively. Autoscaling will be activated on each axis by default. Finally, place four **Numeric Controls** on the front panel labeled **Number of Samples**, **Sampling Frequency**, **Frequency**, and **Amplitude**, respectively. By default, the representation for these Numeric Controls is the double-precision floating-point data type **DBL**. Pop up on **Number of Samples** and change its data type to Long Integer by selecting **Representation>>I32**. Pop up on each of the other Numeric Controls and make sure that (its default) **Representation>>DBL** is selected. Finally, save this VI under the name **Waveform Simulator** in **YourName\Chapter 3**.

Switch to the block diagram and write Mathscript-based code that evaluates Equations [1] through [3]. To produce the two arrays describing a time-varying sine wave's $x$- and $y$-coordinates, which we call $t$ and $x$, respectively, the required code within the Mathscript Node is as follows.

$delta\_t = 1/f\_s$

$start = 0$

$step = delta\_t$

$stop = (N-1)*delta\_t$

$t = start:step:stop$

$x = A*\sin(2*pi*f*t)$

As you code, remember that the characters are case-sensitive. Pop up on each output and select **Choose Data Type>>1D-Array>>DBL 1D**. The **To Double Precision Float**, which is necessary because Mathscript only recognizes double-precision floating-point numerics, is found in **Functions>>Programming>> Numeric>>Conversion**.

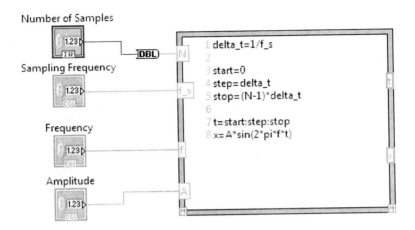

## 3.4 CREATING AN XY CLUSTER

To determine what type of input the XY Graph requires, activate its Help Window

by placing the mouse cursor over the XY Graph's icon terminal  and pressing *<Ctrl+H>*. The Window will appear as shown.

This Help Window indicates that to plot a single waveform, the XY Graph expects a cluster as input. This cluster is to contain the waveform's *x* and *y* arrays and is formed using the **Bundle** icon. Let's call this bundle of a waveform's *x* and *y* arrays its *"xy cluster."*

For future reference, the Help Window also tells us that $N$ waveforms can be plotted simultaneously on an XY Graph. In this situation, the data are passed to the XY Graph's terminal in the form of an "*array of clusters*." To form this object, one must first form the *xy* cluster for each of the $N$ waveforms as above. Then, using the yet-to-be-studied **Build Array** icon, an $N$-element array is constructed with each element being a particular waveform's *xy* cluster.

Refer to your block diagram. Here you will find the XY Graph's icon terminal 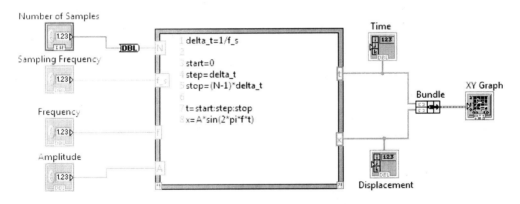. We note that, similar to the Waveform Graph, this terminal has a cluster data-type indicator (in its bottom-center), but unlike the Waveform Graph, its border is brown (rather than pink). The brown color denotes that, in the most general case of multiple waveform plotting, this data input is an array of clusters (rather than a single cluster).

Complete the block diagram wiring as shown below. Use the **Bundle** icon (found in **Functions>>Programming>>Cluster**) to form the sine wave's *xy* cluster. Note that when this cluster is wired to the XY Graph's icon terminal, the icon's border changes from brown (array of clusters) to pink (cluster). This transformation reflects the fact that your request for single waveform plotting requires the input of only a single cluster, rather than the array of clusters necessary for the more complex multiwaveform plots. The **Time** and **Displacement** array indicators can be created by popping up on the appropriate wires and selecting **Create>>Indicator**.

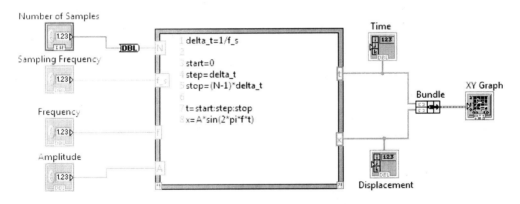

## 3.5 RUNNING THE VI

Return to the front panel, position the **Time** and **Displacement** indicators nicely and then save your work. Set **Number of Samples** and **Sampling Frequency** equal to *100* and *1000*, respectively, and **Frequency** and **Amplitude** to *50* and *5*, respectively. These choices for the parameters simulate sampling a 50-Hz sine wave for 0.1 seconds (100 samples at a rate of 1000 samples per second), and so should result in outputting five cycles of the sine wave. Run your program, and if all goes well, you will see the five cycles. Marvel at the waveform you have created and reflect on how easy it would be to create other shapes by simply changing the equations within the block diagram's Mathscript Node (which is exactly what you will do in a few moments).

## 3.6 MATHSCRIPT INTERACTIVE WINDOW

An invaluable tool while developing Mathscript-based code is the *Mathscript Interactive Window*. Let's take a moment to learn how to use this tool. With a VI open or else in the **Getting Started** window, open the Mathscript Interactive Window by selecting **Tools>>Mathscript Window**...When the window appears, concentrate your attention on its left side. In the lower left you will find the *Command Window* box where Mathscript commands can be entered using the keyboard. Once a command is loaded into the Command Window, it can be executed by pressing *<Enter>* on your keyboard, and the result of this executed command is displayed directly above in the *Output Window*.

Let's test drive the Mathscript Interactive Window by executing commands similar to the ones just used in constructing **Waveform Simulator**. In the Command Window, type *t=0:0.01:0.99*, then press *<Enter>*. In the Output Window, you will first see a copy of the command that you typed, followed by the result of its execution, which in this case is an array of 100 equally spaced *t*-values ranging from $t = 0$ to $t = 0.99$.

Next, in the Command Window, enter (i.e., type the command then press <*Enter*>) *x*=5**sin(2**pi**4**t)*. In the Output Window you will find the sine wave with frequency 4 and amplitude 5 sampled at the 100 equally spaced *t*-values created by the previous command.

Finally, to observe this sine wave, enter *plot(t,x)* in the Command Window. A window with a plot of *x* versus *t* will then appear as shown below, displaying four cycles of the given sine wave.

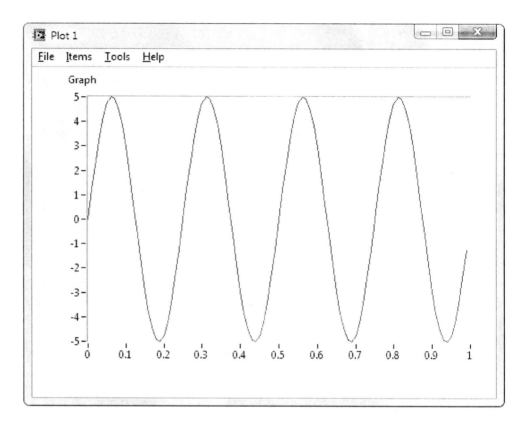

Next, consider a slightly more complicated task: Evaluate the expression $x = t \sin(t)$ at the three times $t = 1, 2, 3$. On the block diagram of **Waveform Simulator**, we multiplied a sine function by a scalar $A$; here, we are multiplying the sine function by a variable $t$. The expected three $x$-values, of course, are $1 \times \sin(1) = 0.841$, $2 \times \sin(2) = 1.819$, and $3 \times \sin(3) = 0.423$, where the argument of the sine function is assumed to be in radians. From the programming we just completed, it would seem that the correct commands to execute this task are:

$$t = 1:1:3$$
$$x = t * \sin(t)$$

In the Command Window, enter $t = 1:1:3$, followed by $x = t*sin(t)$. In the Output Window, somewhat surprisingly, you will receive an error message related to the incompatibility of the matrices involved in the request operation. To understand the problem here, we must delve a bit deeper into the syntax of the Mathscript language. In executing the commands you input, Mathscript produces the following two row vectors:

$$t = [1 \quad 2 \quad 3]$$

$$\sin(t) = [\sin(1) \quad \sin(2) \quad \sin(3)] = [0.841 \quad 0.909 \quad 0.141]$$

Then, in the syntax of Mathscript, the command $t*\sin(t)$ requests that these two row vectors be multiplied together as matrices. That is, the command is interpreted as the following matrix multiplication operation:

$$[1 \quad 2 \quad 3] \times [0.841 \quad 0.909 \quad 0.141]$$

As you know, multiplication of two $1 \times 3$ matrices is an invalid operation, hence the error message in the Output Window. To demonstrate a valid matrix operation (which is not the task we are trying to accomplish), try entering the command $x=t*transpose(sin(t))$. The operation executed will then be the following valid matrix multiplication:

$$x = [1 \ 2 \ 3] \times \begin{bmatrix} 0.841 \\ 0.909 \\ 0.141 \end{bmatrix} = (1)(0.841) + (2)(0.909) + (3)(0.141) = 3.08$$

Instead of the above matrix multiplication operation, we want Mathscript to take the 3-element row vectors $t$ and $\sin(t)$ and form a 3-element row vector, whose $i^{th}$ element is the product of the $i^{th}$ elements of original row vectors. This operation is called *element-wise multiplication* and its Mathscript operator symbol is a period followed by an asterisk (.*). In the Command Window, enter $t=1{:}1{:}3$, followed by $x=t.*sin(t)$. Remember to include the all-important period before the asterisk. You will find that the commands accomplish our originally stated task of evaluating $x = t\sin(t)$. Similarly, element-wise division and power operations can be executed by the operators ./ and .^, respectively.

The Mathscript Interactive Window is also a handy source of help regarding use of Mathscript functions. These (over 500) functions are cataloged into roughly 30 "classes" (e.g., basic mathematical functions, statistics, digital signal processing). To see a list of all of the classes, enter *help classes* in the Command Window. To then see a list of all of the functions in, say, the *dsp (digital signal processing)* class, enter *help dsp* in the Command Window. There you will find the available dsp functions, which include many that perform complex filtering, waveform generation, and spectral analysis operations.

To get some experience with a Mathscript function, let's try using Periodic Signal Generator, which is called *gensig* for short. Enter *help gensig* in the Command Window. The help information that appears states that the gensig command generates two arrays called $x$ and $t$, where $t$ is an array of time values starting at $t = 0$ and stopping at $t = duration$ with the time step $\Delta t = interval$, and $x$ is an array of associated periodic signal values that has the functional form and period given by the parameters *type* and *period*, respectively. The possible values for *type* are *'sin'*, *'square'*, and *'pulse'*. The syntax of the command is

$$[x, t] = gensig(type, period, duration, interval)$$

OK, let's try using *gensig* to generate the same signal that we produced a few moments ago, which was four cycles of a sine wave with frequency and amplitude of 4 and 5, respectively. For a sine-wave frequency of 4, the *period* is $1/4 = 0.25$. For the time variable, we want the *duration* and *interval* to be 0.99 and 0.01, respectively. So, in the Command Window, enter *[x, t] = gensig('sin', 0.25, 0.99, 0.01)*, and then you will find both the *x* and *t* arrays listed in the Output Window. To see a plot of these variables, enter *plot(t,x)* in the Command Window. In the plot window that appears, you should see the expected signal, except the amplitude is 1, rather than the desired 5. To remedy this problem, enter *x=5\*x* in the Command Window, followed by *plot(t,x)*.

As a demonstration of the power of the *gensig* command, produce a square wave by simply changing the command entered in the Command Window to *[x, t] = gensig('square', 0.25, 0.99, 0.01)*. Next, produce a periodic pulse with the command *[x, t] = gensig('pulse', 0.25, 0.99, 0.01)*.

Another useful Mathscript function is *linspace(a,b,N)*, which generates *N* equally spaced samples between the lower value *a* and upper value *b*. Show that the command *t=linspace(0, 0.99, 100)* generates the same array of values as the command *t=0:0.01:0.99*. Also, show that a 100-element array, whose elements all equal 5, can be generated by *t=linspace(5, 5, 100)*. Such an array could describe, for example, the DC voltage output of a battery at 100 equally spaced times. In the program you will soon write, we will call such a constant-valued array a *DC level* waveform.

## 3.7 ADDING SHAPE OPTIONS TO WAVEFORM SIMULATOR

We are now positioned to turn **Waveform Simulator** into a versatile data simulator VI that can output a variety of digitized waveforms. In particular, we will program this VI to create five possible waveform shapes—*Sine, Cosine, DC Level, Square,* and *User-Defined*. So that a user can select the desired waveform, let's add a fifth control called **Shape** to the front panel of **Waveform Simulator**.

## 3.8 THE ENUMERATED TYPE CONTROL

For **Shape**, we will use an *Enumerated Type Control* (nicknamed an *Enum*). Go to **Controls>>Modern>>Ring & Enum**, select an **Enum**, then place it on the front panel and label it **Shape**. Use the ⯑ to perform any necessary resizing and/or positioning.

In **Controls>>Modern>>Ring & Emun**, you will find several types of "*ring*" style controls, each of which is useful when a programmer needs to present the user with a catalog of options. The **Enum** is a special type of ring control that links a sequence of text messages with a series of associated integers. By default, the data type of these integers is unsigned 16-bit (**U16**). When the user selects a particular text message on the front panel control, its associated integer is passed to a terminal on the block diagram. In our present circumstance, the **Enum** control will provide a handy method for selecting a desired waveform shape from among the available options.

The Enum is programmed with its available options as follows. Pop up on the **Enum** and select **Edit Items**... In the dialog box that appears, type *Sine <Enter> Cosine <Enter> DC Level <Enter> Square <Enter> User-Defined*, then click the **OK** button. If you press *<Enter>* after typing *User-Defined*, you will accidentally create a sixth option, which be erased by clicking on the **Delete** button.

On the front panel, use the 🖑 to view the text message (Sine, Cosine, DC Level, Square, User-Defined) for each of the Enum's five programmed options. You can resize the Enum using the ⬉. To make visible the integer associated with each of these text messages, pop up on the **Enum** and select **Visible>>Digital Display**. After viewing the Enum's Digital Display, hide it again (by selecting **Visible>>Digital Display**) and then save your work.

## 3.9 FINISHING THE BLOCK DIAGRAM

Switch to the block diagram and there you will find the **Shape** icon terminal, which is blue, denoting the (**U16**) integer data type emanating from this Enum control. Pop up on the border of the Mathscript Node, select **Add Input** and label it **s**, then complete the code as shown below.

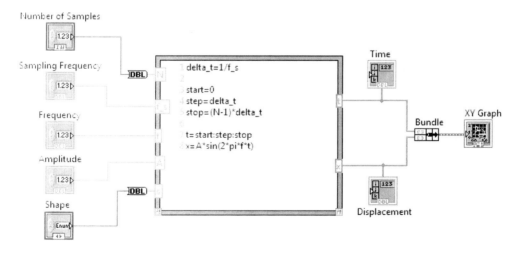

Finally, program the Mathscript Node to produce the desired waveform alternatives. To accommodate the five possible options, we will implement Mathscript's *if-else statement* syntax. For this type of statement, a logical condition is tested (in our situation, the equivalence of *s* to, say, zero via the command $s == 0$; note that two equal signs denote a test of logical equivalence whose outcome is either a Boolean TRUE or FALSE), and if the condition is TRUE, a given command (or list of commands) is executed. The code we require is as follows. For the User-Defined option, we have chosen the sum of two particular sinusoids as its default. A user can then modify this choice to produce any desired waveform shape.

$delta\_t = 1/f\_s$

$start = 0$
$step = delta\_t$
$stop = (N-1) * delta\_t$
$t = start : step : stop$

$if\ s == 0$
   $x = A * \sin(2 * pi * f * t)$
$elseif\ s == 1$
   $x = A * \cos(2 * pi * f * t)$
$elseif\ s == 2$
   $x = linspace(A,\ A,\ N)$
$elseif\ s == 3$
   $[x,\ t] = gensig\ ('square',\ 1/f,\ stop,\ step)$
   $x = A * x$
$elseif\ s == 4$
   $x = 4.0 * \sin(2 * pi * 100 * t) + 6.0 * \cos(2 * pi * 200 * t)$
$end$

Program the Mathscript Node with the above code.

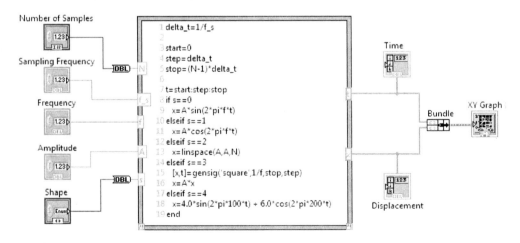

## 3.10 RUNNING THE VI

Now return to the front panel and save your work. Set **Number of Samples** and **Sampling Frequency** as *100* and *1000*, respectively, and **Frequency**, **Amplitude**, and **Shape** as *100*, *5*, and *Sine*, respectively. Run your program, assuming that the **Time**-axis is calibrated in seconds. Then, with this choice of parameters, neighboring data samples are separated by a time interval of 0.001 second, and we are simulating an experiment that acquires 100 data samples over the time span from $t = 0.000$ to $t = 0.099$ second. Thus, for a 100 Hz sine-wave signal whose period is 0.01 seconds, we expect to observe ten of its cycles on the XY Graph. Each cycle is described by a succession of ten samples, as can be seen by popping up on the **Plot Legend** and selecting **Point Style>>Solid Dot**.

Next, without changing **Frequency**, **Amplitude**, and **Shape**, set **Number of Samples** and **Sampling Frequency** to *100* and *10000*, respectively, and then run your program. With this choice of parameters, neighboring data samples are separated by a time interval of 0.0001 second, so now the 100 samples are acquired over the time span from $t = 0.0000$ to $t = 0.0099$ second. Thus, for the 100 Hz sine-wave signal with period of 0.01 seconds, we expect to observe only one of its cycles on the XY Graph, where this single cycle is well-resolved because it is described by a succession of 100 samples.

Finally, run your VI with the **Shape** control set to some of its other available options and enjoy the results.

## 3.11 CONTROL AND INDICATOR CLUSTERS

Let's take this opportunity to add to your bag of LabVIEW programming tricks. We will explore a "clustering" technique that improves the user-friendliness of a VI, both for a user operating its front panel as well as a programmer (including yourself) trying to decipher its block diagram.

We first note that the Numeric Controls on our VI's front panel can be organized into two categories: *Digitizing Parameters* (**Number of Samples, Sampling Frequency**) and *Waveform Parameters* (**Frequency, Amplitude, Shape**). Likewise, both **Time** and **Displacement** Array Indicators are related to the category of *Waveform Output*. This organizational structure within your program can be made explicit by grouping these controls and indicators on the front panel and block diagram. This feat will be accomplished by creating a cluster associated with each of the three categories through the following steps. In **Controls>>Modern>>Array, Matrix & Cluster**, obtain a **Cluster** shell and place it in an empty region of the front panel near the Numeric Controls. Label this **Cluster** shell **Digitizing Parameters**. Similarly, place another **Cluster** shell near the Numeric Controls and label it **Waveform Parameters**. Finally, place a third **Cluster** shell near the Array Indicators and label it **Waveform Output**.

Now, using the , drag the **Number of Samples** and **Sampling Frequency** controls (in that order) into the interior of the Cluster shell labeled **Digitizing Parameters**. If needed, the Cluster shell can be resized at its edges using the ↖, or (more simply) by popping up on its border and selecting **AutoSizing>>Size to Fit** in the pop-up menu (if you pop up within the interior of the Cluster shell, a Controls palette will appear, rather than the desired pop-up menu).

Likewise, drag the **Frequency** control, then the **Amplitude** control, and then the **Shape** control into the interior of the **Waveform Parameters** Cluster shell, and the **Time** and **Displacement** indicators (in that order) into the **Waveform Output** Cluster shell.

Finally, using the ↖, arrange the front-panel objects in some pleasing pattern.

The controls or indicators, which we will generically call *elements*, within a cluster are ordered. This ordering is not based on the position of the elements within the Cluster shell, but initially is determined by the order in which the elements were placed in the Cluster shell during programming. The first element placed in the Cluster shell is indexed zero, the second element is indexed one, and so on. Thus, for example, if you followed the directions above, the **Number of Samples** and **Sampling Frequency** controls within the **Digitizing Parameters** control cluster are indexed 0 and 1, respectively. To verify this ordering, pop up on the border of the **Digitizing Parameters** control cluster and select **Reorder Controls In Cluster**...

The front panel will change its appearance as shown next.

At the bottom right corner of each cluster element, its current *cluster index* is given in the white box. The mouse cursor has morphed into the *cluster order cursor* ⬚, which can be used to change the indexing of the cluster elements by clicking on their associated black boxes. Once the new index assignments have been made, they can be secured by clicking on the **Confirm** button ✓ or reverted to the original settings with the **Cancel** button ✗. You might wish to experiment with this tool for a few moments. Then, after assigning **Number of Samples** and **Sampling Frequency** as the index 0 and 1 elements, respectively, click on the **Confirm** button.

Switch to the block diagram and select **Edit>>Remove Broken Wires**. From **Functions>>Programming>> Cluster, Class, & Variant**, select **Unbundle By Name** ➡ and place it near **Digitizing Parameters** icon terminal. Wire this icon terminal to the left-hand input terminal of the ➡ and when the cluster-wire connection is made, the **Unbundle By Name** icon will morph so that its output (on its right side) contains the name of, and gives access to, the index-zero element in the cluster ➡ Number of Samples. Using the ⬚, resize (at its bottom-center) **Unbundle By Name** so that it has two outputs,

➡ Number of Samples
  Sampling Frequency

thus giving access to both elements . The particular element appearing in a given output terminal can be changed either by clicking on it with the ⬚, or by popping up on it and choosing **Select Item**. The output terminal ordering you select for **Unbundle By Name** does not have to be the same ordering as the cluster wired to its input: For example, the top output terminal does not have to correspond to the input cluster's index-zero element. Complete the wiring as shown below.

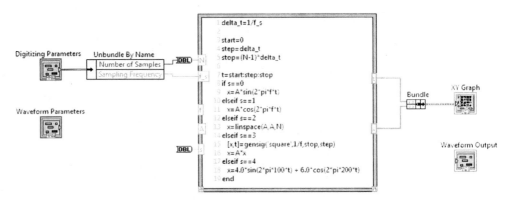

In a similar fashion, use an **Unbundle By Name** to wire the elements within the **Waveform Parameters** cluster to the *f*, *A*, and *s* inputs of the Mathscript Node. Finally, since the XY Graph requires a cluster to be formed containing the *t* and *x* arrays, wire the output of the already present **Bundle** icon to the **Waveform Output** cluster's icon terminal.

As you can see, clustering enhances the readability of a block diagram, especially due to the labeling which appears within the **Unbundle By Name** icons.

Return to the front panel. Run your VI to check that it functions as expected, then save your work.

Finally, save the customized front-panel objects that you have created so that you can reuse them in future programs. To save the **Digitizing Parameters** control cluster, pop up on the border of this cluster and select **Advanced>>Customize** . . . in the pop-up menu.

A *Control Editor* window, which looks similar to the front panel of a VI, will open with your custom-made **Digitizing Parameters** control cluster on it.

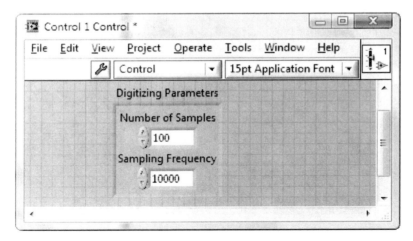

After making sure that the **Control** option (as opposed to **Type Def.** and **Strict Type Def.**) is selected in the toolbar's **Control Type** pull-down menu, use **File>>Save**

to first create a folder named **Controls** within the **YourName** folder, then save your control cluster under the name **Digitizing Parameters** in **YourName\Controls**. The extension **.ctl** will be automatically appended to the control's name. In a similar manner, save your **Waveform Parameters** control cluster and **Waveform Output** indicator cluster under the names **Waveform Parameters** and **Waveform Output** in **YourName\Controls**.

## 3.12 CREATING AN ICON USING THE ICON EDITOR

A well-written LabVIEW program is hierarchical in nature. It consists of a top-level VI whose front panel accepts inputs and displays outputs, while its block diagram is constructed from lower-level *subVIs*. These subVIs, which are analogous to a subroutine in a text-based language such as C, may call even lower-level subVIs, which in turn may call still lower-level subVIs. Just as there is no limit to the level of subroutine layering in a C program, there is no limit to the layers of subVIs used in a LabVIEW program. This modular approach to programming makes programs easy to read and debug.

Your program's subVIs may either be taken from LabVIEW's extensive libraries of built-in icons (found in the Functions Palette) or be custom written by you. It is this latter point that we now wish to focus our attention upon. The importance of what you are about to learn is this: Any VI that you write can then be used as a subVI in the block diagram of a higher-level VI.

To use a program as a subVI, it must have an icon to represent it in the block diagram of the higher-level ("calling") VI. There are two steps in creating this icon: designing its appearance and assigning its connectors. Let's learn these skills by creating an icon for our **Waveform Simulator** program.

## 3.13 ICON DESIGN

First, here is the procedure for designing the icon's appearance. Position the mouse cursor (it doesn't matter which tool it is manifesting at the moment) over the *icon pane* in the upper right-hand corner of the front panel.

Pop up on the icon pane and select **Edit Icon**...from the pop-up menu.

The *Icon Editor* window, shown in the following illustration, will appear.

Within this window, you will find a default picture of the icon and a palette of tools that may be used to redesign its appearance. The tools in the palette, some which may be familiar to you from computer-based drawing programs, have the following functions:

| | | |
|---|---|---|
| Pencil | | Draws and erases pixel by pixel. |
| Line | | Draws straight lines. Press <*Shift*> to restrict drawing to horizontal, vertical, and diagonal lines. |
| Dropper | | Copies foreground color from an element in the icon. Press <*Shift*> to copy background color. |
| Fill Bucket | | Fills an outlined area with the foreground color. |
| Rectangle | | Draws a rectangular border in the foreground color. Double-click on this tool to frame the icon in the foreground color. |
| Filled Rectangle | | Draws a rectangle bordered with the foreground color and filled with the background color. Double-click to frame the icon in the foreground color and fill it with the background color. |
| Select | | Selects an area of the icon for moving, cloning, deleting, or other changes. |
| Text | | Enters text into the icon design. Double-click to access font selection dialog window. |
| Color | | Displays the current foreground and background colors. Click on each to get a palette from which to choose new colors. |

You may wish to explore the use of these tools and create a sophisticated design for your icon (within the constraint of its $32 \times 32$ pixel size). The following step-by-step description will result in an icon with only rudimentary features.

Place the mouse cursor over the **Filled Rectangle** tool, and then double-click.

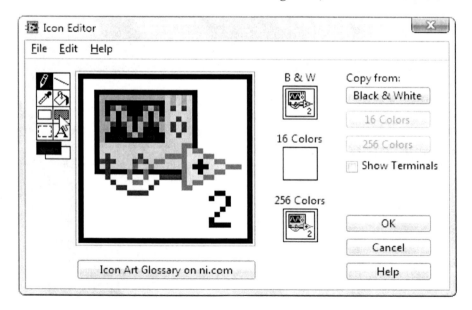

This action will frame the icon in the foreground color (default value is black) and fill it with the background color (white by default). You now have a framed blank canvas on which to create your icon design.

Next, use the **Text** tool to enter *Wave Sim* in the icon's interior. Then click on the **OK** button to save the icon design.

The Icon Editor window will close, returning you to **Waveform Simulator**'s front panel, with your new icon design now in the icon pane.

## 3.14 CONNECTOR ASSIGNMENT

Second, here's how to assign the icon's connectors. Pop up on the icon pane and select **Show Connector**.

A connector will now appear in the icon pane consisting of a pattern of rectangular boxes. By default, LabVIEW chooses the $4 \times 2 \times 2 \times 4$ pattern of boxes shown in the icon pane below. We now wish to associate each of our program's inputs and outputs to a particular box in this pattern.

When we use **Waveform Simulator** as a subVI in the future, two quantities will be passed to it—the **Digitizing Parameters** and **Waveform Parameters** control clusters—and one quantity will be taken from it—the **Waveform Output** indicator cluster. So, with the convention that inputs are on the left and outputs are on the right, we will assign two of the leftmost boxes on the $4 \times 2 \times 2 \times 4$ pattern as the **Digitizing Parameters** and **Waveform Parameters** input terminals and one of the rightmost boxes as the **Waveform Output** output terminal.

Alternately, you may wish to override LabVIEW's default $4 \times 2 \times 2 \times 4$ connector pattern and choose a pattern more appropriate to the present VI, say, a connector with only two inputs and one output. To change the connector pattern, pop up on the icon pane and choose the desired connector from the **Patterns** palette.

If the exact input-output pattern you desire isn't directly on the palette, you can create it by first selecting a related pattern and then applying some operation from the menu such as **Flip Horizontal**. Or, you can create your own pattern of inputs and outputs by selecting **Add Terminal** and/or **Remove Terminal** in the pop-up menu (you might take a few moments to explore the operation of these two commands).

In this book, I will always use LabVIEW's default $4 \times 2 \times 2 \times 4$ connector pattern. Many LabVIEW experts advocate this approach because it promotes uniformity between subVI icons and, as in our present case, leaves some unused terminals that may be handy if the VI is subsequently expanded. Feel free to deviate from this policy if some pattern from the **Patterns** palette better pleases you.

Now assign the **Digitizing Parameters** control cluster to one of the pattern's left-hand terminals through the following steps. Place the mouse cursor over one of the left-hand terminals (the cursor will appear as the Wiring Tool), then click.

The terminal turns dark as shown below.

Click on the **Digitizing Parameters** control cluster. A moving dashed-line marquee will then frame the control and the selected terminal will be colored, indicating the data type of the control to which it has been assigned.

In a similar manner, assign another left-hand terminal to the **Waveform Parameters** control cluster and one of the right-hand terminals to the **Waveform Output** indicator cluster, respectively. The icon pane will then appear as below.

End the assignment procedure by returning the icon design to the icon pane by selecting **Show Icon** from the pop-up menu.

To demonstrate the success of your icon creation, place the mouse cursor over the icon pane and activate the Help Window (via **Help>>Show Context Help**).

Congratulations, you've now completed your first custom-made VI. Using **File>>Save**, store this final version of **Waveform Simulator** in **YourName\Chapter 3**. In later chapters, we'll find out how to use a custom-made VI as a subVI in a higher-level program.

## DO IT YOURSELF

The displacement $x$ of an amplitude-modulated (AM) wave obeys the following relation

$$x = A\,[1+\sin\,(2\pi\,f_{\mathrm{mod}}t)]\,\sin\,(2\pi\,f_{\mathrm{sig}}t) \qquad [4]$$

where $A$ is the amplitude, and $f_{\mathrm{mod}}$ and $f_{\mathrm{sig}}$ are the modulation and signal frequencies, respectively. Write a Mathscript-based program called **AM Wave** that produces and plots an $N$- element array of displacement values for this AM wave. Remember to use Mathscript element-wise operators when appropriate. The front panel and connector assignment for **AM Wave** should appear as shown next.

(Continued)

Run your program with $A = 1$ and some sensible choice of values for $f_{mod}$ and $f_{sig}$. For typical AM wave, $f_{sig} \gg f_{mod}$, so you might try $f_{sig} = 10\, f_{mod}$.

## Problems

1. A unity-amplitude 1-Hz square wave $y(t)$ can be approximated by the following sum of sine waves:

$$y(t) \approx \frac{4}{\pi}\left\{\sin[2\pi t] + \frac{1}{3}\sin[6\pi t] + \frac{1}{5}\sin[10\pi t] + \frac{1}{7}\sin[14\pi t]\right\}$$

[5]

   Write a VI named **Square Wave** that uses a Mathscript Node to evaluate this expression for $y$ in the range $0 \le t \le 3$ seconds, and then plot $y$ versus $t$ on an XY Graph.

2. In polar coordinates $(r, \theta)$, a particular spiral is defined by the relation $r = \theta$, where $0 \le \theta \le 6\pi$. Write a VI called **Spiral**, which uses a Mathscript Node to describe this spiral in terms of Cartesian coordinates $(x, y)$, and then displays the spiral on an XY Graph. You may find the Mathscript function *pol2cart* useful.

3. Open a Mathscript Interactive Window, then type *help roots* in the Command Window. After learning the syntax of the Mathscript command *roots*, use it in the Mathscript Interactive Window to find the two roots of the second-order polynomial $x^2 + x - 1 = 0$. To check that you used *roots* correctly, you can compare your result with that found by using the quadratic formula.

4. Write a Mathscript-based VI called **Noisy Sine**, which generates three cycles of a unity-amplitude 100-Hz sine wave with added random "Gaussian" noise and then plots this waveform on an XY Graph. Calculate $N$ samples of the sine wave at equally spaced times from $t = 0$ to $t = 0.03$ s. Then add a unique "Gaussian" random number to each of these sine-wave samples to create the noisy waveform. A collection of Gaussian random numbers is distributed about zero with probabilities given by a Gaussian ("bell") curve with a standard deviation of $\sigma$. This distribution accurately models the noise in many experimental situations. In your VI, use the Mathscript command *randn(1,N)*, which produce a single row vector of $N$ Gaussian random numbers with $\sigma = 1$.

5. It is easy to show analytically that the integral $\int_0^1 5x^4\, dx$ equals exactly 1. Open a Mathscript Interactive Window, then type *help trapz* in the Command Window. After learning the syntax of the Mathscript command *trapz*, write a VI called **Numerical Integral** that uses *trapz* within a Mathscript Node to calculate the integral $\int_0^1 5x^4\, dx$ numerically and then displays the result in a front-panel indicator with **Digits of precision** equal to 5. In order for your numerical result to be correct to five decimal places (i.e., the first five decimal places are all zero), how many $x$-values must be supplied to *trapz* within the integration range $x = 0$ to $x = 1$?

6. Quite commonly, a projectile lands at a level that is different than that from which it is launched. As shown below, consider a projectile launched with speed $v$ (m/s) at angle

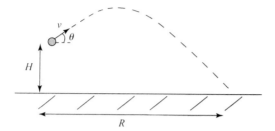

Fig. 3.1  Problem 6.

θ from the horizontal at a height $H$ above the ground and define range $R$ to be the horizontal distance it travels during its flight.

For this situation, it can be shown that

$$R = \frac{v^2}{2g}\left[\sin(2\theta) + \sqrt{\sin^2(2\theta) + \frac{8gH}{v^2}\cos^2\theta}\right]$$

[6]

where θ is positive and negative when above and below the horizontal, respectively.

(a)  Write a Mathscript-based program called **Projectile** that, given $v$ and $H$, calculates $R$ for a sequence of 101 equally spaced angles in the range from $\theta = -90°$ to $\theta = +90°$ and then displays $R$ versus $\theta$ on an XY Graph as well as in an indicator cluster. The front panel of **Projectile** is shown next.

(b)    At a track and field meet, a shot putter launches a shot (heavy metal ball) with speed $v = 13.5$ m/s. Run **Projectile** with this launch speed and assuming the shot lands at the same level from which it was launched. Use the resulting XY Graph and output cluster to demonstrate the well-known fact that the maximum range results if the launch angle equals $+45°$.

(c)    In reality, a shot putter launches the shot from a height close to his or her head. With $v = 13.5$ m/s and $H = 2.1$ m, run **Projectile** and then determine the launch angle that produces the maximum range.

(d)    Finally, if the shot putter launched the shot with $v = 13.5$ m/s from a cliff of height $H = 100$ m, run **Projectile** and then determine the launch angle that produces the maximum range.

7.    Write a VI called **Waveform Simulator (Express)**, which implements one of LabVIEW's easy-to-use Express VIs to generate a waveform. Place a **Waveform Graph** on the front panel, then switch to the block diagram and place **Simulate Signal** there, which is found in **Functions>>Express>>Input**. Immediately after placing this Express VI on the block diagram, its dialog window will open. In this window, choose **Sample per second (Hz)** and **Number of samples** to be *1000* and *100*, respectively, then click the **OK** button near the bottom of the window. When you are returned to the block diagram, complete the code shown below. You will find that a new type of wire called *dynamic data type* emanates from the Express VI's **Sine** output and that the **Waveform Graph** automatically adapts to this format.

(a)    Return to the front panel, and run **Waveform Simulator (Express)** with **Frequency** and **Amplitude** equal to *50* and *5*, respectively. Note that the Waveform Graph's *x*-axis is automatically scaled correctly, a benefit of the dynamic data type wire.

(b)    On the block diagram, pop up on the **Simulate Signal** icon and select **Properties**. The Express VI's dialog will reopen. Select **Signal type>>Sawtooth**, then click the **OK** button. Rerun your VI to view the new waveform shape.

8.    In this problem, you will use the Mathscript command *[p, s] = polyfit(x, y, n)* to fit a given set of data to a second-order polynomial function of the form $y = a_0 + a_1 x + a_2 x^2$, where $a_0$, $a_1$, and $a_2$ are constants. A description of this Mathscript command can be

found by typing *help polyfit* in the Mathscript Interactive Window. Assume that a quantity *y* is measured as a function of another quantity *x*, resulting in the following data set.

| *x* | 1.0 | 2.0 | 3.0 | 4.0 | 5.0 | 6.0 |
|-----|-----|-----|-----|-----|-----|-----|
| *y* | 11.0 | 15.1 | 19.8 | 25.1 | 31.0 | 37.5 |

Construct a VI called **Polynomial Fit (Mathscript)** as follows: Enter the given *y* versus *x* data as row vectors in a Mathscript Node. The *y* row vector, for example, is created with the command $y = [11.0\ 15.1\ 19.8\ 25.1\ 31.0\ 37.5]$. Then use the command $[p, s] = polyfit(x, y, n)$ to fit this given data to a second-order polynomial. The values of $a_2$, $a_1$, and $a_0$ will be given as the first, second, and third elements of the row vector *p*, which can be found by the commands $a2 = p(1)$, $a1 = p(2)$, and $a0 = p(3)$ (the integer in parentheses is the index of the desired element of *p*; unlike LabVIEW arrays, Mathscript indexing begins with *1*, rather than *0*). Display the resulting values of $a_2$, $a_1$, and $a_0$ on the front panel.

# Chapter 4 ▪ ▪ ▪ ▪ ▪ ▪ ▪ ▪ ▪ ▪ ▪ ▪ ▪ ▪ ▪ ▪ ▪ ▪ ▪ ▪ ▪ ▪ ▪

# Data Acquistion using DAQ Assistant

Outfitted with the skills developed in the previous three chapters, you are now equipped to explore one of LabVIEW's most powerful capabilities, computer-controlled data acquisition and generation. In this chapter, you will learn how LabVIEW allows a scientific investigator to reach out through a PC to monitor, as well as create, events in the outside world. In particular, you will carry out this investigation by writing programs based on *DAQ Assistant*, an easy-to-use *Express VI* that controls the operation of a multifunction data acquisition (DAQ) device.

## 4.1 DATA ACQUISITION VIs

LabVIEW's built-in data acquisition functions are designed to operate all of the features on any DAQ device manufactured by National Instruments (NI), the company that also produces LabVIEW. These features include analog-to-digital conversion, digital-to-analog conversion, event timing, pulse counting, and digital input/output operations. Once a NI DAQ device is connected to your computer (e.g., plugged into a PCI expansion slot or attached via a USB cable), LabVIEW offers you both a low-level and high-level approach to control its operation.

In the low-level approach, one develops programs based on the *DAQmx VIs*, which are found in **Functions>>Measure I/O>>NI-DAQmx**. Each of the DAQmx VIs executes a key step (such as setting the sampling rate or storing acquired data in a memory buffer) necessary in performing a complete data acquisition or data generation process. By properly configuring these software building blocks on a block diagram, a LabVIEW programmer can fully control every operation involved in the data acquisition process and so write programs that provide the ultimate in flexibility and power to the user. This control, of course, comes at the price of added complexity to the block-diagram programming.

At the other extreme, LabVIEW offers the programmer a suite of high-level functions that perform common measurement tasks such as communicating with an instrument, storing acquired data in a file, and performing Fast Fourier Transform analysis.

These functions, called *Express VIs*, are found in **Functions>>Express**. Express VIs are straightforward to use and perform many sophisticated tasks automatically, but have limited capabilities. If your programming requirements are simple, you may often find that an Express VI will do your job quite nicely.

For the high-level method of controlling the data acquisition or generation process, LabVIEW offers an Express VI called **DAQ Assistant**. In this chapter, you will explore the fundamentals of controlling a DAQ device by using DAQ Assistant to perform three basic DAQ device operations—analog-to-digital conversion (acquiring a digitized representation of an incoming analog voltage), digital-to-analog conversion (generating an outgoing analog signal from an sequence of digitized numbers), and digital port control (setting and reading the port's HIGH/LOW state). In LabVIEWspeak, these three operations are called *Analog Input (AI)*, *Analog Output (AO)*, and *Digital Input/Output (DIO)*, respectively. Later, in Chapter 11, we will revisit this topic and learn to control a DAQ device using the more advanced DAQmx VIs.

## 4.2 DATA ACQUISITION HARDWARE

National Instruments, based in Austin, Texas, offers a wide variety of computer-based data acquisition devices. The operations these devices are designed to perform are Analog Input, Analog Output, Counter Input (event timing, pulse counting), Counter Output (pulse generation) and Digital Input/Output. Some of the available DAQ devices are quite specialized (for instance, only designed to process digital signals), while others are multipurpose, performing some or all of the above-mentioned operations. There are high-speed devices, which digitize incoming signals at rates over $10^9$ samples per second (S/s), and more garden-variety systems with maximum sampling frequencies on the order of $10^6$ S/s. Digital resolution varies from 8 to 24 bits, and each device connects to a PC through one of about ten possible ways (including the PCI and USB interfaces).

In this chapter, I will use two particular multifunction DAQ devices—the PCI-6251 and the USB-6009—to illustrate the generic features of the particular NI data acquisition product that is present in your system. A brief description of each of these representative devices follows.

**PCI-6251:** This data acquisition board is one of NI's popular multifunction *M Series* DAQ devices. It resides within a computer, plugged into a PCI expansion slot. The 6251 has 16 analog input channels that can perform 16-bit analog-to-digital conversion operations at rates up to $1.25 \times 10^6$ S/s. Also, there are two 16-bit analog output channels, which can update output voltages at a maximum rate of $2.8 \times 10^6$ S/s. Additionally, this DAQ device has 24 digital I/O ports and two 32-bit counters/timers.

**USB-6009:** This low-cost multifunction DAQ device provides basic data acquisition functionality for applications such as simple data logging, portable measurements, and academic lab experiments. It is a small stand-alone unit that connects to your computer via a USB port. The 6009 has eight analog input channels that can perform 14-bit analog-to-digital conversions operations at rates up to $4.8 \times 10^4$ S/s. Also, there are two 12-bit analog output channels, which can update output voltages at a maximum rate of 150 S/s. Additionally, the device has 12 digital I/O ports and a 32-bit counter (with no time-measurement capabilities).

Most of the exercises in this chapter are designed so that they can be completed by the owner of a USB-6009. In the case of hardware-timed waveform generation, however, a device similar to the PCI-6251 is needed.

A conduit carrying signals to and from the outside world must somehow be connected to a DAQ device. For example, among its various capabilities, a PCI-6251 has

## NI PCI/PCIe/PXI/PXIe-6251

| | | | |
|---|---|---|---|
| AI 0 | 68 | 34 | AI 8 |
| AI GND | 67 | 33 | AI 1 |
| AI 9 | 66 | 32 | AI GND |
| AI 2 | 65 | 31 | AI 10 |
| AI GND | 64 | 30 | AI 3 |
| AI 11 | 63 | 29 | AI GND |
| AI SENSE | 62 | 28 | AI 4 |
| AI 12 | 61 | 27 | AI GND |
| AI 5 | 60 | 26 | AI 13 |
| AI GND | 59 | 25 | AI 6 |
| AI 14 | 58 | 24 | AI GND |
| AI 7 | 57 | 23 | AI 15 |
| AI GND | 56 | 22 | AO 0 |
| AO GND | 55 | 21 | AO 1 |
| AO GND | 54 | 20 | APFI 0 |
| D GND | 53 | 19 | P0.4 |
| P0.0 | 52 | 18 | D GND |
| P0.5 | 51 | 17 | P0.1 |
| D GND | 50 | 16 | P0.6 |
| P0.2 | 49 | 15 | D GND |
| P0.7 | 48 | 14 | +5 V |
| P0.3 | 47 | 13 | D GND |
| PFI 11/P2.3 | 46 | 12 | D GND |
| PFI 10/P2.2 | 45 | 11 | PFI 0/P1.0 |
| D GND | 44 | 10 | PFI 1/P1.1 |
| PFI 2/P1.2 | 43 | 9 | D GND |
| PFI 3/P1.3 | 42 | 8 | +5 V |
| PFI 4/P1.4 | 41 | 7 | D GND |
| PFI 13/P2.5 | 40 | 6 | PFI 5/P1.5 |
| PFI 15/P2.7 | 39 | 5 | PFI 6/P1.6 |
| PFI 7/P1.7 | 38 | 4 | D GND |
| PFI 8/P2.0 | 37 | 3 | PFI 9/P2.1 |
| D GND | 36 | 2 | PFI 12/P2.4 |
| D GND | 35 | 1 | PFI 14/P2.6 |

## NI USB-6009

| | | | |
|---|---|---|---|
| GND | 1 | 17 | P0.0 |
| AI 0/AI 0+ | 2 | 18 | P0.1 |
| AI 4/AI 0− | 3 | 19 | P0.2 |
| GND | 4 | 20 | P0.3 |
| AI 1/AI 1+ | 5 | 21 | P0.4 |
| AI 5/AI 1− | 6 | 22 | P0.5 |
| GND | 7 | 23 | P0.6 |
| AI 2/AI 2+ | 8 | 24 | P0.7 |
| AI 6/AI 2− | 9 | 25 | P1.0 |
| GND | 10 | 26 | P1.1 |
| AI 3/AI 3+ | 11 | 27 | P1.2 |
| AI 7/AI 3− | 12 | 28 | P1.3 |
| GND | 13 | 29 | PFI 0 |
| AO 0 | 14 | 30 | +2.5 V |
| AO 1 | 15 | 31 | +5 V |
| GND | 16 | 32 | GND |

16 analog input channels, each standing ready to digitize an incoming analog signal. So how does one get the meaning-laden analog signals from an experiment to the DAQ unit's channel inputs? Typically, an experimenter connects wires from his or her experiment to appropriate terminals on an *I/O Terminal Block* or *NI ELVIS* (these items may be purchased from National Instruments). An I/O Terminal Block has screw (or BNC) terminals to which the experimental wires can be attached, while NI ELVIS wire connections are made on an electronic breadboard. A cable is then connected from the I/O Terminal Block (or NI ELVIS) to a connector at the end of the DAQ device. This connector has (depending on the particular device) 50, 68, or 100 pins, where each pin is associated with a particular function of the DAQ device. The PCI-6251 employs the above-described connection method. Alternately, some DAQ devices are configured with an on-board terminal block to which wires from an experiment are directly attached. The USB-6009 is often configured in this scheme. Regardless of connection method, the function of each DAQ-device pin can be determined from the device's pinout diagram. The pinout diagrams for the PCI-6251 and USB-6009 are reproduced on p. 127.

## 4.3 ANALOG INPUT MODES

In regard to analog input (AI) operations, the above pinout diagrams can be deciphered by noting that the PCI-6251 has 16 AI channels, while the USB-6009 has eight. Through software settings, these inputs can be configured to operate in two distinct analog input modes: *single-ended* and *differential*.

In the single-ended mode, each available analog input pin is an AI channel, with all of these channels referenced to the same common ground. You may choose this common-ground to be the building ground, termed the *referenced singled-ended (RSE)* mode, or supply your own ground-level at the AI SENSE pin, called the *nonreferenced singled-ended (NRSE)* mode. As an example, when operated in the referenced single-ended input mode, the pin assignments for the AI channels on the DAQ devices cited above are given in Table 4.1.

For the NRSE mode on the PCI-6251, the pin assignment of each channel is the same as given in Table 4.1; however, a ground level must be applied to pin 62 (AI SENSE).

**Table 4.1**    Channel Assignments for Referenced Single-Ended (RSE) Input Mode

| Channel | 0 | 1 | 2 | 3 | 4 | 5 | 6 | 7 | 8 | 9 | 10 | 11 | 12 | 13 | 14 | 15 |
|---|---|---|---|---|---|---|---|---|---|---|---|---|---|---|---|---|
| PCI-6251 pin (GND is pin 67, 32, 64, 29, 27, 59, 24, 56) | 68 | 33 | 65 | 30 | 28 | 60 | 25 | 57 | 34 | 66 | 31 | 63 | 61 | 26 | 58 | 23 |
| USB-6009 pin (GND is pin 1, 4, 7, 10) | 2 | 5 | 8 | 11 | 3 | 6 | 9 | 12 | | | | | | | | |

The NRSE mode is not available on the USB-6009, hence the absence of an AI SENSE pin on its pinout diagram.

In differential input mode, available analog-input pins are paired to form eight (PCI-6251) or four (USB-6009) independent AI channels, each of which is sensitive only to the voltage difference between paired pins. On the PCI-6251, differential channel 0 is AI 0 paired with AI 8, differential channel 1 is AI 1 paired with AI 9, and so on. In this "differential amplifier" configuration, noise-pickup by the AI channels is suppressed. Thus, if it is not a problem in your situation to halve the number of available measurement channels, the differential input mode is the preferred *modus operandi*. Again, as an example, the pin assignments for the differential input mode AI channels for our representative boards are given in Table 4.2.

## 4.4 RANGE AND RESOLUTION

When measuring analog input signals with a DAQ device, there are several issues to keep in mind. First, a DAQ device can only measure voltages that fall within an allowed *range*. An accepted range extends from a minimum voltage $V_{min}$ to a maximum voltage $V_{max}$ and a NI DAQ device typically offers the choice of several ranges (such as 0V to +10V, −10V to +10V, and −5V to +5V) with the choice being made by a software setting. Check the spec sheet for your particular DAQ device to find the available ranges. Next, the DAQ device's analog-to-digital converter represents the analog voltage level being sampled with a binary number of $n$-bits, which places a limit on the sampling *resolution*, i.e., the smallest detectable voltage difference $\Delta V$. That is, the converter divides the measurable voltage

**Table 4.2**   Channel Assignments for Differential Input Mode

| INTERFACE BOARD | CHANNEL NUMBER | POSITIVE INPUT PIN | NEGATIVE INPUT PIN |
|---|---|---|---|
| PCI-6251 | 0 | 68 | 34 |
| | 1 | 33 | 66 |
| | 2 | 65 | 31 |
| | 3 | 30 | 63 |
| | 4 | 28 | 61 |
| | 5 | 60 | 26 |
| | 6 | 25 | 58 |
| | 7 | 57 | 23 |
| USB-6009 | 0 | 2 | 3 |
| | 1 | 5 | 6 |
| | 2 | 8 | 9 |
| | 3 | 11 | 12 |

span $V_{span} \equiv V_{max} - V_{min}$ ($V_{span}$ = 20 Volts when the measurable range is from $V_{min} = -10$ V to $V_{max} = +10$ V) into $2^n$ divisions, and so the resulting resolution of the original analog input signal will be

$$\Delta V = \frac{V_{span}}{2^n} \qquad [1]$$

For the typical value of $V_{span}$ = 20 Volts, the resolution provided by a 14-bit (USB-6009) and 16-bit (PCI-6251) DAQ device is 1 mV and 0.3 mV, respectively.

## 4.5 SAMPLING FREQUENCY AND THE ALIASING EFFECT

Another important issue peculiar to digitized data is related to the sampling frequency $f_s$, where $f_s$ is the rate at which analog-to-digital conversions take place. The sampling frequency places an upper limit on the range of frequencies allowed within the original analog signal, if the digitizing process is to result in a faithful representation of the input. This upper threshold is called the *Nyquist frequency* $f_{nyquist}$, which we will show equals one-half of the sampling frequency, that is, $f_{nyquist} = f_s/2$. The physical significance of the Nyquist frequency is this: For a given sampling rate $f_s$, an input sine wave of frequency $f_{nyquist}$ will be sampled in the minimal manner necessary to represent the sinusoid's peaks and valleys—just twice per cycle. If the input frequency exceeds this threshold, the sampling rate becomes insufficient and the analog-to-digital conversion process becomes inaccurate in a way described next.

Fairly often, the bandwidth limitation placed on a "to-be-digitized" input analog signal does not cause you—the experimentalist—much of a problem. For example, if dealing with audio-produced electrical signals, you might know on physical grounds that the frequency content of the analog input is bracketed by 20 and 20,000 Hz. Then, with a sampling rate of at least 40 kHz, this input can be properly acquired. Alternately, an analog signal may have passed through an amplifier, which behaves as a low-pass filter due to its finite bandwidth response. In this case, the sampling rate must be twice the maximum frequency passed by the amplifier. If no natural frequency bracketing exists in your experiment, then you must impose a high-frequency cutoff by placing a low-pass filter in your data-gathering circuitry. Given the available sampling frequency $f_s$, the filter's components then are chosen so that frequencies higher than $f_s/2$ are not passed.

What happens if a frequency higher than the Nyquist limit accidentally is input to your digitizer? Something much worse than you might expect. In a process called *aliasing*, that too-high frequency appears falsely as a lower-than-Nyquist frequency when processed by the digitizer. This phenomenon, which is peculiar to discrete sampling, is illustrated here.

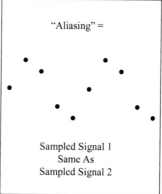

**Fig. 4.1** **Alaising Effect when signal with frequency greater than the Nyquist frequency is digitally sampled.**

Quantitatively, assume a sine wave with (high) frequency $f$ is digitally sampled at $N$ discrete times $t_i = i\Delta t$, where $i = 0,1,2,\ldots, N\!-\!1$. Then, the sampling frequency and Nyquist frequency are $f_s = 1/\Delta t$ and $f_{nyquist} = f_s/2$, respectively. If $f > f_{nyquist}$, it can be shown that at every digitizing times $t_i$, the displacement of the sine wave $\sin(2\pi ft_i)$ is equal to plus-or-minus the displacement of a sine wave with the "alias frequency" $f_{alias}$, where $0 \le f_{alias} \le f_{nyquist}$. By applying the condition $\sin(2\pi ft_i) = \pm\sin(2\pi f_{alias}t_i)$ at every $t_i$ (see Problem 1), one finds that

$$f_{alias} = |f - nf_s| \quad \text{(Aliasing Condition)} \qquad [2]$$

where $n = 1,2,3\ldots$

Demonstrate the aliasing effect by opening **Waveform Simulator** (located in **YourName\ Chapter 3**) and programming it to produce a 100 Hz sine wave of amplitude 5. That is, in the **Waveform Parameters** control cluster, set **Frequency Amplitude**, and **Type** equal to *100*, *5*, and *Sine*, respectively. Then, in the **Digitizing Parameters** control cluster, taking **Number of Samples** and **Sampling Frequency** to be *20* and *2000*, respectively, run this VI to demonstrate what a digitized sinusoid with frequency less than the Nyquist frequency (in this case, $f_{nyquist}$ = 1000 Hz) looks like. Note that one cycle occurs over the course of 0.01 second, so the frequency of the wave is 1/0.01 s = 100 Hz, as expected. Now, with the **Digitizing Parameters** unchanged, program **Waveform Simulator** to produce a 2100 Hz sine wave, and then rerun the VI. You will find that, when digitized, this "higher-than-Nyquist" frequency wrongly appears as a 100 Hz sine wave with period of 0.01 second, consistent with the prediction of Equation [2] with $n = 1$, that is, $f_{alias} = |2100\,\text{Hz} - (1)(2000\,\text{Hz})| = 100\,\text{Hz}$. Predict what will happen if you program **Waveform Simulator** with some other "higher-than-Nyquist" frequencies such as 1800 Hz and 4100 Hz, then see if your expectations are realized via use of **Waveform Simulator**.

## 4.6 MEASUREMENT & AUTOMATION EXPLORER (MAX)

To carry out the exercises in this chapter, you must have a National Instruments data acquisition device connected to your computer, and its driver software (called *NI-DAQmx*) must be correctly installed. To verify that these conditions are met, we will use the handy utility *Measurement & Automation Explorer*, which is nicknamed *MAX*.

To open MAX, either select **Tools>>Measurement & Automation Explorer...**(if you have an open VI or **Getting Started** window), or else double-click on MAX's desktop icon (if available). After MAX opens, double-click on **Devices and Interfaces** in the **Configuration** box. This action will command MAX to determine all of the data acquisition devices present within your computing system.

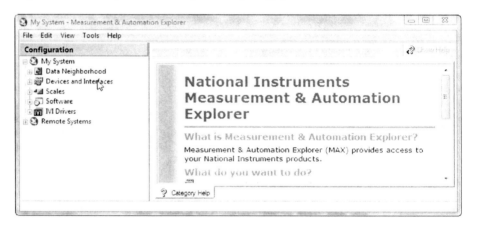

MAX will return the findings of its device survey in hierarchical tree fashion as shown next. We are interested in DAQ devices, so next double-click on **NI-DAQmx Devices**.

All of the DAQ devices that MAX found will then be listed. For the system used below, a *NI PCI-6251* device was found to be connected to the computer. Note that the driver software has (automatically) given this device the shorthand name *Dev1*. The DAQ device on your system may have a different shorthand name. When you go through this procedure and for some reason a device that you believe is connected to your computer does not appear on the **NI-DAQmx Devices** list, try selecting **View>>Refresh** (or press *<F5>* on your keyboard) to command MAX to perform its device survey again. If the device in question doesn't appear then, there is something wrong with the device's connection that must be repaired.

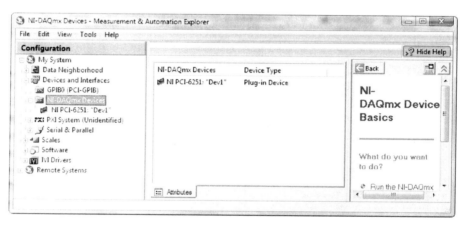

To verify that your DAQ device is properly functioning, pop up on its name and select **Self-Test** from the pop-up menu. Alternately, you can click on the **Self-Test** button near the top of the window.

A brief test of your device's functionalities will be performed, and, ideally, you will receive a dialog-box message stating that all is well.

In a moment, we will program your DAQ device to digitize an analog voltage signal applied at its input. Before we execute this analog input operation, we must secure the connection from the source of analog voltage to the correct pins on your DAQ device. Let's perform this AI operation in differential mode, using differential channel 0, called *ai0* for short. For a DAQ device with a total of eight AI channels, the differential positive (ai0+) and negative (ai0−) channels are pins AI 0 and AI 4, respectively. For a DAQ device with 16 AI channels, ai0+ and ai0− are pins AI 0 and AI 8, respectively. To determine the location of these pins, pop up on the name of your DAQ device and select **Device Pinouts** from the pop-up menu.

A window will appear with the pinout diagram for your device.

Connect a voltage difference of (approximately) +5 Volts to the differential analog input channel 0 of your DAQ device. For example, on a PCI-6251 board in the differential mode, connect +5 Volt to pin 68 and GND to pin 34, while on a USB-6009, attach +5 Volt to pin 2 and GND to pin 3. On an NI ELVIS breadboard, the pins are labeled AI 0+ and AI 0− (or ACH0+ and ACH0−) in the Analog Input Signals section.

Next, pop up on the name of your DAQ device and select **Test Panels...**(or press the **Test Panels...**button). A diagnostic window will appear, which will allow you to test the functioning of your set-up. With the mouse cursor, select the **Analog Input** tab, then enter the settings as shown below. In the **Channel Name** box, the text *Dev1/ai0* selects (differential) AI channel 0 on the DAQ device with shorthand name *Dev1* (on your Test Panel, use the shorthand name of your particular DAQ device, which may be different than *Dev1*).

Then, press the **Start** button. If the connections from the voltage source to your DAQ device are properly made, the digitized value of (approximately) +5 Volts obtained by the DAQ device will appear in the **Value** box. Until you press the **Stop** button, this measurement will be repeated several times per second and the results plotted on the chart.

After pressing the **Stop** button, you may want to take a moment to explore the diagnostic features available under the other tabs (e.g., Analog Output).

## 4.7 SIMPLE ANALOG INPUT OPERATION ON A DC VOLTAGE

Let's turn your computer into a DC voltmeter by building a VI that carries out the same Analog Input operation that we just ran using MAX. For this exercise, keep the 5 V voltage difference connected to differential channel ai0 of your DAQ device.

Construct the front panel shown below in which an acquired voltage is displayed in the **Numeric Indicator** labeled **Voltage**. Through Equation [1], you can determine the resolution $\Delta V$ of the analog-to-digital converter on your DAQ device and use this information to select the proper number of **Digits of precision** on the **Voltage** indicator (using the pop-menu's **Display Format...** option). For example, if $\Delta V = 0.3$ mV, **Digits of precision** should be *4*. Using **File>>Save**, first create a new folder named **Chapter 4** within the **YourName** folder, then save this VI under the name **DC Voltmeter (Express)** in **YourName\Chapter 4**.

Switch to the block diagram. There, place **DAQ Assistant**, found in **Functions>> Express>>Input**.

A moment after you place DAQ Assistant on the block diagram, a dialog window called **Create New Express Task...** will open automatically. DAQ Assistant can be configured to perform any of the data acquisition operations within the capabilities of your DAQ device. In this dialog window, you will select the particular operation that you would like DAQ Assistant to perform for this VI. Since we want to digitize ("acquire") an incoming analog voltage, click on **Acquire Signals**.

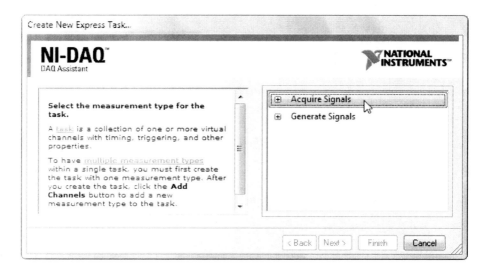

A list of options under **Acquire Signals** will appear in hierarchal tree fashion. Click on **Analog Input**.

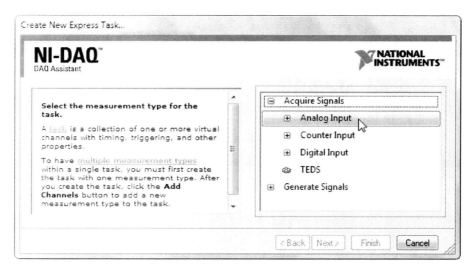

Then, among the measurement types available under **Analog Input**, select **Voltage**.

Finally, the available AI channels on your DAQ device (called *physical channels*) will be listed. From this list, select channel **ai0**. Then click on the **Finish** button. From now on, our shorthand for such a sequence of selections will be **Acquire Signals>>Analog Input>>Voltage>>ai0**.

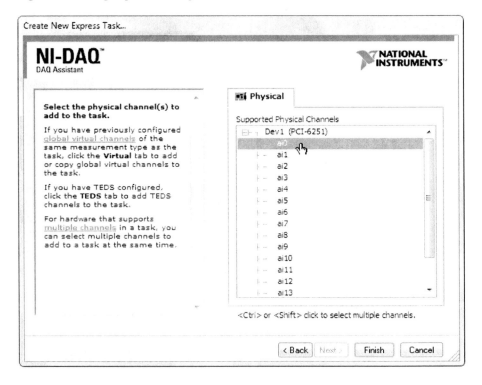

Above, we selected a particular data acquisition operation (analog input voltage measurement) to be performed on a particular physical channel (ai0). This combination of selections becomes part of the definition of a newly created *task*. The **DAQ Assistant** dialog window will now appear automatically. In this window we will configure data-taking details which will complete the task's definition.

The **DAQ Assistant** dialog window opens with the **Express Task** tab selected and the (automatically given default) name of our task—**Voltage**—is highlighted. Under the **Settings** tab, enter the values shown in the following diagram. Under **Timing Settings**, select **Acquisition Mode>> 1 Sample (On Demand)**. In this mode, DAQ Assistant will acquire (only) one voltage reading every time it is executed.

To familiarize yourself with some of the available information within this dialog window, click on the **Connection Diagram** tab. There (after possibly having to input the I/O Terminal Block used by your system) you will find a helpful list and diagram outlining the proper DAQ device pins to use when connecting to differential channel ai0.

Return to the original window by clicking on the **Express Task** tab. Select **Display Type>>Chart**, then click on the **Run** button near the upper left corner of the window. DAQ Assistant will execute repeatedly, acquiring one voltage sample per execution, and the sequence of acquired voltage readings is plotted on the chart. If this plot consists of a sequence of values, each (approximately) equal to +5 Volts, then you have successfully configured DAQ Assistant for our intended task.

Click **OK** near the bottom right corner to finalize this configuration process. When the dialog window disappears, **DAQ Assistant** will appear on your block diagram as an *expandable node*, with inputs and outputs appropriate for the selections that you made within the DAQ Assistant dialog window. In an expandable node, some of its terminals are "unexpanded" unnamed arrows (inward direction for inputs, outward directed for outputs), while other terminals are "expanded" in their own named band near the bottom of the icon. Below, only the **data** output terminal is expanded. This terminal reports the acquired voltage reading each time DAQ Assistant executes.

To expand all of DAQ Assistant's terminals, place the ⬉ at the bottom center of the expandable node. The tool will morph into a resizing handle, which you can click and drag downward until the node has expanded to its full extent.

When you release the mouse button, all of the terminals will appear expanded.

Code the block diagram as shown next. Here, the While Loop iterates once every 0.1 seconds until the **Stop Button** is clicked. With each iteration, DAQ Assistant acquires a single voltage reading from channel ai0, and this value is displayed on the front panel **Voltage** indicator.

There are a few subtle features of this block diagram, which we will pause for a moment to discuss. First, the **Wait (ms)** icon is included within the While Loop so that, when executing, this program does not monopolize your computer's resources. Because DAQ Assistant alone executes very quickly, without the delay produced by Wait (ms), the loop would iterate at such a high rate that your processor would have to devote most of its efforts to running this program, leaving little time for the other operations for which it is responsible.

Second, look closely and you'll find that the wire connecting the **data** output of DAQ Assistant and Voltage's icon terminal is dark blue and banded, a form that we have not encountered up to this point. The format of this wire is called *dynamic data type*, a data type used by Express VIs. In addition to the data associated with a signal (in our present case, a single digitized voltage value), a dynamic data type wire includes attributes of the signal such as its name and time stamp (i.e., the date and time that the data was acquired). Note that, because of the polymorphic nature of a **Numeric Indicator**, we are able to wire the dynamic data type wire directly to Voltage's icon terminal. A red coercion dot appears, though, indicating that only the data value is being passed to the icon terminal (and all additional information about this signal's attributes is being ignored). Later in this chapter, we will see that these attributes can be quite useful when plotting an acquired signal on a Waveform Graph.

Finally, all LabVIEW data acquisition icons (including DAQ Assistant) have an *error cluster* input and output which you can elect to use for runtime error reporting. Include this useful feature in your VI by popping up on the **error out** terminal and selecting **Create>>Indicator**.

Return to the front panel, where you will now find an **error out** cluster included. The error cluster consists of three elements: **status** (Boolean TRUE if there is an error, FALSE if no error), **code** (integer identifying the error), and **source** (descriptive text identifying the function within which the error occurred). When an error does occur, its integer code can be deciphered by popping up on **code** and selecting **Explain Error**.

Use the ⬎ to position all the front-panel objects nicely. Then, with the 5 V voltage difference connected to differential channel ai0, run **DC Voltmeter (Express)**. Does it read the input voltage correctly? Save your work on this VI.

## 4.8 DIGITAL OSCILLOSCOPE

In the previous exercise, you configured DAQ Assistant so that with each execution it acquired a single voltage reading of the analog signal at a channel input. By putting this icon within a repetitive loop and controlling the time per loop iteration with Wait (ms), you wrote **DC Voltmeter (Express)**, a program that, with the simple addition of some data storage and plotting capability, one might consider using to sample and display *N* data samples of a time-varying waveform. The last sentence outlines the schematic for one of the most useful laboratory monitoring systems—the *digital oscilloscope*. Perhaps you've already thought of some cool things to do with this idea and are ready to go back and start to make the needed modifications to **DC Voltmeter (Express)**...but, unfortunately, Wait (ms) is an Achilles' heel in this plan. Wait (ms) measures time by accessing a clock within your computer, which has an accuracy of one millisecond. Thus, by using Wait (ms) to mark time in the data-taking process, the moment of each voltage digitization will be determined with an uncertainty on the order of 0.001 second. For very low-frequency inputs (say, less than 10 Hz), this level of uncertainty in *x*-axis values might possibly be acceptable for the envisioned oscilloscope's *Voltage* versus *Time* output, but it will be useless over the higher range of frequencies that you'd want an oscilloscope to operate.

Thankfully, a much better clock, with precision on the order of microseconds (or, in the case of the USB-6009, 20 $\mu$s), exists on a National Instruments DAQ device. And, even better, LabVIEW provides access and control of this clock through DAQ Assistant. Let's explore this feature of DAQ Assistant as we write a useful digital oscilloscope program. The program we wish to write, called **Digital Oscilloscope (Express)**, will acquire $N$ equally spaced voltage-samples of a time-varying analog input, and then quickly plot this array of data values. By repeating this process over and over, we'll achieve a real-time display of the waveform input.

Open a blank VI and, using **File>>Save**, store it in **YourName\Chapter 4** under the name **Digital Oscilloscope (Express)**. On the front panel of **Digital Oscilloscope (Express)**, place a **Stop Button** and a **Waveform Graph** with its $x$- and $y$-axis labeled **Time** and **Voltage**, respectively. If you'd like, hide the Stop Button's label by toggling it off via **Visible Items>>Label**. Also, in the Controls Palette, choose **Select a Control…**, and then navigate to the **YourName\Controls** folder in the **Look in:** box. There, select a **Digitizing Parameters.ctl** control cluster (by either double-clicking or highlighting, then pressing the **OK** button), then place this control cluster on the front panel. (If you haven't created and saved **Digitizing Parameters.ctl** previously, refer to section 3.11 "**Control and Indicator Clusters**" and create this control cluster now.)

Switch to the block diagram. Place a **While Loop** there and then place a **DAQ Assistant** (from **Functions >>Express>>Input**) icon within the loop. When the **Create New Express Task…** dialog window opens automatically, select **Acquire Signals>>Analog Input>>Voltage>>ai0**. Then, when the **DAQ Assistant** dialog window appears, configure it as shown below. Note that we have set **Acquisition Mode** to **N Samples**, and that in this mode the settings for **Samples to Read** and **Rate (Hz)** are activated. These settings were disabled in the **1 Sample (On Demand)** mode we used for **DC Voltmeter (Express)**.

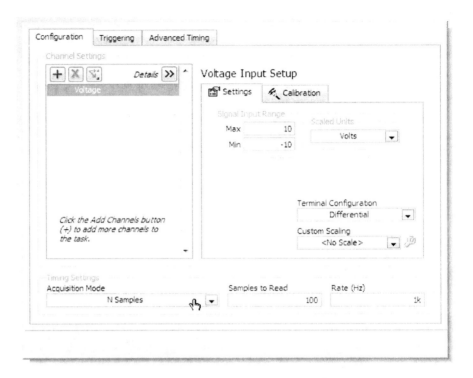

The **N Samples** acquisition mode is a *buffered hardware-timed analog input* operation. In this process, a time-dependent analog signal is digitized at $N$ equally spaced times. The time interval between neighboring samples $\Delta t = 1/f_s$, where $f_s$ is the sampling frequency, is very accurately controlled by a hardware clock on the DAQ device. As the data samples are acquired, they are first placed in a memory buffer on the DAQ device, and then transferred from there into the computer's memory. This method assures that no data samples are lost when the computer multitasks (i.e., switches among its various chores) during the data-taking process. Under the dialog window's **Timing Settings**, **Samples to Read** and **Rate (Hz)** are used to select the values for $N$ and $f_s$, respectively.

Once you have programmed the appropriate values in the DAQ Assistant dialog window, close it by pressing the **OK** button. If later you wish to reopen this dialog window, you can do so by either popping up on **DAQ Assistant** and selecting **Properties** in the pop-up menu, or simply placing the ⬏ over **DAQ Assistant** and double-clicking.

After you are returned to the block diagram, observe all of the available input and output terminals for DAQ Assistant when configured in the **N Sample** mode by resizing the DAQ Assistant icon so that all of its terminals are expanded.

We could now proceed directly to wiring, but some convoluted wire paths would result from the present ordering of the DAQ Assistant terminals. Let's first change the terminal ordering to something more desirable. Pop up on the (top) **data** terminal, and choose **Select Input/Output>>number of samples**. DAQ Assistant will then appear as shown next.

Through a similar procedure, order the terminals as shown below and then resize DAQ Assistant so that **error in**, **task out**, and **timeout (sec)** are no longer expanded.

Finally, complete the block diagram as shown next. When **Unbundle By Name** (found in **Functions>>Programming>>Cluster, Class, & Variant**) is first placed on the block diagram, it will only have one output terminal. Use the ⬧ to make both the **Number of Samples** and **Sampling Frequency** terminals visible, then wire these outputs to DAQ Assistant's **number of samples** and **rate** inputs, respectively. Also, create the **error out** cluster by popping up on DAQ Assistant's **error out** terminal and selecting **Create>>Indicator**. Note that when you connect DAQ Assistant's **data** output to the Waveform Graph's terminal, the resulting wire is of the dynamic data type.

Besides being wired to the While Loop's conditional terminal ◎, the **Stop Button** is wired to DAQ Assistant's **stop (T)** input for the following reason. Within DAQ Assistant several operations are carried out that either start up or shut down the requested data acquisition task. Thus, these particular operations need only be called during the first or last of the multitude of While Loop iterations. However, by default, the **stop (T)** input of DAQ Assistant is TRUE, causing these start-up and shut-down (collectively termed *"overhead"*) operations to be executed each time **DAQ Assistant** executes. Since your data-taking system is not able to digitize the incoming stream of real-time data during overhead operations, needlessly including them increases the program's *dead time*, the percentage of each iteration period during which the system is not sensitive to the incoming data. With increasing dead time, more and more cycles of the incoming repetitive data will stream past your system's input undetected. In certain situations, such as when collecting a large number of data-cycles with the intent of adding them together so as to average out random noise, a programming inefficiency that causes you to miss a significant percentage of the incoming data-cycles can easily extend the completion time of your experiment by minutes (or sometimes even hours). By wiring **stop (T)** to the **Stop Button**, this input will be FALSE during all but the last While Loop iteration. Then, DAQ Assistant will perform its start-up operations only during the first iteration and its shut-down operations during the last, thus enhancing the performance of this program.

Return to the front panel, neaten the arrangement of objects using the , and then save your work.

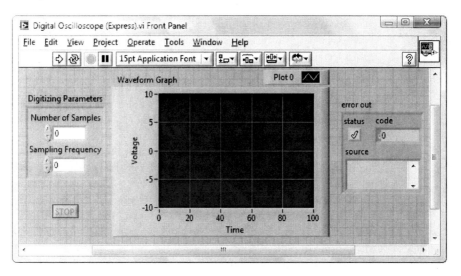

Let's test drive **Digital Oscilloscope (Express)**. Adjust the settings on a function generator so that it is outputting an analog sine wave with amplitude less than 10 V and a frequency of about 100 Hz. Using the appropriate method for your particular system, connect the positive and negative (ground) outputs from the function generator to the channel inputs ai0+ and ai0− on your DAQ device. Then, set **Number of Samples** and **Sampling Frequency** equal to *100* and *1000*, respectively, and run **Digital Oscilloscope (Express)**. If all goes well, you will see a plot of about ten cycles of the 100 Hz sine wave.

Note that, without any coding effort on our part, the **Plot Legend** is labeled with the task name **Voltage** and the *x*-axis is properly scaled to reflect the time increment $\Delta t$ between successive sample acquisitions. These automatic actions are a benefit of the dynamic data type format used to pass the acquired data from DAQ Assistant to the Waveform Graph. Attribute information about this data is included within the dynamic data type wire and is used by the Waveform Graph to accomplish these automatic actions.

Be impressed by the relative ease with which you wrote this program, which acquires and plots a succession of digitized sine waves. However, I'm sure that one flaw in this computer-based instrument is apparent to you. Most likely, the sine wave trace you are observing does not appear stationary but instead is either moving to the right or to the left. And, by slightly changing the function generator's sine-wave frequency, the plotted sine wave can be made to move first one direction, then the other. Why is this so?

In your program, DAQ Assistant is configured so that with each execution, it outputs an array of *N* equally spaced data samples. Let's call this array of *N* samples a *trace*. As you know from coding the block diagram, the sine-wave image viewed on the Waveform Graph is not just a single trace, but is actually a sequence of many traces, where a new trace is generated and plotted with each iteration of the While Loop. Since nothing in your code forces each trace to begin at an equivalent point on the repetitive incoming waveform, the successive traces are displaced from each other along the *x*-axis, resulting in the undesirable situation of a moving image or, worse, a jumbled mess. The issue we are broaching here is called *triggering* and, for **Digital Oscilloscope (Express)** to be a useful program, we must build-in this capability.

In a commercially available oscilloscope, triggering is accomplished in the following manner. The input signal is monitored by an analog *"level-crossing"* circuit. The purpose of this circuit is to determine each time the incoming signal passes through a specified voltage level and then immediately trigger the scope's data acquisition process. By using knobs on the scope's front panel, the oscilloscope user sets the circuit's threshold level and specifies whether acquisition should be initiated when the level is passed through starting from above (*negative*, or *falling*, *slope*) or starting from below (*positive*, or *rising*, *slope*).

Many National Instruments DAQ devices are capable of performing the analog level-crossing triggering procedure just described. For example, the PCI-6251 has this capability; however, the lower-cost USB-6009 does not. So that all readers can include triggering in their **Digital Oscilloscope (Express)** VI, we'll implement an alternate mode—*digital edge triggering*—in our program because this mode is available on all NI DAQ devices.

A digital signal can be in one of two possible states, termed HIGH and LOW. For the digital ports on NI DAQ devices, these states conform to the *transistor-transistor logic (TTL) standard* with HIGH (close to) 5 Volts and LOW (close to) 0 Volts. As a digital signal alternates between its two states, the transition from LOW to HIGH is called the *rising edge*, while the HIGH to LOW transition is the *falling edge*. To execute a data acquisition operation under digital edge triggering, one connects a digital signal to the appropriate pins on a DAQ device. Then, by choosing the proper software settings, the desired data acquisition operation can be begun ("triggered") whenever the digital signal has a rising, or alternately, falling edge.

Stop **Digital Oscilloscope (Express)**, if it's running, then configure it for digital edge triggering as follows. On the block diagram, double-click on **DAQ Assistant**. When the DAQ Assistant dialog window opens, click on the **Triggering** tab, and select **Trigger Type>>Digital Edge** and **Edge>>Rising**. **Trigger Source** will identify the possible DAQ device pins at which the digital signal can be applied; choose one of the available pins. In the window below, pin **PFI0** is chosen on a PCI-6251, which from the pinout diagram for this device corresponds to pin 11. The ground for the digital signal would also need to be connected to the digital GND pin 12. Once you have made your selection, click on the **OK** button.

That's it. **Digital Oscilloscope (Express)** is now configured for digital edge triggering. Return to the front panel and save your work.

Besides producing a sine wave with frequency $f$ at its *main* output, your function generator creates a TTL digital signal of the same frequency $f$, which is available at its *sync* (sometimes called *TTL* or *trigger out*) output. The digital edges of this TTL signal coincide in time with particular points on the sine wave's cycle. For example, on many generator models, the rising edge of the TTL signal occurs when the sine wave is at its peak. Thus, the sync output provides a convenient digital signal with which one can control the acquisition of sine-wave data (i.e., always begin at the same equivalent point on the cycle) via digital edge triggering.

Connect the sync output from your function generator to the digital triggering pins on your DAQ device. For example, sync's positive and negative (ground) terminals would connect to PFI0 (pin 29) and GND (pin 32), respectively, on a USB-6009 device.

On a PCI-6251, sync's positive and negative terminal connect to PFI0 (pin 11) and GND (pin 12), respectively. On an NI ELVIS breadboard, the relevant pins are labeled PFI 0 (or TRIGGER) and GROUND.

On the front panel, set **Number of Samples** and **Sampling Frequency** equal to *100* and *1000*, respectively, and set the sine-wave frequency on your function generator to approximately 100 Hz. Run **Digital Oscilloscope (Express)**. If all goes well, you will see a "stationary" plot of about ten cycles of the 100Hz sine wave.

On the function generator, set the frequency $f$ of the input sine-wave signal to several values within the range $0 \leq f \leq 500$ Hz. You should find that frequencies of signals in this frequency range are faithfully reproduced by **Digital Oscilloscope (Express)**, but that sine-wave shape becomes distorted as the signal frequency approaches 500 Hz (due to the paucity of sampled points per sine-wave cycle at these higher input frequencies).

We have set the sampling frequency equal to $f_s = 1000$ Hz, thus the Nyquist frequency is $f_{nyquist} = f_s/2 = 500$ Hz. For input frequencies greater than the Nyquist frequency, that is, $f > 500$ Hz, we expect the aliasing effect. Demonstrate that for $f = 1100$ Hz, the digitized sine wave appears to have a frequency of 100 (rather than 1100) Hz. Use Equation [2] to predict two other input frequencies $f$ which will appear as a 100 Hz sine wave when digitized due to aliasing, then demonstrate that your predictions are correct using **Digital Oscilloscope (Express)**.

## 4.9 ANALOG OUTPUT

Multifunction DAQ devices manufactured by National Instruments typically have two Analog Output (AO) channels. Each AO channel can perform $n$-bit digital-to-analog conversion operations at speeds up to a *maximum update rate* given in samples per second (S/s). The possible analog voltages produced fall within a range from $V_{min}$ to $V_{max}$. Within this voltage span $V_{span} \equiv V_{max} - V_{min}$, the output analog voltage can be one of $2^n$ possible values. Thus, the resolution $\Delta V$ with which a voltage can be produced (i.e., the smallest difference between one possible output voltage and the next) is $\Delta V = V_{span}/2^n$. As a typical example, if $n = 16$ and $V_{span} = 20$ Volts, $\Delta V = 0.3$ mV. The specifications for the PCI-6251 and USB-6009 are listed in Table 4.3.

The voltage difference associated with Analog Output channel *ao0* is output between pins AO 0 and AO GND, while for channel *ao1* the output is between pins

Table 4.3

| DEVICE | NUMBER OF AO CHANNELS | AO VOLTAGE RANGE (MAXIMUM)* | BITS | UPDATE RATE (MAXIMUM) |
|---|---|---|---|---|
| USB-6009 | 2 | 0 to +5 Volts | 12 | 150 S/s |
| PCI-6251 | 2 | −10 to +10 Volts | 16 | 2.8 MS/s |

* Other ranges are software-selectable.

**Table 4.4**

| BOARD | AO 0 (PIN) | AO 1 (PIN) | AO GND (PIN) |
|---|---|---|---|
| USB-6009 | 14 | 15 | 13, 16 |
| PCI-6251 | 22 | 21 | 54, 55 |

AO 1 and AO GND. On our two representative DAQ devices, the relevant pins are given in Table 4.4.

On an NI ELVIS breadboard, AO 0 and AO 1 are labeled AO 0 and AO 1 (or DAC0 and DAC1), respectively, in the Analog Outputs section. AO GND is labeled GROUND.

## 4.10 DC VOLTAGE SOURCE

To familiarize ourselves with the AO process, let's start by writing a program that simply outputs a requested voltage value. On a blank VI, place a **Numeric Control** labeled **Voltage** and a **Stop Button**. If you'd like, hide the Stop Button's label (toggle it off via **Visible Items>>Label**) and make the Voltage control's **Digits of precision** appropriate for the resolution of your particular DAQ device. Save this VI in the **YourName\Chapter 4** folder under the name **DC Voltage Source (Express)**.

Switch to the block diagram and place **DAQ Assistant** within a While Loop there. When the **Create New Express Task...** window opens, make the following sequence of selections: **Generate Signals>>Analog Output>>Voltage>>ao0**, and then press **Finish**.

When the **DAQ Assistant** dialog window opens, make the selections shown in the following diagram, then click on the **OK** button. Note we are programming $V_{min}$ and $V_{max}$ to be 0 and +5 Volts, respectively (values appropriate for all DAQ devices, including the USB-6009). With the **Generation Mode>>1 Sample (On Demand)** option selected, the DAQ device will update the voltage output at channel ao0 every time that DAQ Assistant is executed in your program. This method of updating the AO channel is called *software timing*.

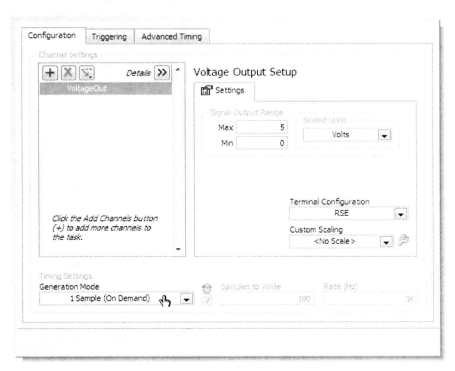

Upon return to the block diagram, use the ↖ to expand all of the terminals of DAQ Assistant when programmed for a **1 Sample (On Demand)** Analog Output operation.

Then, resize DAQ Assistant so that only the **data**, **stop**, and **error out** terminals are expanded, and in the order shown next. Complete the block diagram, remembering the front-panel **error out** cluster indicator can be created by popping up on the **error out** terminal and selecting **Create>>Indicator**. The red coercion dot indicates that the DBL value from **Voltage** is converted to the dynamic data type required at the **data** input to DAQ Assistant. When this code runs, the While Loop will execute once every 10 ms, until the **Stop Button** is clicked. With each iteration, DAQ Assistant will instruct the DAQ device to output a voltage difference between pins AO 0 and AO GND equal to the current value of the **Voltage** control.

Switch to the front panel and arrange the objects as you wish.

Connect the positive and negative (ground) input of a voltmeter (or oscillo-scope in auto triggering mode) to the AO 0 and AO GND pins of your DAQ device, respectively.

Then, set the **Voltage** control to some value within the range 0 to +5 Volts, and press the **Run** button. The voltmeter should inform you that the requested voltage difference is being produced by your DAQ device. With the VI still running, set **Voltage** to a few other values within the 0 to +5 Volt range to convince yourself that the VI is functioning properly.

Next, let's purposely cause a runtime error in this program and see what happens. Remembering that we programmed the DAQ device to output voltages only within the range from 0 to +5 Volt (by choosing $V_{min} = 0$ and $V_{max} = +5$ V in the DAQ Assistant dialog window), set **Voltage** equal to a value outside of this range, say, 6 Volts.

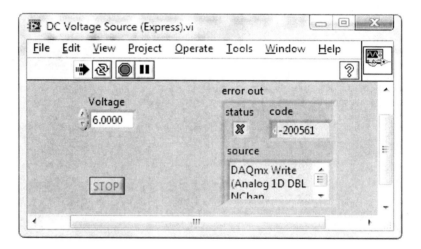

The **error out** cluster indicator informs us that an error has occur with the ✖ in its **status** box, which corresponds to a Boolean value of TRUE. The **source** box indicates where within the program the error occurs, and **code** identifies the cause of the error via an integer code. To decipher this integer code, pop up on the code box and select **Explain Error**. A dialog window will appear, describing the source of the error, which in the present case was due to the requested voltage being "too large or too small."

Note that your program continues to run, despite this error, and if you set **Voltage** back to within the acceptable range, the VI will function correctly as if nothing out-of-sorts had ever happened. In some programming situations, the "continue to run as if nothing is wrong" behavior you observed here is fine, but more commonly one wants a program to respond in some appropriate way (such as to cease operation) if an error occurs. Appropriate response to an error is called *error handling*.

Let's build error handling into **DC Voltage Source (Express)** by making the VI cease operation when an error occurs. Click the **Stop Button**, if the VI is still running, then switch to the block diagram so that we can add the desired error-handling code.

Presently, there is only one way to stop our VI. When the **Stop Button** is clicked, a TRUE Boolean value is passed to the ⊙, which causes the While Loop to stop iterating. Consider the following alternate way to stop the program. When an error occurs, the **status** element within the **error out** cluster becomes TRUE. By passing this value to the ⊙, we could stop the While Loop. You can include both of these two possible program-stopping methods in your VI with the following coding.

Place an **Unbundle By Name** (found in **Functions>>Programming>>Cluster, Class, & Variant**) on the block diagram and wire it to the **error out** terminal of DAQ Assistant. Once wired to the **error out** terminal, **Unbundle By Name** can be resized to have output terminals for all three of the elements with the **error out** cluster (status, code, source), if desired. However, since we only need **status** in our present program, keep **Unbundle By Name** sized to a single terminal. The element associated with this terminal can be selected by popping up on the right (output) side of **Unbundle By Name** and using **Select Item**.

On your block diagram, allow a TRUE Boolean value from either the **Stop Button** or the error cluster's **status** element to cease While Loop iteration by including an **Or** icon (found in **Functions>>Programming>>Boolean**), as shown next.

Return to the front panel and save your work. Run the VI with various values for **Voltage**. You will now find that when **Voltage** is set to a value outside of the range 0 to +5 Volts, the program ceases operation.

When you are finished, run **DC Voltage Source (Express)** one more time with **Voltage** set to *0*, so the channel ao0 is left in the zero-voltage state.

## 4.11  SOFTWARE-TIMED SINE WAVE GENERATOR

By continually updating an Analog Output channel with appropriately chosen voltage values, one can turn a DAQ device into a waveform generator with a time-dependent voltage output of any desired shape (e.g., sine wave). To obtain an output voltage whose time-dependence accurately represents that desired shape, one must assure that the AO channel changes ("updates") to particular values at precisely the right moments. The most accurate method for timing these required updates involves use of a high-frequency hardware clock. Such clocks are contained internally on DAQ devices such as the PCI-6251 and can be used to create high-quality analog waveforms. Unfortunately, hardware clocks are not included on less expensive DAQ devices such as the USB-6009, and so only the (much) less accurate method of software-timed waveform generation is possible on these devices. Let's first write a program that creates an analog sine wave using software timing. Owners of any DAQ device will be able to run this program on their system. After that, you will have the option of writing a VI that implements hardware-timed AO operations, if your DAQ device possesses this capability.

With a few quick edits, we can transform **DC Voltage Source (Express)** into a software-timed sine-wave generator. In this program, DAQ Assistant is configured in the **1 Sample (On Demand)** mode and so outputs a given voltage level each time the While Loop iterates. To obtain a sine-wave output, we will arrange for the sequence of voltage levels fed to DAQ Assistant over the course of several iterations to follow a sine-wave pattern. **Wait (ms)** will control the time per iteration of a While Loop, and so will determine the rate at which the sine wave is generated, that is, its frequency. We want to make this VI compatible with the capabilities of a USB-6009, which has the following two limitations: First, the USB-6009's maximum AO update rate is 150 S/s, which means that the fastest time interval over which an AO channel's voltage level can be changed is 1/150 second, or about 7 ms. Thus, we will program **Wait (ms)** to wait 10 ms. If you have a device capable of faster update, you can (slightly) improve this VI by programming **Wait (ms)** to wait 1 ms, which is the shortest allowable time to wait when using this icon. Second, AO channels on the USB-6009 can only produce voltages in the range from 0 to +5 Volts. Hence, we will create a sine wave with amplitude of 1 Volt, which oscillates about the 2-Volt (rather than 0-Volt) level.

Let's use our previous work on **DC Voltage Source (Express)** to jump start our development of this sine-wave generator program. With **DC Voltage Source (Express)** open, select **File>>Save As...** In the dialog window that appears, choose **Copy>>Substitute copy for original**, and then press the **Continue** button. In the next window that appears, navigate to the **YourName\Chapter 4** folder in the **Save in:** box, name this copied VI **Sine Wave Generator (Software-Timed)** in the **File name:** box, and press the **OK** button. The original file **DC Voltage Source (Express)** will be closed (and safely saved in **YourName\Chapter 4**), while the newly created file **Sine Wave Generator (Software-Timed)** is open and ready for you to modify.

On the front panel, change the Numeric Control's label from **Voltage** to **Samples per Cycle**. Then, modify the block diagram to appear as shown below. The **Mathscript Node** (found in **Functions>>Programming>>Structures**) has two inputs ($i$ and $N$) and one output (data). Pop up on the **data** output and select **Choose Data Type>Scalar>>DBL**. As the iteration terminal increments, the code within the Mathscript Node evaluates the sine function at $N$ equally spaced arguments in each sine-wave cycle. The spacing in radians between adjacent arguments is $2\pi/N$, where $N$ is the number of samples per cycle. The icon **To Double Precision Float** wired to the (32-bit integer) iteration terminal is necessary because a Mathscript Node only recognizes floating-point numerics. This icon is found in **Functions>>Programming>>Numeric>>Conversion**.

Connect the positive and negative (ground) input of an oscilloscope to the AO 0 and AO GND pins of your DAQ device, respectively.

Switch to the front panel and save your work.

You can now observe how well a digitally produced sine wave approximates the (continuous) sine function. Run your VI with **Samples per Cycle** set to *10*, so that each sine cycle is represented by ten voltage levels. Given that each While Loop iteration takes 10 ms, the period of the observed sine wave should be (10 ms/update) × (10 updates) = 100 ms, making its frequency 10 Hz. Observe the 10 Hz sine wave on an oscilloscope. Although the underlying sine pattern is apparent, the ten voltage levels used to form this waveform are too well-resolved to give the appearance of a continuous function.

160

Try increasing **Samples per Cycle** to *20*, then *30*, and so on. Since the While Loop iteration time remains constant, as you increase **Samples per Cycle**, the frequency of the output waveform will decrease. You will need to adjust your oscilloscope's *Time/Div* control appropriately to view a complete cycle of the waveform. At what value for **Samples per Cycle** does the output waveform start to take on a continuous appearance?

## 4.12 HARDWARE-TIMED WAVEFORM GENERATOR

If your DAQ device supports hardware-timed analog output operations (PCI-6251 does, USB-6009 doesn't), try building the following high-quality waveform generator. On the front panel of a blank VI, place a **Stop Button** and an **XY Graph** with the *x*- and *y*-axis labeled **Time** and **Voltage**, respectively. Then, save this VI under the name **Waveform Generator (Express)** in **YourName\Chapter 4**.

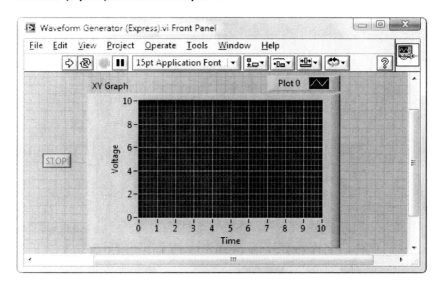

Switch to the block diagram. Place a **While Loop** there and then a **DAQ Assistant** within it. When the **Create New Express Task…** window opens, make the following sequence of selections: **Generate Signals>>Analog Output>>Voltage>>ao0**, and then press **Finish**. When the **DAQ Assistant** dialog window opens, make the selections shown in the following diagram. Be sure to uncheck the **Use Waveform Timing** box (located next to the 🖑 in the illustration). Note we are programming $V_{min}$ and $V_{max}$ to be −10 and +10 Volts, respectively. In the **Generation Mode>>Continuous** mode, a 1D array consisting of a sequence of voltage values is first written into a memory buffer associated with DAQ Assistant, then the DAQ device outputs this sequence of voltages at channel AO 0 in a circular fashion (i.e., after the last array value is output, the next output is the first array value). When all of the selections are properly made, click on the **OK** button.

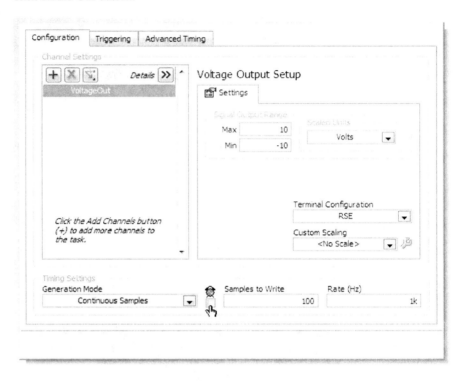

Upon return to the block diagram, expand the terminals of DAQ Assistant, and order them as shown next using **Select Input/Output** in each terminal's pop-up menu.

The plan for our VI is this: Use the **Waveform Simulator** program written in Chapter 3 to generate a 1D array of voltage values that describes a desired waveform. Then, pass this 1D array to the **data** input of **DAQ Assistant**, which initiates the process of outputting the voltage waveform at channel ao0.

## 4.13 PLACING A CUSTOM-MADE VI ON A BLOCK DIAGRAM

First, we must place **Waveform Simulator** (as a subVI) on our block diagram. Here's how to do that. On the Functions Palette, choose the **Select a VI...** subpalette.

The **Select the VI to Open** dialog window will appear. In the **Look in:** box, navigate to the **YourName\Chapter 3** folder. Once the file list for this folder is displayed, double-click on **Waveform Simulator** or highlight it, then click on the **OK** button in the dialog box.

You will then be returned to the block diagram, where you can place your custom-written icon in a convenient position. Then, with the ✎, create **Digitizing Parameters** and **Waveform Parameters** control clusters by popping up on these terminals and using **Create>>Control**.

## 4.14 COMPLETING AND EXECUTING WAVEFORM GENERATOR (EXPRESS)

Finally, complete the block diagram as shown next. When you wire Unbundle By Name's **Displacement** output to DAQ Assistant's **data** input, the **Convert to Dynamic Data** icon will be inserted into the wire automatically. This icon converts the 1D floating-point numeric array produced by **Waveform Simulator** to the dynamic data type used by Express VIs such as DAQ Assistant. If you ever need to place **Convert to Dynamic Data** on a block diagram manually, it is found in **Functions>>Express>>Signal Manipulation**.

Return to the front panel, tidy it up, and then save your work.

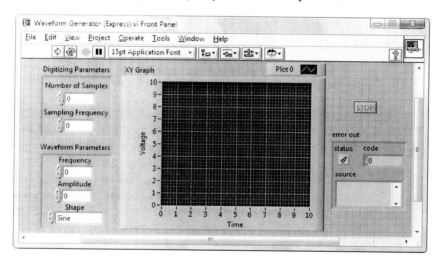

Here's how to program your VI to output a continuous waveform. Using the **Digitizing Parameters** and **Waveform Parameters** control clusters, chose values that instruct **Waveform Simulator** to produce one (or, possibly, a few) complete cycle(s) of the desired periodic waveform. This software-representation of the waveform's cycle will then be written into a memory buffer by **DAQ Assistant**. The array of waveform values will then be output sequentially (in a circular fashion) by your DAQ device at a rate given by the **Sampling Frequency** until a TRUE value is sent to DAQ Assistant's **stop** input.

Test-drive your VI by using it to output a 100 Hz sine wave of amplitude 5 Volts. First, in the **Waveform Parameters** control cluster, program **Frequency**, **Amplitude**, and **Shape** to be *100*, *5*, and *Sine*, respectively. Next, assuming 100 points are required to well-define each cycle of the sine wave, the sampling frequency should be 100 times larger than the sine wave's frequency. Thus, for **Waveform Simulator** to create one complete cycle, in the **Digitizing Parameters** control cluster, let **Number of Samples** and **Sampling Frequency** be *100* and *10000*, respectively.

Run your VI. The software waveform produced by **Waveform Simulator** will be displayed on the **XY Graph** for you to view. Using an oscilloscope, observe the voltage waveform output at channel ao0. Is it a (quasi) continuous sine wave with a 100 Hz frequency and 5 V amplitude?

With your VI still running and keeping all other parameters the same, change **Frequency** (in the **Waveform Parameters** control cluster) to *150*. From the XY Graph, you will find that **Waveform Simulator** produces 1½ cycles of this waveform. Using your knowledge of the circular fashion in which the DAQ device sequences through the memory buffer values, can you explain the discontinuities that you observe in the voltage waveform being output at channel ao0?

With the VI still running, try changing **Frequency** to *200, 400, 500,* and *1000*. With each of these choices, **Waveform Simulator** will produce an integer number of complete cycles of the waveform and so produce voltage waveform outputs with no discontinuities. Why does the waveform output at channel ao0 become less well-defined for the higher frequencies? Also, you might enjoy changing **Shape** and **Amplitude** as the VI is running.

Finally, this VI has one constraint, which reflects a limitation the DAQ Assistant Express VI. Namely, DAQ Assistant sets the DAQ device's sampling frequency during its first execution within the program (i.e., during the first iteration of the While Loop) and this value cannot be changed as the program continues to run. You can verify this point as follows: Start the VI with **Frequency**, **Amplitude**, and **Shape** equal to *100*, *5*, and *Sine*, respectively, and **Number of Samples** and **Sampling Frequency** equal to *100* and *10000*, respectively. **Waveform Simulator** will create one cycle of a 100 Hz sine wave, and you will observe a 100 Hz sine wave at ao0. Now, with the VI continuing to run, change **Frequency** and **Sampling Frequency** to *1000* and *100000*, respectively. **Waveform Simulator** will create 100 points that represent one cycle of a 1000 Hz sine wave if output at a rate of 100 kHz, but you will observe a 100 Hz sine-wave output at channel ao0 because the DAQ device's sampling frequency did not change from its initial value of 10 kHz.

## 4.15 MODIFIED WAVEFORM GENERATOR

Above, we discovered the following properties of DAQ Assistant when placed in an iterative loop: After being set during this icon's initial execution, the value of **Sampling Frequency** remains fixed. However, the array read into its **data** input can be changed during any execution of the icon, allowing the parameters **Frequency**, **Amplitude**, **Shape**, and **Number of Samples** to be adjusted as the iterative program runs. Consider the following approach for increasing the flexibility of **Waveform Generator (Express)**: Set **Sampling Frequency** to a large value. Then, during each iteration as the program runs, determine an appropriate data array based on the current **Waveform Parameters** settings and input this array to DAQ Assistant. Let's arbitrarily define the "appropriate data array" to be four cycles of the waveform described by the **Waveform Parameters** settings.

If interested, with **Waveform Generator (Express)** open, use **File>>Save As...** to create **Waveform Generator (Modified Express)** and save it in **YourName\Chapter 4**. Then delete the **Digitizing Parameters** control cluster so that the front panel is as shown next.

On the block diagram, hardwire **Sampling Frequency** to, say, 100000 Hz. Additionally, using the value of **Frequency** input on the front panel, the **Number of Samples** value required to produce four cycles of the waveform is determined and input to **Waveform Simulator** as shown in this block diagram.

Run **Waveform Generator (Modified Express)**. It will produce high-quality voltage waveforms of any frequencies up to a few kilohertz.

## DO IT YOURSELF

In this exercise, you will write a VI that turns an LED on or off. A Digital Input/Output (DIO) pin on an NI DAQ device is called a *line*. A typical DAQ device has two or three *ports*, where each port is composed of eight *lines*. A particular line is specified as *P port:line*, so, for example, P0.3 is line 3 of port 0. Locate the DIO line P0.0 on your DAQ device, and then construct the hardware circuit shown in Figure 4.2. In this circuit, when P0.0 goes HIGH, it supplies +5 Volts to this circuit, illuminating the LED.

**Fig. 4.2**   **Hardware circuit for Do It Yourself project. DIO line P0.0 of DAQ device connected to (current-limiting) resistor in series with LED.**

Write a VI, which can illuminate the LED as follows:

Place a Boolean switch (e.g., **Push Button**) on the front panel of a new VI. Then on the block diagram, place **DAQ Assistant** and configure it as **Generate Signals>>Digital Output>>Line Output>> port0/line0** and **Generation Mode>>1 Sample (On Demand)**. With this configuration, when a Boolean TRUE (FALSE) is passed to DAQ Assistant's **data** input, P0.0 will go HIGH (LOW). Build the following block diagram. Since a DAQ device has *N* DIO lines, **data** is formatted to accept an array of Boolean values with up to *N* elements. Hence, the **Build Array** icon (found in **Functions>>Programming>>Array**) is necessary to convert the Boolean switch's single Boolean value to a Boolean array with one element.

Run the VI several times and show that you can toggle the LED in your hardware circuit on and off via the setting of the Boolean switch.

Next, construct the hardware circuit shown in Figure 4.3 with two LEDs, which we will call *A* and *B*. Then write a While Loop–based VI called **Alternating LEDs**. When this program runs, during even iterations of its While Loop, *A* is lit and *B* is dark; during odd iterations, *B* is lit and *A* is dark. The LEDs alternately light until a front-panel **Stop Button** is clicked.

**Fig. 4.3** Hardware circuits controlled by Alternating LEDs VI.

In constructing **Alternating LEDs**, besides **Build Array**, you may find the following icons useful: **Select** and **Equal To 0?** in **Functions>>Programming>>Comparison** as well as **Quotient & Remainder** in **Functions>>Programming>>Numeric**.

   **For NI ELVIS Users:** In its normal functioning, first-generation versions of NI ELVIS block access to the port 0 DIO lines because these lines are used by the NI ELVIS to communicate with its associated DAQ device. In order to obtain control of these DIO lines, you must disable this communication by setting NI ELVIS's front panel **Communications** switch to **Bypass** (as opposed to **Normal**), then run a VI called **NI ELVIS-Enable Communications Bypass.vi**, which found in **Functions>>Instrument I/O>>Instrument Drivers>>NI ELVIS>>Low Level NI ELVIS VIs>>Digital I/O**. Construct the following block diagram and run it to obtain control of the port 0 DIO lines.

The port 0 lines will then appear on the NI ELVIS breadboard at the pins labeled DI 0 through DI 7 in the DIGITAL I/O section.

## Problems

1.  Writing the aliasing condition as $f = \pm f_{alias} + n\, f_s$, show that $\sin(2\pi f t_i) = \pm\sin(2\pi f_{alias} t_i)$ at all $t_i = i\Delta t$.
2.  Observe the resolution $\Delta V$ of an analog input signal digitized by your DAQ device and verify that your observation is consistent with the prediction of Equation [1]. Attach a slowly varying analog signal (e.g., 1 Hz triangle wave) and the sync output from a function generator to the analog input channel and digital triggering pins programmed into **Digital Oscilloscope (Express)**. Run **Digital Oscilloscope (Express)** with a fast sampling rate and fairly small number of samples so that the input signal is effectively constant over the trace that you obtain. Stop the VI, and then, by changing the $y$-axis scaling, zoom in on the acquired data (alternately, you can use a **Probe** to view the numerical values of the data samples). Under "high magnification," you should find that the data samples of this "effectively constant" input signal are distributed in discrete levels (due to electronic noise). Measure the spacing $\Delta V$ between two of these adjacent levels. Does your value for $\Delta V$ agree with the prediction of Equation [1]?

3. By passing an analog signal through a low-pass filter prior to inputting it to a digitizer, the aliasing effect can be suppressed. To demonstrate this procedure, attach an analog sine-wave signal and the sync output from a function generator to the analog input channel and digital triggering pins programmed into **Digital Oscilloscope (Express)**.

   On the front panel of **Digital Oscilloscope (Express)**, set **Number of Samples** and **Sampling Frequency** equal to *100* and *1000*, respectively, and set the sine-wave frequency *f* on your function generator to approximately 100 Hz. Run **Digital Oscilloscope (Express)**. If all goes well, you will see a "stationary" plot of about ten cycles of the 100 Hz sine wave. Turn off autoscaling on the *y*-axis.

   Next, set *f* on your function generator to approximately 1900 Hz, and then fine-tune *f* until **Digital Oscilloscope (Express)** displays an aliased waveform of about 100 Hz. Use Equation [2] to explain why the input with *f* = 1900 Hz appears as a 100 Hz sine wave when digitized, due to aliasing.

   Finally, without changing the frequency setting on your function generator, pass the *f* = 1900 Hz sine-wave voltage through the low-pass op-amp filter shown in Figure 4.4 with *R* = 15 kΩ and *C* = 0.1 *μ*F, prior to inputting it to the DAQ device. By what factor is the amplitude of the aliased signal attenuated in comparison to the unfiltered situation? What is the 3dB-frequency $f_{3dB} = 1/(2\pi RC)$ of the low-pass filter? If you wanted the aliased signal to be attenuated even further, what would you change? [This process can be optimized by using a low-pass filter with a steeper roll-off than the simple (first-order) circuit used here.]

**Fig. 4.4    Problem 3.**

4. Build a two-channel digital oscilloscope. With **Digital Oscilloscope (Express)** open, use **File>>Save As…**to create a new VI named **Dual Digital Oscilloscope (Express)**. On the block diagram, you must reprogram DAQ Assistant to read two channels rather than just one. Open DAQ Assistant and add the new channel, then close the dialog window. Your VI is now complete.

   Attach a voltage waveform and the sync out digital signal from a function generator to the two AI channels of your DAQ device programmed on the block diagram. Additionally, attach the sync out digital signal to the PFI0 channel of your DAQ device Run **Dual Digital Oscilloscope (Express)** and verify that it simultaneously displays traces of the two inputs.

5. Thermocouples are widely used as temperature sensors. A thermocouple is constructed by joining the ends of two dissimilar metals, for example, a copper and a constantan wire for a Type T thermocouple. This junction produces a millivolt-level voltage, which

has a well-documented temperature dependence, where the temperature is measured relative to a "cold junction" reference temperature. Conveniently, this cold junction can be provided by a compact electronic device called a Cold Junction Compensator (CJC), which effectively makes the reference temperature equal to 0 °C.

Connect a thermocouple to a CJC and then connect the plus and minus output of the CJC to the AI+ and AI− pins of your DAQ device. Using DAQ Assistant, write a program called **Thermocouple Thermometer (Express)** that, every 250 ms until a Stop Button is clicked, reads the thermocouple voltage, converts this value to the corresponding temperature in Celsius, and then displays this temperature in a front-panel indicator. Program DAQ Assistant as follows: **Acquire Signals>>Analog Input>>Temperature>> Thermocouple>>ai0**, followed by **Scaled Units>>deg C, CJC Source>>Constant**, and **CJC Value>>0**, and **Acquisition Mode>>1 Sample (On Demand)**. Program **Thermocouple Type** for your particular type of thermocouple (e.g., T).

Run **Thermocouple Thermometer (Express)** and use it to measure room temperature as well as the temperature of your skin.

6. If your DAQ device supports hardware-timed analog output operations, write a program called **Bode Magnitude Plot**, which performs the following process: Apply an AC voltage with frequency $f$ and 1 Volt rms amplitude to the input of a circuit, measure the resulting rms voltage amplitude $V_{out}$ at the circuit's output, and then calculate the gain (in decibel) $G = 20\log(V_{out}/V_{in})$, where $V_{in} = 1$ $V_{rms}$. Construct the VI such that this process is repeated for frequencies in the range $f = 10$ Hz to $f = 2010$ Hz in 25 Hz increments, and the resulting data is displayed on a plot of $G$ versus $\log(f)$. Build your VI according to these guidelines:

The workhorse of this VI is formed by configuring three Express VIs as shown below. Here **Simulate Signal** (found in **Functions>>Express>>Input**) is used to create a sine wave of frequency $f$ and 1 Volt rms amplitude, which is passed to a **DAQ Assistant** configured to generate this waveform at an analog output channel. Assuming this sine-wave signal is applied at the input of a circuit, the second **DAQ Assistant** is configured to read 500 samples of the response at the circuit's output at a sampling frequency of $50f$ (i.e., ten cycles of the output will be read and made available at the **data** terminal). Finally, these ten cycles are supplied to **AC & DC Estimator.vi** (found in **Functions>>Signal Processing>>Signal Operation**). Note in the dialog window for Simulate Signal, select **Samples per second (Hz)>>40000** and check **Integer number of samples**. For the analog output DAQ Assistant, select **Generation Mode>>Continuous** and check **Use Waveform Timing**. For the analog input DAQ Assistant, select **Acquisition Mode>>N Samples**.

**Fig. 4.5    Problem 6.**

Place an XY Graph on the front panel and supply it with bundled arrays of *f* and *G* from the block diagram. To obtain a log scale on the *X*-axis, pop up on the XY Graph and select **X Scale>>Mapping>>Logarithmic**.

When completed, use **Bode Magnitude Plot** to obtain a Bode plot on the *RC* circuit shown in Figure 4.5, where R = 4.7 kΩ and C = 0.1 μF.

7. If your DAQ device supports hardware-timed analog output operations, write a program called **Music Box**, which plays the scale of A major with each note sounding for 500 ms. The eight notes in this scale have the following frequencies in Hertz: 440.0, 493.9, 554.4, 587.3, 659.3, 740.0, 830.6, 880.0. These frequencies can be stored either in a front-panel **Array Control** or in a block-diagram **Array Constant** (created by placing a **Numeric Control** within an **Array Constant** shell). On the block diagram, input each frequency *f* one at a time into a For Loop that contains the subdiagram shown below. Here, the **Simulate Signal** Express VI (found in **Functions>>Express>>Input**) creates an array representing an integer number of cycles of the desired sine wave with frequency *f*, which is then output as a real voltage waveform at an AO channel by DAQ Assistant. When programming **Simulate Signal**, select **Samples per second (Hz)>>100000, Number of samples>>1000** (and uncheck **Automatic**). Check **Integer number of cycles**.

To play the scale, attach an amplifier and speaker to the AO channel such as shown in Figure 4.6, and then run **Music Box**.

**Fig. 4.6  Problem 7.**

# Chapter 5 ▪ ▪ ▪ ▪ ▪ ▪ ▪ ▪ ▪ ▪ ▪ ▪ ▪ ▪ ▪ ▪ ▪ ▪ ▪ ▪ ▪ ▪

# Data Files and Character Strings

The outcome of any experiment is embodied in the data it generates. In computer-controlled experimentation, then, it is important to have the ability to save experimental data in disk files for future analysis. Hopefully, these data are stored in a convenient format so that they can be easily read by a user-written program or a commercially available software package that analyzes and displays the experimental results.

Computers commonly communicate alphanumeric data using the *American Standard Code for Information Interchange* (ASCII). In this coding scheme, seven bits of a byte are used to represent a character, while the byte's eighth "parity" bit is used for error checking when the data is received by a reader. The ASCII set of $2^7 = 128$ distinct states is used to represent all of the keyboard's alphanumeric characters as well as nonprintable control characters such as carriage return (CR) and line feed (LF). In the Extended ASCII set, all eight bits of the byte are utilized for character coding, yielding 256 distinct states.

A sequence of ASCII characters is called a *string*. Character strings, of course, can be used to represent text messages. However, strings also are useful in passing commands and data to and from the computer and stand-alone instruments. Additionally, as you will find in this chapter, strings can be used to store numerical data in files on a computer disk.

The closest approximation to a universally readable data file format is the *ASCII text file*. Storing data in this manner has the following advantages. Your files will be read accurately by computers of all manufacturers. Additionally, the files can be viewed with a word processor and, if desired, easily cut and pasted into a document for report generation. Finally, your data will be easy to import into commercially available data analysis software packages. A large number of these application programs prefer that your file is in tab-delimited ASCII text, or what is commonly called the *spreadsheet format*. In this format, tabs separate columns and end of line (EOL) characters separate rows, as shown next.

```
0.00 →  0.4258¶
1.00 →  0.3073¶                          → = Tab
2.00 →  0.9453¶                          ¶ = EOL
3.00 →  0.9640¶
4.00 →  0.9517¶
```

Opening this file using a spreadsheet program such as Microsoft Excel yields:

|   | A | B | C |
|---|---|---|---|
| 1 | 0.00 | 0.4258 | |
| 2 | 1.00 | 0.3073 | |
| 3 | 2.00 | 0.9453 | |
| 4 | 3.00 | 0.9640 | |
| 5 | 4.00 | 0.9517 | |
| 6 | | | |

Alternately, numerical data may be stored as a *binary-format byte stream file*. This file format is simply a bit-for-bit image of the data that reside in your computer's memory. Thus, when data are exchanged between a binary file and computer memory, little or no data conversion is required, so you get maximum performance. Also, binary data files provide the most memory-efficient storage method for numerical data. To demonstrate this fact, consider the number of bytes necessary to store the integer *54321*. Since this number is less than $2^{16} = 65536$, it will require only two bytes of memory in a binary-format file. However, in an ASCII text file, this number will occupy 5 bytes of memory, one byte for each of the characters *5, 4, 3, 2,* and *1*. The disadvantage of binary files is their lack of portability. They cannot be viewed by a word processor and they cannot be read by any program without a detailed knowledge of the file's format.

LabVIEW contains built-in VIs that facilitate data storage and retrieval using either the ASCII text or binary file format. If ease of use and compatibility with a spreadsheet application program are of most concern, use the ASCII-based file storage. If, on the other hand, efficient memory use and high speed are desired, a binary file is the best choice.

In this chapter, you will learn how to store data in an ASCII-based spreadsheet file. You will write a program that generates data using your custom-made **Waveform Simulator** as a subVI and then store these data by implementing icons from **Functions>>Programming>>File I/O**.

## 5.1 STORING DATA IN A SPREADSHEET-FORMATTED FILE

Here's the plan for our data-storage VI: Use the **Waveform Simulator** program created in Chapter 3 to generate some data, and then store these data in a spreadsheet-formatted data file, which is saved on your computer's hard drive.

Start by opening a new VI. Using **File>>Save**, create a new folder called **Chapter 5** within the **YourName** folder, then save this VI under the name **Spreadsheet Storage** in **YourName\Chapter 5**.

Now switch to the block diagram. First, you must place **Waveform Simulator** (as a subVI) on your block diagram. In summary, here's how to do that (you may wish to review section 4.13 "**Placing a Custom-Made VI on a Block Diagram,**" which illustrates the steps in the process). On the Functions Palette, choose the **Select a VI...** subpalette. The **Select the VI to Open** dialog window will appear. In the **Look in:** box, navigate to the **YourName\Chapter 3** folder. Once the file list for this folder is displayed, double-click on **Waveform Simulator**, or highlight it, then click on the **OK** button in the dialog box. You will then be returned to the block diagram, where you can place the custom-written icon in a convenient position.

Next, with the 🖑, create **Digitizing Parameters** and **Waveform Parameters** control clusters on the front panel by popping up on these terminals and using **Create>>Control**.

Then, place a **Write To Spreadsheet File.vi** icon (found in **Functions>> Programming>>File I/O)** on the diagram.

## 5.2 STORING A ONE-DIMENSIONAL DATA ARRAY

To understand how the **Write To Spreadsheet File.vi** icon functions, view its Help Window, as shown below. Here, you will find that this VI has the potential to function in numerous ways. Also, as is generically true for Help Windows, the default values for the icon's various inputs are given in parentheses. In its most basic operating mode, one simply provides a 1D array of numerical values to Write To Spreadsheet File.vi's **1D data** input. The VI then prompts the user for a file name and stores the values under this title using the spreadsheet format. By operating **Write To Spreadsheet File.vi** in this elemental way, we will gain insight into the necessity of its other available options.

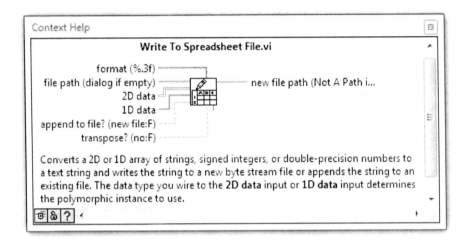

Place **Unbundle By Name** (found in **Functions>>Programming>>Cluster, Class, & Variant**) on your diagram, then wire it to the **Waveform Output** terminal of Waveform Simulator. Using the ▶, expand **Unbundle By Name** so that output terminals for both elements of the cluster (**Time** and **Displacement**) are accessible, then wire the **Displacement** terminal to the Write To Spreadsheet File.vi's **1D data** input.

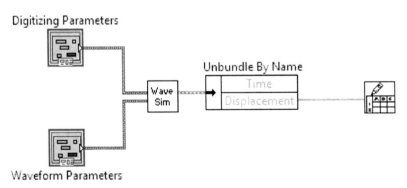

Now return to the front panel, tidy up the object arrangement, then save your work. Program the inputs to create a small number (such as *10*) of displacement values for a 100 Hz sine wave of amplitude 4.0 with a sampling frequency of 1000 Hz. Then run your program.

A dialog box called **Choose file to write** will appear. Navigate to the **YourName\ Chapter 5** folder in the **Save in:** box, name the file **Sine Wave Data** in the **File name:** box, then click on the **OK** button. The program will then complete execution by creating the requested spreadsheet file in **YourName\Chapter 5**.

Use a word processor program to open the ASCII-based **Sine Wave Data** spread-sheet file (in the Open dialog window, you may have to choose the file type as **All Files** in order to see **Sine Wave Data** listed). Shown next is what I found by viewing this data file in Microsoft Word (the appearance of your file may vary slightly due to the particular tab settings of your word processor). Here I have activated Word's **Show¶** command, which allows one to see the file's nonprinting characters, such as Tab (→) and EOL (¶). If your word processor has a similar option, use it to view these usually hidden characters.

We see that the ten sine-wave values are delimited by tabs and that the sequence concludes with an EOL. From our knowledge of the spreadsheet formatting convention, we conclude that a spreadsheet application program will interpret this string as a row of data, placing these ten numerical values in a sequence of columns contained within a single row. Close **Sine Wave Data** with the word processor, but don't save it (else the program will possibly embed its own characteristic formatting statements within the file).

If available, use a spreadsheet application program to read your **Sine Wave Data** file. To do this, your application program may give you the option of using an **Open** or an **Import** command. Either will work. You may have to tell the program that you're read-ing in a text file (as opposed to binary, Excel, etc.). Also you may encounter a dialog window in which you must tell the program that your data is tab-delimited. Next, I show the result of reading the **Sine Wave Data** file using Microsoft Excel. As expected, the ten numerical values appear in ten sequential columns of a single row.

| | A | B | C | D | E | F | G | H | I | J | K |
|---|---|---|---|---|---|---|---|---|---|---|---|
| 1 | 0 | 2.351 | 3.804 | 3.804 | 2.351 | 0 | -2.351 | -3.804 | -3.804 | -2.351 | |
| 2 | | | | | | | | | | | |
| 3 | | | | | | | | | | | |

## 5.3 TRANSPOSE OPTION

If you have a bit of experience with spreadsheet applications, the above illustration will cause some concern. In using such programs to generate plots, perform curve fitting, and carry out other useful data analysis operations, the array of values for a particular quantity (e.g., a sine-wave displacement) is expected to reside in a single column, with the array's elements indexed by the row numbers. This organizational scheme for data is often called *column-major order*. We have seen that, in its default setting, the **Write To Spreadsheet File.vi** icon does not function in a manner consistent with this convention. However, once this problem has been identified, it is trivial to remedy by consulting the icon's Help Window. Here you will find that the VI possesses a **transpose?** input, whose default value is FALSE. By wiring a TRUE **Boolean Constant** to this input, the file will be recorded in a "column-like" manner, rather than "row-like." Perform this wiring on your block diagram. An easy way to obtain the **Boolean Constant** is to pop up on the **transpose?** input and use **Create>>Constant**. The value of the **Boolean Constant** can be toggled from the FALSE (default) to the TRUE state by using the 🖑.

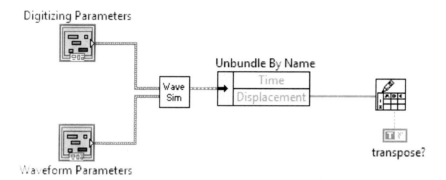

Run your program to produce a spreadsheet file of ten sine-wave values. You can either reuse the file name **Sine Wave Data** (by opening this name in the **Chose file to Write** dialog box, assuming the file is initially closed) or else invent a new name. View this file using a word processing program. As shown next (using Microsoft Word), we see that each data value is separated from its neighbor by an EOL character. Thus, when read by a spreadsheet application, this array of values will be placed in a single column.

| Sine Wave Data - Microsoft Word |
| --- |
| 0.000¶ |
| 2.351¶ |
| 3.804¶ |
| 3.804¶ |
| 2.351¶ |
| 0.000¶ |
| -2.351¶ |
| -3.804¶ |
| -3.804¶ |
| -2.351¶ |

Page: 1 of 1    Words: 10    100%

If available, open the data file in a spreadsheet application. Here (using Microsoft Excel), we see that our expectation of column-like data is fulfilled.

| | A | B | C | D | E | F | G | H | I | J | K |
|---|---|---|---|---|---|---|---|---|---|---|---|
| 1 | 0 | | | | | | | | | | |
| 2 | 2.351 | | | | | | | | | | |
| 3 | 3.804 | | | | | | | | | | |
| 4 | 3.804 | | | | | | | | | | |
| 5 | 2.351 | | | | | | | | | | |
| 6 | 0 | | | | | | | | | | |
| 7 | -2.351 | | | | | | | | | | |
| 8 | -3.804 | | | | | | | | | | |
| 9 | -3.804 | | | | | | | | | | |
| 10 | -2.351 | | | | | | | | | | |
| 11 | | | | | | | | | | | |

## 5.4 STORING A TWO-DIMENSIONAL DATA ARRAY

To analyze and/or plot **Waveform Simulator**'s sine-wave data using a spreadsheet application, one would need to import both its **Time** and **Displacement** arrays into two spreadsheet columns. Let's see how, by implementing the **2D data** input of the **Write To Spreadsheet.vi**, it is easy to modify our **Spreadsheet Storage** program to produce the appropriate data file. We start with **Time** and **Displacement**, the two 1D array outputs from **Waveform Simulator**. Each of these 1D arrays has $N$ elements, where $N$ is determined by the value of the **Number of Samples** control. We now need to splice these two 1D arrays together to form a 2D array with the $N$ values of **Time** and **Displacement** in the 2D array's first and second row, respectively. Mathematicians would call this object a 2-by-$N$ matrix; in LabVIEW it is called a 2D array.

To construct this 2D array, you will use an icon called **Build Array**. However, first, you must open up some room for **Build Array** between **Unbundle By Name** and **Write To Spreadsheet File.vi**. To widen the gap between these two icons, click on the wire that connects them, and then delete it. Next, place the ↳ (it will appear as a cross in the empty diagram region) as shown below.

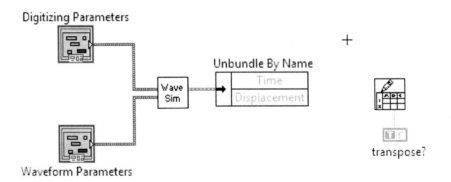

Click, and then drag cursor until a dotted rectangle frames both the **Write To Spreadsheet File.vi** and the **Boolean Constant** icon.

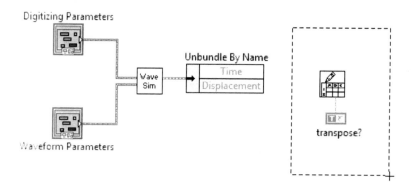

When you release the mouse button, both icons will be highlighted with a marquee. You can then move them to the right as one object by either using the keyboard's <*Right Arrow*> key, dragging with the Positioning Tool while depressing the <*Shift*> key (allowing only horizontal motion), or simply dragging with the ⬎.

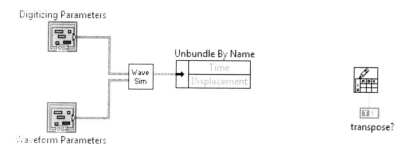

Now select the **Build Array** icon from **Functions>>Programming>>Array**. When you put this icon on the block diagram, it will initially appear with only one input.

183

Place the ⬚ at the bottom-middle of the icon until it morphs into a resizing handle. Resize the icon so that it can accommodate two inputs.

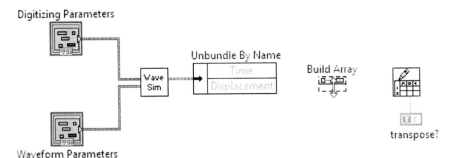

Then wire Unbundle By Name's **Time** and **Displacement** outputs to the top and bottom inputs of **Build Array**, respectively. **Build Array** will splice these inputs together to form the required two-dimensional array.

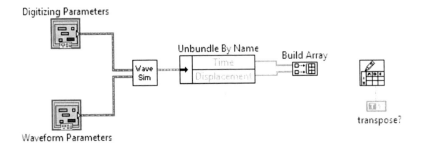

Wire the output of **Build Array** to the **2D data** input of Write To Spreadsheet File. vi. Note that this connection appears as a double wire, the LabVIEW designation for a 2D array.

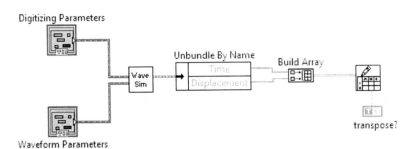

Return to the front panel and run your program with a small value such as *10* for **Number of Samples**. Then view your resulting data file using a word processor. It should appear like this.

This formatting appears correct for importing the **Time** and **Displacement** data into two adjacent columns of a spreadsheet sheet. To verify this expectation, open your file with a spreadsheet application program. It should appear as below. If you're familiar with the operation of the spreadsheet application, you may enjoy plotting column B versus column A or curve fitting the column B data to a sine function.

|    | A     | B      | C |
|----|-------|--------|---|
| 1  | 0     | 0      |   |
| 2  | 0.001 | 2.351  |   |
| 3  | 0.002 | 3.804  |   |
| 4  | 0.003 | 3.804  |   |
| 5  | 0.004 | 2.351  |   |
| 6  | 0.005 | 0      |   |
| 7  | 0.006 | -2.351 |   |
| 8  | 0.007 | -3.804 |   |
| 9  | 0.008 | -3.804 |   |
| 10 | 0.009 | -2.351 |   |
| 11 |       |        |   |

## 5.5 CONTROLLING THE FORMAT OF STORED DATA

Now let's explore some of **Write To Spreadsheet File.vi**'s other options. First, you may have noticed that, when viewing your data file, each of its numerical values was stored with a precision of three digits to the right of the decimal point. Is this the precision of the numerical values produced by the **Waveform Simulator** VI?

Try this test: Open **Waveform Simulator** by double-clicking on its icon within **Spreadsheet Storage**'s block diagram. Pop up on the **Time** and **Displacement** front-panel indicators, select **Display Format...**, and then increase **Digits of precision** from its default value to something large like *10* and disable the **Hide trailing zeros** option. You may then want to resize the indicators so that all of the new digits are visible. Run the VI with some appropriate choices for the **Digitizing Parameters** and **Waveform Parameters** inputs. You will then find that the array indicators display values of the precision that you requested. Now, with this modification in **Waveform Simulator**, rerun **Spreadsheet Storage**, and then view the new **Sine Wave Data** file it produces. You will find that the stored values there still have a precision of three digits. Thus, we surmise that **Waveform Simulator** is not influencing the formatting of values within **Sine Wave Data**. In fact, independent of its **Digits of precision** setting, the **Waveform Simulator** VI calculates highly accurate floating-point values, which it outputs to connecting wires. The **Digits of precision** parameter merely selects how many decimal places of these highly accurate values are to be displayed on the VI's front-panel indicator.

Since we have eliminated **Waveform Simulator** as a suspect, we must look elsewhere within **Spreadsheet Storage** to find where the decision is being made to truncate highly precise **Waveform Simulator**–produced values in the thousandths place upon storage in the file. A quick glance at Write To Spreadsheet.vi's Help window identifies that the decision is made at this icon's **format** input. By leaving this input unwired, we have instructed the icon to store numerical values in the default format of *%.3f*. To understand this instruction, we need to know that the **format** string obeys the following syntax:

% [WidthString] [.PrecisionString] ConversionCharacter

Here, optional features are enclosed in square brackets. Each syntax element, and how it affects the form of the stored number within the data file, is explained in the following table.

**Table 5.1**

| | |
|---|---|
| % | Character that denotes the beginning of a format specification. |
| **WidthString** (Optional) | Integer specifying the total number of ASCII characters to be used to represent the stored number. If your specified value of **WidthString** exceeds the number of characters actually required to represent a particular number, the excess portion of the string will be padded with spaces. If **WidthString** is not specified (or the stored number requires more characters than the specified **WidthString**), the numeric string will expand to as long as necessary to represent the stored number. |

*(Continued)*

**Table 5.1** *(Continued)*

| | |
|---|---|
| **.PrecisionString** (Optional) | Period (.) followed by an integer specifying the number of digits to the right of the decimal point in the stored number. If **.PrecisionString** is not specified, six digits to the right of the decimal point are stored. If only the period (.) appears and **PrecisionString** is missing or zero, no digits to the right of the decimal point are stored. |
| **ConversionCharacter** | Single character that specifies in what manner the number is to be stored with the following code:<br>d    decimal integer<br>x    hex integer<br>o    octal integer<br>f    floating point with fractional format<br>e    floating point with scientific notation |

Thus we see that *%.3f* instructs **Write To Spreadsheet File.vi** to store numbers in fractional floating-point format with each number's total width auto-adjusted under the restriction of maintaining three digits to the right of the decimal point.

Place a **String Constant** on the block diagram, define it to be *%8.5f*, and then wire it to Write To Spreadsheet File.vi's **format** input. The easiest way to create the **String Constant**, of course, is to pop up on the **format** input, then select **Create>>Constant**. You can also find this icon in **Functions>>Programming>>String**.

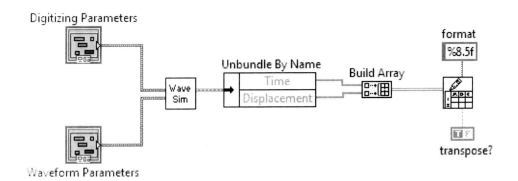

Run your program and store the data in a file. View the file using a word processor. Do the file's numerical values appear in the format that you specified? You might try exploring the effect of changing the **ConversionCharacter** to *d* (decimal integer) or *e* (floating point with scientific notation).

## 5.6 THE PATH CONSTANT AND PLATFORM PORTABILITY

Next, let's look more closely at the act of naming the data file. We note that, in some cases, it would be desirable to hardwire the name of a data file into our program code. Glancing back at the Write To Spreadsheet File.vi's Help Window, we find that this option is available via the **file path (dialog if empty)** input. Previously, we have left this input unwired; thus the default **dialog if empty** (i.e., a dialog box prompting us for a file name) has been operational. Now we will take the alternate approach of programming-in a specific file name.

As you probably know, the *path* to a particular file in your computer system is specified by a hierarchical directory structure. The path is typically denoted by a **drivename**, followed by **directory names** (commonly called **folders**), followed by the **filename** itself. In Windows, these various levels of the hierarchy are separated by backslashes (\); on the Macintosh, forward slashes (/) are used. Thus, on a Windows machine with a hard drive named **C:**, the file **Sine Wave Data** within the directory **Users\Account\Desktop\ YourName\Chapter 5** has the path **C:\Users\Account\Desktop\YourName\Chapter 5\Sine Wave Data**. On a Mac with a hard drive named **Macintosh HD**, a similarly named file on the **Desktop** has the path **Macintosh HD/Users/Account/Desktop/YourName/Chapter 5/Sine Wave Data**. To hardwire a specific file name into a program, one might anticipate that the path appropriate to your computing platform (Windows or Macintosh) is simply inserted into a **String Constant** and then this object is wired to the **file path** input of the Write To Spreadsheet File.vi icon. However, in deference to the issue of *platform portability*, this expectation is slightly modified.

Platform portability is one of the outstanding features of LabVIEW programming. This means, for example, that you can write a LabVIEW program called **Widget** on a Windows-based system, then send it (electronically or on a disk) to a colleague for use in a Macintosh-based laboratory. Once your colleague has copied the VI onto his or her Mac system, **Widget** can be opened from within LabVIEW running on the Macintosh system. While opening it, LabVIEW detects that **Widget** was written on another platform and recompiles it to use the correct instructions for the local processor. As seen above, one small part of this recompilation must include translating any hardwired path names that are contained within **Widget** into the form appropriate for the new system. To facilitate this translation process, a special LabVIEW object called the **Path Constant** (found in **Functions>>Programming>>File I/O>>File Constants**) is used for the specification of path names. By dedicating the **Path Constant** to the sole use of enclosing path names, these special strings are distinguished from other character strings when the program is being ported.

By popping up on the **file path** input of Write To Spreadsheet File.vi, place a **Path Constant** on your block diagram, then enter a path appropriate to your computing platform into it. Here, I show the resulting diagram on my Windows system.

Save your work, and then run this program. Verify that it performs as expected by viewing the resulting data file with a word processor or spreadsheet application program.

## 5.7 FUNDAMENTAL FILE I/O VIs

**Write To Spreadsheet File.vi** is actually a high-level VI constructed from more fundamental File I/O functions. To gain more control of the file input/output process, let's first use the fundamental File I/O functions to write a program with the same capabilities as **Write To Spreadsheet File.vi**, and then use our deepened understanding to enhance its functionality.

All data storage ("file output") operations must perform the following three steps: *open a file*, *write data to the open file*, *close the file*. Retrieving previously stored data ("file input") is performed in the same three steps, except that data is read from the open file (rather than written to it). In **Functions>>Programming>>File I/O**, the following three icons are available to perform the three required steps of data storage: **Open/ Create/Replace File**, **Write To Text File**, and **Close File**. To understand how to wire these three icons together to accomplish data storage, we will first briefly describe the function of each individual icon.

The Help Window for **Open/Create/Replace File** is shown next. Given a file defined at its **file path** input, the job of this icon is either to open that existing file or to create it (if it doesn't already exist). If nothing is wired to **file path**, a dialog path will appear to input the file's path. Using the given file path as well as other configurational information such as the location within the file of the last I/O operation and degree of user access (e.g., read only), this icon also produces a "reference number" (called *refnum* for short). To pass this needed information to other File I/O icons, the refnum is available at the **refnum out** output terminal. Note that all of the inputs for this icon are optional (required inputs are shown in bold type). Thus, when **Open/Create/Replace File** executes, each input left unwired will take on its default value, which is shown in parentheses.

**Write To Text File**, whose Help Window follows here, performs the actual data storage. Once presented with the open file's *refnum* configurational information at its **file** input, this icon writes the ASCII string at its **text** input into the open file. Additionally, the refnum is available at the **refnum out** output terminal.

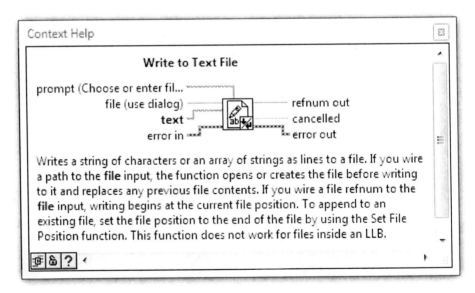

Finally, **Close File**'s Help Window is given below. This icon closes the file specified at its **refnum** input. Also, the file path, which is contained within the input refnum, is made available at the **path** output terminal.

Note that all three of the above icons include error reporting via the error cluster, which is appears at the **error in** and **error out** terminals.

The three-step data storage process is accomplished by wiring these three icons together as follows.

This wiring scheme takes advantage of the principle of LabVIEW programming called *data dependency*. Simply stated, data dependency means that an icon cannot execute until data is available at *all* of its inputs. In this diagram, all of the inputs to **Open/ Create/Replace File** are wired (the **Path Constant** is explicitly wired to **file path**; the other optional inputs are implicitly wired to their default values). So, when this diagram is run, **Open/Create/Replace File** executes immediately. Upon completion, **Open/ Create/Replace File** outputs a refnum at its **refnum out** terminal, which is passed through the wiring to Write to Text File's **file** input. Because of data dependency, **Write to Text File** cannot execute until it receives the refnum from **Open/Create/Replace File.** Then, after **Write to Text File** completes its execution, it passes the refnum from its **refnum out** terminal to Close File's **refnum** input. Only then can **Close File** execute. Thus, through this programming scheme, we are assured that the icons will execute in the desired sequence: **Open/Create/Replace File** followed by **Write to Text File** followed by **Close File**.

Also, the correct manner of chaining together File I/O icons for error reporting is shown above. If an error does occur at one point in the chain, subsequent icons will not execute and the error message will be passed to the **General Error Handler.vi**, which will display the message in a dialog box. **General Error Handler.vi** is found in

**Functions>>Programming>>Dialog & User Interface**. Rather than using this dialog box approach to error reporting, one could alternately pass error information to a front-panel error cluster for viewing (as you did in your Chapter 4 VIs).

With **Spreadsheet Storage** open, select **File>>Save As…**, and then use it to create a new VI named **Spreadsheet Storage (OpenWriteClose)** in the **YourName\Chapter 5** folder. Modify the block diagram to appear as shown below. The required icons are found in **Functions>>Programming>>File I/O** and **Functions>>Programming>>Dialog & User Interface**.

Pop up on the **operation** input of Open/Create/Replace File and select **Create>>Constant**. A block-diagram **Enumerated Constant** ("Enum") will be created. Place the 🖐 atop the **Enum**, click the mouse, then select the **replace or create** option. This option will be secured when you release the mouse button. With the path input at **file path**, this icon will then either create and leave open a new file (if a file with that path does not previously exist), or else open the existing file with that path and instruct the subsequent icons to overwrite ("replace") its existing data with new data. This mode of operation is the same as that programmed into **Write To Spreadsheet File.vi**. Also, note that, by leaving the other inputs (**prompt, access, error in**) to Open/Create/Replace File unwired, these inputs will be automatically programmed to their defaults values (**no prompt, read/write, no error**).

Finally, we need to supply **Write to Text File** with a string containing the 2D array of waveform values in spreadsheet formatting. This string can be created through the use of **Transpose 2D Array** (found in **Functions>>Programming>>Array**) and **Array To Spreadsheet String** (found in **Functions>>Programming>>String**). Complete the block diagram shown next, with the spreadsheet values formatted as *%8.5f*.

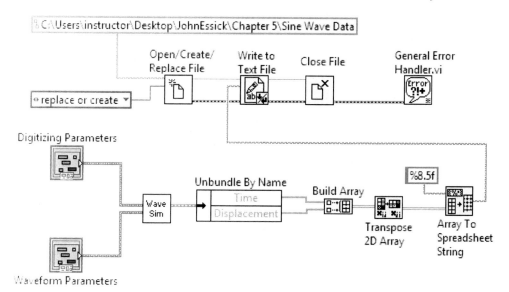

Save your work. Run **Spreadsheet Storage (OpenWriteClose)** with the same choice of values for its front-panel controls as when you ran **Spreadsheet Storage** a few minutes ago. Then use a word processor to verify that the file **Sine Wave Data** appears the same as when it was created using **Spreadsheet Storage** (which implements **Write To Spreadsheet File.vi**).

If interested, place **Write To Spreadsheet File.vi** on a block diagram, and then open this VI by double-clicking on it. Look at its block diagram and open (by double-clicking) its subVI **Write Spreadsheet String.vi**. You will find that **Open/Create/Replace File**, **Write to Text File**, and **Close File** are the basic building-blocks from which **Write To Spreadsheet File.vi** is constructed.

## 5.8 ADDING TEXT LABELS TO SPREADSHEET FILE

In the previous exercise, you created an ASCII text file with two tab-delimited columns of $(x, y)$ data, the first column containing a sequence of **Time** values, the next column containing the associated values of sine-wave **Displacement**. Because of the file format, you were able to use a word processor or spreadsheet application program to view the contents of this file. A desirable addition to this data file would be the ability to provide descriptive text at the top of each column, labeling each experimental quantity. More generally, one might wish to include all sorts of text statements within the file that would provide a complete record of the experimental conditions and choice of parameters under which the data were taken. Let's see how to provide labels at the top of each data column. Once this skill is in place, it's a small step to including any other desired text into the file.

In our text-based method of data storage, the entire set of $(x_i, y_i)$ values is contained within one long ASCII character string. LabVIEW indexes these $N$ values in the range of $i = 0$ to $N - 1$, so the structure of this spreadsheet-formatted string is as follows:

$$x_0 <Tab>\ y_0 <EOL>\ x_1 <Tab>\ y_1 <EOL> \ldots x_{N-1} <Tab>\ y_{N-1} <EOL>$$

One can then attach the labels **Time** and **Displacement** as the initial entries in the first and second columns, respectively, by simply adding these text characters as a spreadsheet-formatted prefix *Time* $<Tab>$ *Displacement* $<EOL>$. The string will then appear as:

$$\text{Time} <Tab>\ \text{Displacement} <EOL>\ x_0 <Tab>\ y_0$$
$$<EOL> x_1 <Tab> y_1 <EOL> \ldots x_{N-1} <Tab>\ y_{N-1} <EOL>$$

Thus, our task is to construct the label prefix, append the long ASCII string containing all the data values, and then store this entire string (termed a *byte stream*) in a file. Let's see how to do this.

The required byte stream is constructed using the **Concatenate Strings** icon, which is found in **Functions>>Programming>>String**. This icon splices together all its input strings into a single output string.

Delete the pink wire connecting **Array To Spreadsheet String** to **Write to Text File**, then place **Concatenate Strings** on your block diagram as shown next. You will find that at first this icon has only two inputs. Position the ⬉ at the icon's bottom center, where ⬉ will morph into a resizing handle, then expand the icon until it has five inputs.

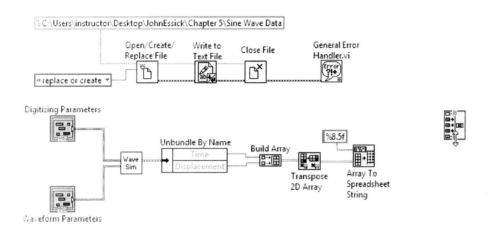

You will now construct the desired label prefix using the top four inputs of **Concatenate Strings**. Wire a **String Constant** (found in **Functions>>Programming>> String** or by popping up on the input) containing the text *Time* and *Displacement* to the first and third input of **Concatenate Strings**, respectively. Obtain a **Tab Constant** and an **End of Line Constant** icon from **Functions>>Programming>>String** and wire them to the second and fourth **Concatenate Strings** inputs, respectively.

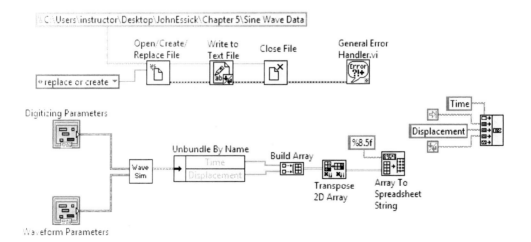

Complete construction of the desired byte stream by wiring Array To Spreadsheet String's **spreadsheet string** output to the fifth input of **Concatenate Strings**. Then, wire Concatenate Strings' **concatenated string** output to Write to Text File's **text** input.

Save your work. Run **Spreadsheet Storage (OpenWriteClose)** with some choice of values for its front-panel controls. Then view the resulting file using a word processor. As shown next, with Microsoft Word I found the desired labels atop each column just as we had planned. You may have to play with your word processor's tab settings to get the labels to align well with the numerical data.

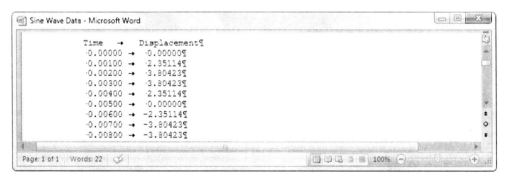

If available, try opening this text-based data file into a spreadsheet application program. In Excel, after a sequence of dialog windows within which I selected the option to begin data import at the file's first line, the data appeared as shown below. If the column labels are not needed, begin the data import at the file's second line. With the data input thusly an experimentalist can then analyze and graph his or her experimental results in Excel.

| | A | B | C |
|---|---|---|---|
| 1 | Time | Displacement | |
| 2 | 0 | 0 | |
| 3 | 0.001 | 2.35114 | |
| 4 | 0.002 | 3.80423 | |
| 5 | 0.003 | 3.80423 | |
| 6 | 0.004 | 2.35114 | |
| 7 | 0.005 | 0 | |
| 8 | 0.006 | -2.35114 | |
| 9 | 0.007 | -3.80423 | |
| 10 | 0.008 | -3.80423 | |
| 11 | 0.009 | -2.35114 | |
| 12 | | | |

## 5.9 BLACKSLASH CODES

To conclude, let's explore a shortcut method for including nondisplayable ASCII characters within a **String Constant**. The trick is to pop up on the **String Constant** icon, and then select '\' **Codes Display** (as opposed to **Normal Display**) from the menu that appears. When this option is activated, LabVIEW will interpret characters immediately following a backslash (\) as a code for nondisplayable characters in a manner listed in Table 5.2. You must use uppercase letters for the hexadecimal characters and lower case letters for the special characters such as Tab and CR.

**Table 5.2**

| CODE | LABVIEW INTERPRETATION |
|---|---|
| \00–\FF | Hexadecimal value of an 8-bit character |
| \b | Backspace (ASCII BS, equivalent to \08) |
| \f | Formfeed (ASCII FF, equivalent to \0C) |
| \n | Line Feed (ASCII LF, equivalent to \0A) |
| \r | Carriage Return (ASCII CR, equivalent to \0D) |
| \t | Tab (ASCII HT, equivalent to \09) |
| \s | Space (equivalent to \20) |
| \\ | Backslash (ASCII \, equivalent to \5C) |

Let's demonstrate the convenience of backslash codes by employing them in the construction of the labeling-string *Time <Tab> Displacement <EOL>*. To build this string, I must tell you that the EOL marker is platform-dependent. The definitions for EOL peculiar to each computing systems are:

Windows: Carriage Return, then Line Feed (\r\n)
Macintosh: Carriage Return (\r)

Thus, by enabling the '\' **Codes Display** option, the entire column-labeling text can be constructed by inserting *Time\tDisplacement\r\n* within a single **String Constant**, assuming we are working on a Windows system. However, if in the future an attempt is made to port this program to another platform, the explicit expression of the EOL character will cause problems. Due to this portability issue, it is good programming practice to take the following slightly modified approach to the string construction: Enclose *Time\tDisplacement* within a single **String Constant**, then concatenate this object with the **End of Line** icon found in **Functions>>Programming>>String**. The **End of Line** icon is a "smart" EOL character. At compile time, it checks which platform is currently being implemented and automatically generates the appropriate EOL character(s), thus infusing the program with portability.

Here is a step-by-step procedure for modifying **Spreadsheet Storage (OpenWriteClose)** to accomplish the above. Start by replacing all of the labeling-string construction code by a **Concatenate Strings** icon with three inputs. Wire an empty **String Constant** to the top input (by popping up and selecting **Create>>Constant**), an **End of Line** icon (from **Functions>>Programming>>String**) to the middle input, and Array To Spreadsheet String's **spreadsheet string** output to the bottom input.

Next, pop up on the empty **String Constant** and enable backslash coding by selecting '\' **Codes Display** from the pop-up menu. Then, using the 🖑, enter *Time\tDisplacement* into the **String Constant**, then press the numeric keypad's *<Enter>*.

When you are finished, you can resize and reposition this object using the ⤡, if needed. Then wire the **Concatenate Strings** output to the Write to Text File.vi's **text** input.

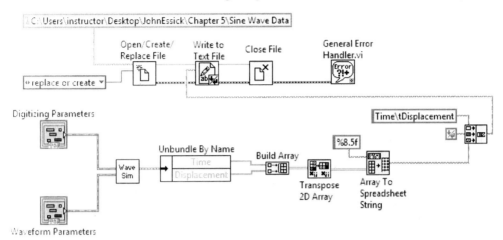

Save your work, and then run the program. Use a word processor to verify that backslash coding produces a data file with the desired column labeling.

## DO IT YOURSELF

Write a VI called **Spreadsheet Read**, which reads and displays data from a two-column spreadsheet-formatted disk file. Name the first and second column of data within the spreadsheet file the *X Array* and *Y Array*, respectively. Construct **Spreadsheet Read** to do the following: Open the desired two-column spreadsheet file, read its contents, and then plot the *Y Array* versus *X Array* on an **XY Graph** as well as display these arrays in an indicator cluster labeled **output cluster**. Design an icon for your VI and assign its connector terminals consistent with the following Help Window. Save this VI in **YourName\Chapter 5**.

When **Spreadsheet Read** is completed, run **Spreadsheet Storage** with its front-panel controls set to the values shown below. After **Spreadsheet Storage** executes, there will be a two-column spreadsheet data file named **Sine Wave Data** in the **YourName\Chapter 5** folder with its X Array and Y Array containing Time and Displacement values for five cycles of a 50 Hz sine wave of amplitude 5.

Next, use **Spreadsheet Read** to read **Sine Wave Data**. After successfully running this VI, the front panel of **Spreadsheet Read** should appear as below.

Here are three helpful tips:

- Use **Read from Spreadsheet File.vi** (found in **Functions>>Programming>>File I/O**) to open and read the contents of **Sine Wave Data**. Leave **file path** unwired so that the user will be prompted for the file name via a dialog box. The data will be output as a 2D array at the **all rows** output.
- A column of a 2D array can be "sliced off" (that is, isolated as a 1D array) using **Index Array** (found in **Functions>>Programming>>Array**). When first placed on a block diagram, **Index Array** has just one **index** input (which appears as a small black box) as shown below at the left. However, when a 2D array is wired to its **n-dimensional array** input, two **index** inputs appear as shown below in the middle. The top and bottom **index** inputs are the row and column indices of the 2D array, respectively. To slice off the first column (indexed as column 0), just wire a **Numeric Constant** of value *0* to the bottom (column) **index** input, as shown below at the right. The top (row) **index** input is left unwired. The **subarray** output of Index Array will then be a 1D array containing the values of column 0 from the input 2D array.

Thus, to slice off both columns of a two-column 2D array, the required subdiagram is as follows.

- Create **output cluster** from the block diagram by popping up on an appropriate block-diagram object and using **Create>>Indicator**. Then, on the front panel, pop up on the index display of each Array Indicator within **output cluster**, select **Visible Items>>Label**, and then enter its appropriate name.

## Problems

1. Write a program called **Read Spreadsheet Text**, which uses the icons **Open/Create/Replace File**, **Read from Text File**, and **Close File** to read the contents of a

spreadsheet-formatted text file and display the resulting string in a front-panel **String Indicator** as well as a decoded numeric representation in a numeric array indicator.

(a)    Create a spreadsheet-formatted text file by running **Spreadsheet Storage** with its front-panel controls set to the values shown below. After **Spreadsheet Storage** executes, there will be a two-column spreadsheet text file with ten rows named **Sine Wave Data** in the **YourName\Chapter 5** folder.

Next, read **Sine Wave Data** by running **Read Spreadsheet Text**. Does the text displayed in the front-panel indicator appear as expected? Pop up on this indicator and select **'\' Codes Display**. Nondisplayable ASCII characters will now be visible in the indicator. Explain the pattern of characters that you now see.

(b)    On the block diagram, use the icon **Spreadsheet String To Array** (found in **Functions>>Programming>>String**) to convert the text string to a 2D double-precision floating-point array with a format consistent with that used in creating **Sine Wave Data**. Display this 2D array in a front-panel indicator. Run **Read Spreadsheet Text**. Is the array of data properly represented in the front-panel array indicator?

2.    A LabVIEW programmer modifies **Spreadsheet Storage** as shown next in an attempt to add column labels to the stored spreadsheet data.

With **Spreadsheet Storage** open, use **File>>Save As...** to create a new VI called **Spreadsheet Storage with Labels** in **YourName\Chapter 5**. Modify the block diagram of this VI so that it replicates the code shown above. Run this program and demonstrate (with the help of a word processor) that it does not work as hoped.

(a) Explain why this code does not accomplish its stated goal.

(b) Fix the block diagram code so that it functions properly. You should just have to change one wire.

(c) Once the code functions properly, explain why this approach to achieving column labeling is less efficient than that used in **Spreadsheet Storage (OpenWriteClose)**.

3. Write a For Loop–based program called **Four Text Files**, which (with no runtime input from a user) creates four separate files named **Text0.txt**, **Text1.txt**, **Text2.txt**, and **Text3.txt**, each containing a unique text message, and stores these files in the **YourName\Chapter 5** folder. Create one file during each iteration of the For Loop. The file's path can be constructed using **Build Path** (found in **Functions>>Programming>>File I/O**) as well as **Format Into String** (found in **Functions>>Programming>>String**. This latter icon can be expanded so it has multiple inputs, for example, one for an **I32** integer and another for the string **.txt**. Make the text messages in **Text0**, **Text1**, **Text2**, and **Text3** be *This is file 0*, *This is file 1*, *This is file 2*, and *This is file 3*, respectively.

4. Write a VI called **Binary Storage**, which stores the 2D array of **Time** and **Displacement** values produced by **Waveform Simulator** in a binary-formatted file named **Sine Wave Data**. Much of

the program you will write is similar to **Spreadsheet Storage (OpenWriteClose)**, so you may want to begin there and create **Binary Storage** using **File>>Save As...**and save it in **YourName\Chapter 5**. The front panel of **Binary Storage** should look as follows.

On the block diagram, implement the required three-step data storage process using **Open/Create/Replace File**, **Write to Binary File**, and **Close File** found in **Functions>>Programming>>File I/O**. Wire a TRUE **Boolean Constant** to Write to Binary File's **prepend array or string size?** input. This selection instructs LabVIEW to begin the binary file with an 8-byte header, which records the number of rows and columns into which the data is arranged.

(a)    Run **Binary Storage** with its front panel controls programmed as shown above. Find the icon for the binary-formatted **Sine Wave Data** file in the **YourName\Chapter 5** folder and determine its size (in kilobytes). The file's exact size is found by right-clicking on the icon, and selecting **Properties**. Given the front-panel input values and binary formatting, explain why **Sine Wave Data** has the size it does.

(b)    Open **Sine Wave Data** with a word processor. Offer a qualitative explanation for what you observe. Close **Sine Wave Data**.

(c)    Run **Spreadsheet Storage** with its front-panel controls programmed with the same values as shown above. Determine the size (in kilobytes) of the spreadsheet-formatted **Sine Wave Data**. You will find that this file is now slightly larger than its previous binary-formatted size. Given the front-panel input values, explain why **Sine Wave Data** has the size it does under spreadsheet formatting.

(d)    By changing a particular parameter on the block diagram of **Spreadsheet Storage**, you can cause spreadsheet-formatted **Sine Wave Data** to be approximately twice as large as binary-formatted **Sine Wave Data**. Predict the block-diagram change needed to result in this file-size doubling. Make this change and verify your prediction.

5.  Create a binary-formatted file called **Sine Wave Data** by running **Binary Storage** (the VI written in Problem 4) with the choice of front-panel control values shown next. In **Binary**

**Storage**, make sure that the Write to Binary File's **prepend array or string size?** input is wired TRUE, so that an 8-byte informational header is inserted at the beginning of the **Sine Wave Data** file.

Your assignment is to write a VI called **Binary Plot**, which reads the binary-encoded **Time** and **Displacement** data within **Sine Wave Data** and then plots these data on an **XY Graph**. When the completed VI is run, its front panel should look as follows.

Here are some guidelines to follow in writing your program:

- Retrieval of data from a file is a three-step (*open, read, close*) process. These three steps are accomplished by configuring the appropriate File I/O icons (found in **Functions>>Programming>>File I/O**) as shown below. The 8-byte header at the beginning of the **Sine Wave Data** file informs **Read from Binary File** about the size of the data file to be read as well as the required form (i.e., the number of rows and columns) of the data array output.

- The fact that the data consists of a 2D array of double-precision floating-point numerics is communicated to **Read from Binary File** by wiring a 2D DBL Array Constant to its **data type** input as shown in the next diagram. This constant inputs no data to **Read from Binary File**, but rather simply specifies the desired data type for the output data. To form this Array Constant, first place an **Array Constant** shell from **Functions>>Programming>>Array** on the block diagram. Then, place a **Numeric Constant** (from **Functions>>Programming>>Numeric**) within the Array Constant shell, pop up on the **Numeric Constant,** and select **Representation>>Double Precision (DBL)**. Finally, pop up on the Array Constant's **index display** and select **Add Dimension**. The icon is now a 2D Array Constant, where the top and bottom index display show the row and column index, respectively.

- The 2D data array emanating from Read from Binary File's **data** output has two rows, where the first and second rows contain the **Time** and **Displacement** values, respectively (note that this 2D array has never been transposed, in contrast to the 2D array involved in the spreadsheet VIs that we wrote). Each of these rows must be sliced off in preparation for plotting on an **XY Graph.** To accomplish this slicing operation, use **Index Array** (from **Functions>>Programming>>Array**) with a **Numeric Constant** of appropriate value wired to its top (row) **index** input. Leave the bottom (column) index input unwired (see the helpful tips section at the end of the Do It Yourself project in this chapter).

6. Write a program called **Palindrome Detector**, which determines whether an input word, phrase, or sentence reads the same backward as forward. As shown below, a **Round LED** indicator is lit on the front panel when the input string is determined to be a palindrome.

Construct your block diagram using icons from **Functions>>Programming>>String** and its subpalette **Additional String Functions**. Before comparing the equality of the input string with its reverse, first change the input to all lowercase characters and also remove all spaces. The spaces can be removed by replacing each space character with an **Empty String Constant**.

When completed, test **Palindrome Detector** with the following inputs: *radar, Able was I ere I saw Elba*.

7. The day of the week for a given (Gregorian calendar) date can be found using the following formula called Zeller's Congruence:

$$day = \text{mod}(I, 7)$$

where

$$I = D + floor\left[(M + 1)2.6\right] + Y + floor\left[\frac{Y}{4}\right] + floor\left[\frac{C}{4}\right] + 5C$$

Here, the *floor [x]* function truncates its argument *x* to the closest lower integer and the modulo function mod(*I*,7) equals the remainder when the integer *I* is divided by 7. The quantities appearing in Zeller's Convergence are defined as follows:

    *day* = day of the week (0 = Saturday, 1 = Sunday,..., 6 = Friday)
    *D* = day of the month
    *M* = month (3 = March, 4 = April, 5 = May,..., 10 = December)
    *C* = century given by *floor* (*year*/100)
    *Y* = year within century given by *mod* (*year*, 100)

For the months of January and February, *M* = 13 and *M* = 14 for the previous year, i.e., *year* → *year* −1.

207

Write a program called **Day of the Week** which implements Zeller's Convergence to find the day of the week for a given date. The front panel of this VI should appear as shown below. The date is input in a **String Control** in the format *mm/dd/yyyy*, and the result displayed as text (e.g., *Thursday*) in a **String Indicator**.

On the block diagram, use icons from **Functions>>Programming>>String** and its subpalettes (e.g., **Match Pattern, Decimal String To Number**) to parse the input string and convert it into numeric representations of M, D, and year. Also, create a seven-element array of strings (e.g., using **Build Array** or a block-diagram **Array** shell containing with a **String Constant**) where the index-zero, index-one,…, index-six elements are the strings Saturday, Sunday,…, Friday, respectively. Then, use **Index Array** to select the correct element from this array to output to **Day of the Week**. Finally, the icon **Select** from **Functions>>Programming>>Comparison** may come in handy.

Verify that your VI works correctly by finding the day of the week for July 4, 1776 (Thursday) and January 1, 2000 (Saturday).

8. Computer-controlled scientific instruments often report their measurement results in the form of an ASCII string. As an example, assume a multimeter sends the string *VOLTS DC +1.345E+02* in which *VOLTS DC* identifies the type of measurement being reported and *+1.345E+02* is the actual measured value. In order for this measured value to be useful within a program (e.g., plotted on a chart, input to a calculation), the string received from the multimeter has to be "parsed" (i.e., the identifier portion separated for the measured value portion) and the measured value portion converted from its ASCII string representation to a numeric format. Write a program called **Parse and Convert Multimeter String**, in which *VOLTS DC +1.345E+02* is input in a front-panel **String Control**. On the block diagram, this string is then parsed appropriately and the measured value portion is converted to the double-precision floating-point format. The needed icons for these operations are found in **Functions>>Programming>>String** and its subpalette **String/Number Conversion**.

When properly functioning, the front panel of **Parse and Convert Multimeter String** should appear as shown below. In the String Control, **"\" Codes Display** is selected so that nondisplayable characters are made visible.

# Chapter 6

## Shift Registers

A looping structure allows one to implement a common requirement in computer programming—repeating the same operation many times. In earlier chapters, we found that LabVIEW provides two such looping structures, the For Loop and the While Loop. We discovered that, through the creation of a tunnel at its border, each of these loop structures possesses a form of memory. By toggling on the tunnel's auto-indexing feature, the loop (once it has completed its full execution) will output an array in which is stored the sequence of values created by the succession of loop iterations. Often, however, another form of memory is needed—that which interconnects successive loop iterations. In this guise, a value created by the previous iteration is transferred for use within the calculations of the present iteration. This form of memory is called a *local variable* and can be accomplished in LabVIEW's looping structures via the creation of a *shift register*.

Shift registers are created by activating a pop-up menu at a loop boundary. For example, you may create a shift register by popping up on a For Loop's border, then selecting **Add Shift Register**.

When you release the mouse button, the shift register will appear. It is comprised of a pair of terminals directly opposite each other on the vertical sides of the loop border.

The right terminal stores a value upon completion of an iteration. This stored value is then shifted to the left terminal for use in calculations during the next iteration. Such a feature is useful, for example, in summing the components of a quantity calculated over the span of the entire set of loop iterations.

You can configure the shift register to remember values from more than one previous iteration. To accomplish this feat, pop up on the shift register's left terminal and select **Add Element** from the menu.

Upon this selection, a second left terminal will appear.

By repeating this operation, you can create as many left terminals as desired.

In the previous illustration, we have created three left terminals that will function as follows. When a subdiagram within the For Loop (not shown) is calculating some quantity during the $i^{th}$ loop iteration, the top left terminal contains the value of this quantity calculated during the $(i-1)^{th}$ iteration, the middle left terminal contains the value from $(i-2)^{th}$ iteration, and the bottom left terminal contains the value from $(i-3)^{th}$ iteration. Over the course of the complete For Loop execution, this set of left terminals will behave as a First In, First Out (FIFO) digital shift register, if that is a familiar concept to you.

In this chapter, you will explore the use of local variables in LabVIEW programming. First, you will use a shift register to integrate a discrete data set numerically. Then, with a shift register that recalls values from two past iterations, you will differentiate this same data set.

## 6.1 POWER FUNCTION SIMULATOR VI

Imagine some physical system for which one of its measurable quantities $y$ (for example, temperature or pressure) is a function of the continuous variable $x$ (e.g., position or time). That is, $y$ is perfectly described by $y = f(x)$, where $f$ is an analytic function. If an experimenter sets out to determine the function $f$, he or she will have to confront this reality: It is impossible to design an experiment that samples the quantity $x$ and its associated $y$ values in a continuous manner. One must instead settle for discretely sampling $x$ a total of $N$ times and recording each associated $y$ value. Thus the experiment will result in a set of $(x, y)$ data samples, where $x = x_0, x_1, x_2,..., x_{N-1}$, and $y = f(x_0), f(x_1), f(x_2),..., f(x_{N-1})$.

In the next exercise, we will study the use of shift registers by writing a program that numerically integrates a discretely sampled data set. In an actual computer-controlled experiment, the discretely sampled data set might be obtained through the use of LabVIEW's built-in data acquisition (DAQ) VIs. Given this set, an integration VI might be used to perform some appropriate analysis on the data. Here, we will use software to generate a simulated set of discretely sampled $(x, y)$ data that follows the power function $y = ax^b$, where $a$ and $b$ are the *prefactor* and *power*, respectively. For our present purposes, it is actually beneficial to use simulated data. In contrast to a real experiment, we will know the exact functional form by which the data was generated, and this form is easy to integrate analytically. Thus we will be able to judge the accuracy of our numerical integration methods by comparison with an analytical result.

We desire a VI that, given a lower limit $x_0$ and upper limit $x_{N-1}$, creates $N$ equally spaced points along an $x$-axis, then produces an associated array of $y = ax^b$ values. Open a blank VI and use **File>>Save** to first create a folder named **Chapter 6** within the **YourName** folder, then save this new program under the name **Power Function Simulator** in **YourName\Chapter 6**. As shown next, on the front panel place two **Cluster** shells (found in **Controls>>Modern>Array, Matrix & Cluster**), then label them **Digitizing Parameters** and **Function Parameters**. Within the **Digitizing Parameters** cluster, place three **Numeric Controls** labeled **Number of Samples, Lower Limit,** and **Upper Limit**; within the **Function Parameters** cluster, place two **Numeric Controls** labeled **Prefactor** and **Power**, respectively. Set the representation of **Number of Samples** to I32. Finally, place an **XY Graph** on the front panel with its $x$- and $y$-axis labeled **x** and **y=a\*x ^ b**, respectively.

As an aid in arranging these front-panel objects into a pleasing pattern, you might enjoy experimenting with the **Alignment Tool** and **Distribution Tool** in the toolbar. To use these tools, you must first use the to highlight the group of objects on which you wish to operate.

Pop up on the icon pane, select **Edit Icon. . .** , and then design a new icon using the Icon Editor. Save your design by pressing the **OK** button.

$N$ equally spaced sampled values of $x$, ranging from $x_0$ (lower limit) to $x_{N-1}$ (upper limit) can be created by determining the necessary spacing between neighboring points $\Delta x$:

$$\Delta x = \frac{x_{N-1} - x_0}{N - 1}$$

Then

$$x_i = x_0 + i\,\Delta x \quad i = 0, 1, 2, \ldots, N-1$$

Once the array of $x$-values is known, then $y_i = f(x_i)$ can be calculated, where (for simplicity) we are taking $f$ to be the power function.

Switch to your block diagram and write code shown on the next page that produces the $x$ and $y$ arrays in the manner described above. Note that the Mathscript operator .^ (period followed by a caret) performs the element-wise power operation. Alternately, the Mathscript command *power* $(x, b)$, which calculates $x^b$, can be used. Remember to pop up on the Mathscript node $x$ and $y$ outputs and select **Choose Data Type>>1D-Array>>DBL 1D**. The data type for the *delta_x* output is **Scalar>>DBL**. Also **output cluster** is easily created by popping up on the output of the associated **Bundle** icon and selecting **Create>>Indicator**.

Return to the front panel and use the ⬐ to position all objects nicely. In **output cluster**, label the top, middle, and bottom Array Indicator as **x array**, **y array**, and **delta_x**, respectively (an Array Indicator's label is made visible by popping up on its index display and

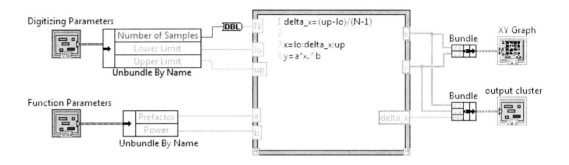

selecting **Visible Items>>Label**). It may be helpful to activate **AutoSizing>>Size to Fit** in the cluster's pop-up menu so that it automatically resizes as you add labeling. If you make mistakes while doing this operation, **Edit>>Undo**, or its keyboard short cut <*Ctrl+Z*>, can come in handy.

Pop up on the icon pane and select **Show Connector**. You can either keep the default 12 terminal pattern, or else pop up on the icon pane again and select another (e.g., two-input, one-output) connector from the **Patterns** palette. Assign the terminals according to the following scheme, then pop up on the icon pane and select **Show Icon**.

Input values for the five controls, then run your program to verify that it works. An example of a successful run is shown below. Using **File>>Save**, store your completed program in **YourName\Chapter 6**, then close its window.

## 6.2 NUMERICAL INTEGRATION USING A SHIFT REGISTER

In the following illustration (Figure 6.1), we have a sequence of $x_i$ values that are spaced apart by a constant step $\Delta x$. The solid curve represents a function $f(x)$ that has known values (shown as solid dots) at each $x_i$. Let $f(x_i) \equiv y_i$.

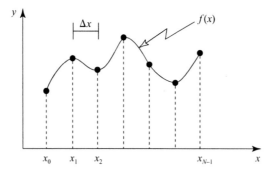

**Fig. 6.1**

We wish to integrate the function $f(x)$ between $x_0$ and $x_{N-1}$. This integral can be interpreted geometrically as the total area enclosed under the $f(x)$ curve between the two endpoints. In the previous illustration, two adjacent dotted lines at the abscissas $x_i$ and $x_{i+1}$ define a columnar-shaped area bounded on top by the curve $f(x)$. We note that the sum of the areas of all such columnar shapes in the region between $x_0$ and $x_{N-1}$ is equal to the integral we want to evaluate. Thus, if we have a general method for determining the area of each columnar shape, we can evaluate the integral.

## 6.3 NUMERICAL INTEGRATION VIA THE TRAPEZOIDAL RULE

The obstacle, of course, to writing a general formula for the area of a particular columnar shape is that each area has a curved top. In the *Trapezoidal Rule* for numerical integration, one assumes that the curve $f(x)$ at the top of each columnar shape can be approximated by a straight line. This approximation becomes better and better as $\Delta x$ is made smaller and smaller.

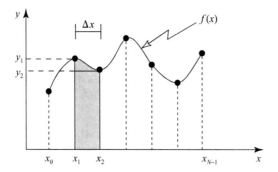

**Fig. 6.2**

Then, for example, the shaded columnar area in Figure 6.2 between $x_1$ and $x_2$ is approximated as the trapezoidal area given by

$$y_2 \, \Delta x + \frac{1}{2} (y_1 - y_2) \Delta x = \frac{1}{2} (y_1 + y_2) \Delta x$$

From this equation, we find that the Trapezoidal Rule is equivalent to assuming that the columnar shape defined by $x_1$ and $x_2$ is approximately a rectangle with height equal to the average of $f(x_1)$ and $f(x_2)$. So,

$$\int_{x_1}^{x_2} f(x) \, dx \approx \left[ \frac{y_1 + y_2}{2} \right] \Delta x$$

where the graphical interpretation of this relation is shown in Figure 6.3.

**Fig. 6.3**

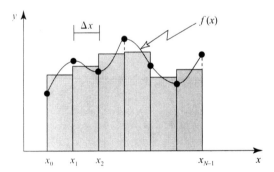

**Fig. 6.4**

Using the Trapezoidal Rule, the approximation for the entire integral then is given in Figure 6.4.

Thus

$$\int_{x_0}^{x_{N-1}} f(x) \, dx \approx \left[ \frac{y_0 + y_1}{2} \right] \Delta x + \left[ \frac{y_1 + y_2}{2} \right] \Delta x + \dots + \left[ \frac{y_{N-2} + y_{N-1}}{2} \right] \Delta x$$

or

$$\int_{x_0}^{x_{N-1}} f(x)\, dx \approx \left[ \frac{1}{2} y_0 + y_1 + y_2 + \dots + y_{N-2} + \frac{1}{2} y_{N-1} \right] \Delta x$$

[1]

In our program, we will find it most convenient to write this expression in the following equivalent way:

$$\int_{x_0}^{x_{N-1}} f(x)\, dx \approx \left[ \{ y_0 + y_1 + y_2 + \dots + y_{N-2} + y_{N-1} \} - \frac{1}{2} \{ y_0 + y_{N-1} \} \right] \Delta x$$

[2]

On the right-hand side (RHS) of Equation [2], the first and second term in the square brackets are the sum of *y* array values and one-half the sum of the *y* array endpoints, respectively.

## 6.4 TRAPEZOIDAL RULE VI

Assume that an experiment has produced a set of *N* equally spaced (*x*, *y*) data points, where the *x*-axis spacing between points is $\Delta x$ and the variation of *y* can be described by some function *f(x)*. Let's write a VI that, given the *y* array and $\Delta x$, numerically integrates *f(x)* between the extremal *x*-axis values of $x_0$ and $x_{N-1}$ using the Trapezoidal Rule.

Construct the front panel below and save this VI under the name **Trapezoidal Rule** in the **YourName\Chapter 6** folder. Provide plenty of **Digits of precision**, say *10*, for the **Value of Integral** indicator. Remember the **y array** input is formed by placing a **Numeric Control** inside of an **Array** shell. Design an icon and assign the connector's terminals consistent with the Help Window shown.

Now switch to the block diagram. Add a For Loop to your diagram and place a shift register on its boundary. To add the shift register, pop up on either of the For Loop's vertical borders and select **Add Shift Register**. You can then reposition the terminal pair using the ▶, if desired.

We are going to use the For Loop to compute the sum of the *y*-array values of our data set. In a text-based language, we would write something along the lines of the following:

*Sum* = 0.0
*For i* = 0, *N* − 1
    *Sum* = *y* (*i*) + *Sum*

To implement this code in LabVIEW, initialize the shift register to zero by wiring a **Numeric Constant** (defined as *0.0*) to its left terminal. Place an **Add** icon within the loop and wire an input and its output to the shift register's left and right terminal, respectively.

Now, we take advantage of a handy LabVIEW convenience: In addition to its (previously studied) ability to build an array at a loop's output, auto-indexing can be used to sequence through an array at a loop's input. First, complete the following wiring, and then I'll explain its meaning. Simply wire the **y array** control's terminal to the remaining **Add** input.

Note that the just-completed wire is thick (denoting an array) outside the loop, then becomes thin (denoting a scalar) inside the loop. Why is this so? Auto-indexing is activated in a For Loop by default, so upon execution, the loop will sequentially input one element from **y array** each time it iterates. That is, on the first loop iteration when ⬜ equals 0, the element of **y array** with index 0 will be input, on the second iteration when ⬜ equals 1, the element of **y array** with index 1 will be input, and so on until the end of the array is reached. As an added convenience, when auto-indexing is enabled on an N-element array entering a For Loop, LabVIEW automatically sets the loop's count terminal to N, thus eliminating the need to wire a value to ⬜.

The previous diagram works like this. The For Loop steps through the complete set of **y array**'s elements, accumulating the sum of these numeric values as it goes. During a particular iteration, the shift register's left terminal provides the sum accumulated up through the last iteration. The newly indexed array element is added to this sum and the result is stored in the shift register's right terminal for use in the next loop iteration. When the For Loop completes its operation, the sum of all of **y array**'s elements is contained in the shift register's right terminal. Thus, the value output from this terminal is the first term within the square brackets on the RHS of Equation [2].

We might flirt with the idea of neglecting the second term within the square brackets on the RHS of Equation [2]. This term is an endpoint correction to the integral expression. Since it involves the sum of just two array elements, it may be small compared to the first term calculated above. If we were to neglect this small correction term, the integral would be determined by simply multiplying the For Loop output by $\Delta x$, as shown below.

With not too much more work, though, we'll be able to write the code necessary to determine the endpoint correction, which, from Equation [2], is given by one-half the sum of the first and last $y$-array elements. We can explicitly include this code on the present diagram, but let's take advantage of the modular nature of LabVIEW programming by tucking this code away within its own VI called **Endpoints**. We will then include **Endpoints** as a subVI on the **Trapezoidal Rule** diagram.

Open a new front panel and save it in **YourName\Chapter 6** under the name **Endpoints**. Place a **Numeric Control** within an **Array** shell on the panel and label it **Array**. Also label a **Numeric Indicator** as **Half of Endpoints Sum**. Pop up on the icon pane, select **Edit Icon...**, and then design an appropriate icon.

Pop up on icon pane again and select **Show Connector**. Assign the terminals in a manner consistent with the following Help Window.

Switch to the block diagram. Here, given the *N*-element **Array**, we need to extract its first (index 0) and last (index *N* − 1) elements. The **Index Array** icon, found in **Functions>>Programming>>Array**, performs this function. Its Help Window is shown next.

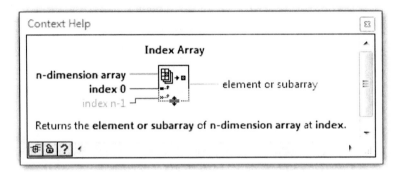

When operating on a two-dimensional (or higher) array, proper configuration of this icon can require some thought. But for our present needs, involving the one-dimensional **Array**, **Index Array** is exceedingly easy to use. One simply wires the 1D array and an integer-containing **Numeric Constant** to the **n-dimensional array** and **index** inputs, respectively. The integer denotes the index of the array element to be extracted. Once extracted, the numeric value of this element appears at the icon's **element** output terminal.

Another related icon that will prove useful is **Array Size**, also found in **Functions>>Programming>>Array**. As shown in its Help Window below, this icon determines the number of elements present within an array. In the one-dimensional case, given an *N*-element 1D array at its input, the icon returns the integer (**132**) *N* at its output.

Complete the following block diagram. The left half of this diagram extracts the index 0 and index $N-1$ elements from **Array**, while the right half adds these two numeric values together then divides by two. Save your work, then close **Endpoints**. The convenient **Decrement** icon is found in **Functions>>Programming>>Numeric** and can be used to construct the integer $N-1$.

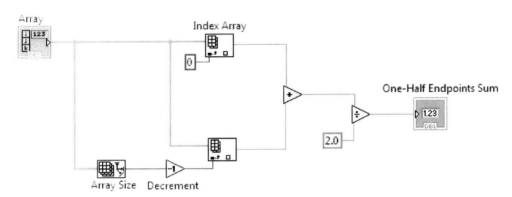

Alternately, with slightly more work, you can find $N-1$ using the **Subtract** icon.

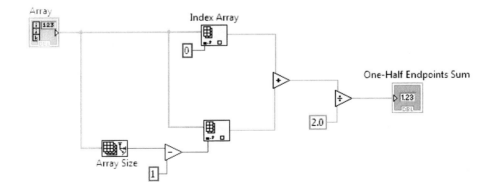

Now return to the **Trapezoidal Rule** block diagram. Using **Functions>>Select a VI...**, include **Endpoints** as a subVI on this diagram in the manner shown below. The program now fully manifests all aspects of Equation [2]. Save your work as you close this VI.

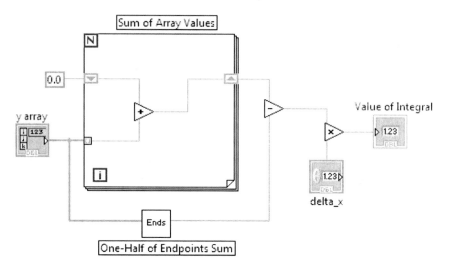

Let's see if the **Trapezoidal Rule** VI really works by supplying it with some known data from **Power Function Simulator**. Construct the following front panel and block diagram called **Trapezoidal Test** and save it in **YourName\Chapter 6**. It's easiest to first code the block diagram, and then use icons' automatic creation feature to make the front-panel objects.

Digitizing Parameters

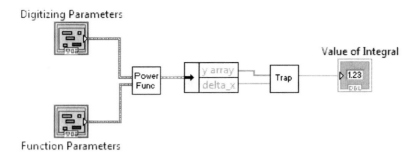

Function Parameters

Value of Integral

Assign the icon pane's connector terminals consistent with the following Help Window.

Let's use **Trapezoidal Test** to evaluate the integral $\int_0^1 5x^4\,dx$ numerically. It is easy to show analytically that this integral is equal to exactly one. Knowing this true value, we'll be able to check the precision of our approximate numerical method.

Enter *0* and *1* on **Trapezoidal Test**'s front-panel **Lower Limit** and **Upper Limit** controls, respectively, and choose **Prefactor** and **Power** to be *5* and *4*, respectively, so that **Power Function Simulator** is programmed to calculate $f(x) = 5x^4$.

The appropriate value for **Number of Samples** is more problematic. A larger value for this parameter makes $\Delta x$ smaller. The smaller you make $\Delta x$, the less the function $f(x)$ will vary across the top of each columnar shape whose area is needed in the calculation of the integral. Thus, the less bent the top of the columnar shape, the more closely it resembles the straight line assumed by the Trapezoidal Rule approximation. So a large value for **Number of Samples** will results in more accuracy for the Trapezoidal Rule approximation, with the drawback that the program will take longer to run. Try running your program with a variety of choices for **Number of Samples** and note the resulting **Value of Integral** in each case. Can you find a good compromise between high precision and acceptable runtime? Remember, when evaluated analytically, the value of the integral is exactly 1.0000000. . . .

## 6.5 CONVERGENCE PROPERTY OF THE TRAPEZOIDAL RULE

To explore the effect of **Number of Samples** in determining the precision of the resulting integral value, write the following program called **Convergence Study (Trap)**. Place an **XY Graph** on a new front panel and label its *x*- and *y*-axis **Number of Samples** and **Value of Integral**, respectively. Save this VI in **YourName\Chapter 6**.

Switch to the block diagram. Our plan is to write code that increments **Number of Samples** from a small to large value, calculates (using **Trapezoidal Test**) and stores the integral value at each step, and then displays the resulting array of values on the **XY Graph**. Produce the diagram shown below. Here, the **For Loop** and **Add** function will produce **Number of Samples** values ranging from *10* to *200* as the loop iterates. The two control clusters are created by popping up on the inputs of **Trapezoidal Test** and selecting **Create>>Control**.

There is one last problem to solve, namely, achieving block-diagram control over the value of **Number of Samples** that is input to **Trapezoidal Test**. As the VI is presently written, the value for **Number of Samples** can only be controlled on the front panel, and this value is transferred to **Trapezoidal Test** within the cluster wire emanating from the **Digitizing Parameters** terminal. However, the icon **Bundle By Name** will allow us to intercept the **Number of Samples** value "locked" within the block-diagram cluster wire and change its value before being passed on to **Trapezoidal Test**'s input.

Here's the procedure for accomplishing this feat. First, delete the cluster wire connecting the **Digitizing Parameters** terminal and **Trapezoidal Test**. Then place **Bundle By Name** (found in **Functions>>Programming>>Cluster, Class, & Variant**) as shown.

Next, wire from the **Digitizing Parameters** terminal to the central terminal of **Bundle By Name**. When this connection is made, the input (left) terminal of **Bundle By Name** will provide access to the first (index-zero) element in this cluster by default. If you want access to a different element, simply pop up on this input and use **Select Item**, or expand **Bundle By Name** to gain access to all the cluster's elements.

Finally, complete the block diagram coding as shown below. The cluster input to **Trapezoidal Test** will now have the value of **Number of Samples** generated by the **Add** icon (rather than the value from the front-panel control).

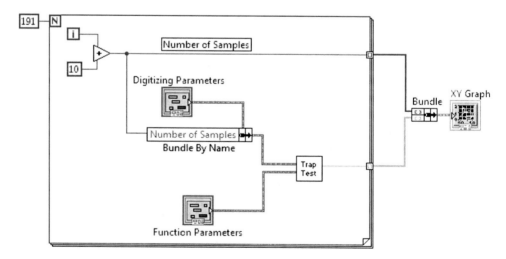

Switch back to the front panel and tidy it up using the 🔖. Save your work.

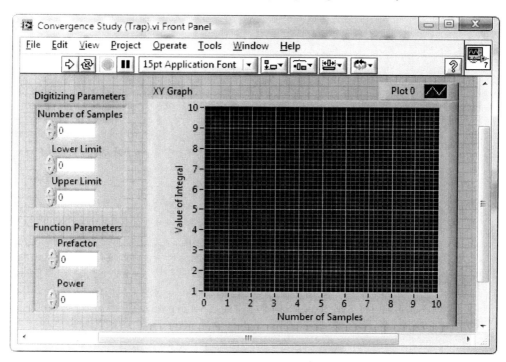

Run **Convergence Study (Trap)** with it programmed to calculate the integral $\int_0^1 5x^4 \, dx$. The value of **Number of Samples** on the front panel will not matter; it is controlled by the block diagram now. You should find that a (diffusely defined) threshold value for **Number of Samples** exists. Increasing **Number of Samples** to values greater than this threshold value only produces incremental enhancement in the accuracy of the integral value. Such a study of the convergence properties of your calculational method would be quite useful in making an intelligent choice for the parameter **Number of Samples**, if you were to use this VI to evaluate a complicated integral numerically.

## 6.6 NUMERICAL DIFFERENTIATION USING A MULTIPLE-TERMINAL SHIFT REGISTER

In the last section, a shift register was used to recall a value created in the loop iteration immediately preceding the current one. In this section, we will explore how this register can be configured to remember values from not just one, but several previous loop iterations. We will explore this feature by writing a program that evaluates the derivative at each data point in a discretely sampled data set.

Assume we are given an equally spaced $N$-element set of $(x, y)$ data such that

$$x_i = x_0 + i\,\Delta x \qquad i = 0, 1, 2, \ldots, N-1$$

$$y_i = f(x_i)$$

where

$$\Delta x = \frac{x_{N-1} - x_0}{N-1}$$

and $f$ is an analytic function. At the $i^{th}$ data point, the derivative can be evaluated numerically using the following expression that involves knowledge of the two neighboring data points:

$$\frac{dy_i}{dx} = \frac{y_{i+1} - y_{i-1}}{2\,\Delta x} \tag{3}$$

In writing a program to evaluate Equation [3], one encounters the following problem: As the elements of an array are sequentially read into a For Loop, during the loop's $i^{th}$ iteration, previously read $y$-values (such as $y_{i-1}$) can be made available in shift registers, but to-be-read values (such as $y_{i+1}$) cannot. To solve this problem, we simply rewrite Equation [3] by letting $i \to i-1$. Then, our expression for the derivative becomes

$$\frac{dy_{i-1}}{dx} = \frac{y_i - y_{i-2}}{2\,\Delta x} \qquad i = 2, 3, 4, \ldots, N-1 \tag{4}$$

In Equation [4], the range of $i$ begins at 2 (i.e., does not include 0 and 1) because $y_{-2}$ and $y_{-1}$ are unknown.

Given a For Loop with the configuration of shift registers shown below, all of the quantities needed to calculate $dy_{i-1}/dx$ via the prescription of Equation [4] will be available during the $i^{th}$ iteration of the loop. Labels, which denote the indices of the array elements appearing in each shift register terminal during this particular loop iteration, are included in the illustration.

Thus, Equation [4] is coded by the following subdiagram:

Initiating the execution of this diagram in compliance with Equation [4], however, presents us with a challenge. Since $y_{-2}$ and $y_{-1}$ do not exist, the derivative of lowest index that can be calculated using Equation [4] is $dy_1/dx$, which corresponds to $i = 2$. To calculate this derivative during the For Loop's first iteration when $\boxed{i}$ equals zero, the top- and bottom-left terminals of the shift register must contain $y_1$ and $y_0$, respectively, while auto-indexing should place $y_2$ at the $y$-array input to the loop. To initialize the left terminals, we simply wire the desired values to them from outside the loop. In the case of the $y$ array, we must somehow slice off its first two values so that the $y_2$ value in the original $N$-element array becomes the index-zero element in the newly formed array. The new array then will have $N - 2$ elements.

Let's write a VI called **Extract First Two** that manipulates the $y$ array as desired above. Construct the following front panel, which accepts **Array** as an input and outputs its $i = 0$ and $i = 1$ elements to the **Index-Zero Element** and **Index-One Element** Numeric Indicators, respectively. In addition, **Extract First Two** outputs a new **Sliced-Off Array** that is formed by deleting the index-zero and index-one elements from the input **Array**. Package all three outputs in a cluster as shown. Save the VI in **YourName\Chapter 6**.

Design an icon for the icon pane and assign the connector terminals consistent with the following Help Window.

Then write the following block diagram, save your work, and close the VI.

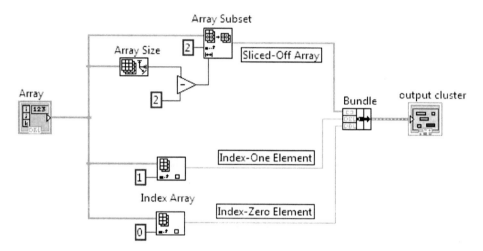

Here, we have used the **Array Subset** icon, found in **Functions>> Programming>>Array**. Its Help Window is reproduced below. This VI receives an array as input and sets aside a given number (= **length**) of its sequential elements, starting at a prescribed index (= **index**). The VI then outputs this subset of elements as a new array at **subarray**. When using this icon, always keep in mind that arrays (as all other LabVIEW structures) are *zero-indexed*. That is, the first element has an index of zero, the second has index of one, and so on.

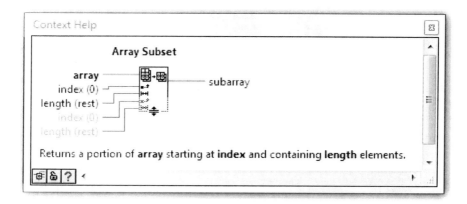

Now we are prepared to write a VI that performs differentiation. This VI will evaluate the derivative numerically at each point, except at the endpoints, when given an *N*-element discretely sampled data set supplied by **Power Function Simulator**. The calculated derivative will be displayed on an **XY Graph** as well as in an Array Indicator labeled **derivative**.

Place an **XY Graph** with its *x*- and *y*-axis labeled as **x** and **Derivative**, respectively. Then use **File>>Save** to store this VI under the name **Power Function Derivative** in **YourName\ Chapter 6**. Switch to the block diagram and complete the following code. This diagram will compute the derivative as an array of *N*–2 elements whose first and last elements are $dy_1/dx$ and $dy_{N-2}/dx$, respectively, where we are using the indexing appropriate to the original *y* array input from **Power Function Simulator**. The **Digitizing Parameters** and **Function Parameters** control clusters are made by first placing **Power Function Simulator** on the block diagram and implementing the pop-up menu automatic creation feature.

Finally, we need to slice the first and last elements off of the **x array** output from **Power Function Simulator**, so that its indexing properly matches an *x*-value with the correct **derivative** array element. Use **Array Subset** to perform this slicing operation, as shown below. Then, complete bundling the resulting **x** and **derivative** arrays, and wire this cluster to the **XY Graph** terminal. Pop up on the cluster wire and create the **output cluster**.

Return to the front panel and tidy up. Within **output cluster**, label the top and bottom array indicators as **x** and **derivative**, respectively, then save your work.

To judge whether your VI is coded properly, program its front panel so that **Power Function Simulator** will create the function $f(x) = 5x^2$. Take **Number of Samples**, **Lower Limit**, and **Upper Limit** to be *100*, *0*, and *1*, respectively. Then run **Power Function Derivative**. Is the slope of the curve on your **XY Graph** as anticipated?

Next, use your VI to calculate the derivative of $f(x) = 5x$. Is the result as expected? You may want to scale the *y*-axis manually here.

## 6.7 MODULARITY AND AUTOMATIC SUB VI CREATION

One of the hallmarks of a good LabVIEW program is its *modularity*. In such a program, the top-level VI acts as a supervisor, coordinating the execution of its collection of sub-VIs. In this approach to programming, a subVI is written to carry out each component task required for the successful completion of the greater program's overall mission. Modularity means that each subVI can be executed as a "stand-alone" program (i.e., independent of the top-level VI), simplifying the programmer's inevitable job of program debugging. Additionally, since subVI are designed to perform well-defined tasks, many times they can be reused as components in other top-level programs.

In the exercise we have just finished, **Power Function Derivative** is the top-level VI, while **Power Function Simulator** and **Extract First Two** are its subVIs. When viewed from this perspective, a programmer might suddenly wish that **Power Function Derivative** had been written with an increased level of modularity. For example, in the upper left region of the block diagram, several icons accomplish the subtask of snipping off the **x array**'s two endpoints. Thus this group of icons performs a well-defined task and is an excellent candidate for becoming a subVI.

When such a realization is made, here's an editing trick that is most helpful. First, use the ▶ to select the group of block-diagram objects that you wish to include within a subVI.

When you release the mouse button, this group will be highlighted with a marquee. Then, in the **Edit** pull-down window, select **Edit>>Create SubVI**. The highlighted group will be automatically converted into a subVI with a default icon as shown below.

To view this subVI, double-click on its icon. The front panel will open and you will find the new subVI's automatically created controls and indicators with default labels. The subVI itself will also have a default name.

You can then use the $A$ to create more accurate labels. Also a name can be given to the subVI using **File>>Save** and the subVI's icon can be designed by popping up on the icon pane and selecting **Edit Icon....** You will find that the terminals have been automatically associated with the controls and indicators, but these associations can be changed through the icon pane's pop-menu, if you wish.

Carry out these suggested procedures to create the front panel of a sub VI called **End-Less Array** as shown next.

When you switch to the block diagram, you will find code that you selected. Upon closing this subVI, **Power Function Derivative** will appear as shown below.

One then might suddenly realize that the entire central region of **Power Function Derivative** is the generic code required to calculate the numerical derivative of any function, and so might be of future use when working with functions other than the power function. This central region can be easily morphed into a VI called **Numerical Derivative** by first selecting it as shown next.

Then, using **Edit>>Create SubVI**, the **Numerical Derivative** VI shown below results after some work with the  .

The block diagram of **Power Function Derivative** then appears as shown below.

Digitizing Parameters

XY Graph

Power Func

Num Deriv

Function Parameters

output cluster

## DO IT YOURSELF

The factorial of integer $N$ is defined to be $N! \equiv 1 \times 2 \times 3 \times \ldots \times (N - 1) \times N$ with $0! \equiv 1$.

(a) Write a VI called **Factorial** that, given $N$, calculates $N!$. To check that your VI works, show that $10! = 3,628,800$ and $15! = 1,307,674,368,000$. If your program does not give the correct answer for both of these cases, figure out the problem within your code and fix it. **Factorial** should be able to produce correct values up to about $170!$

(b) Stirling's Approximation states that, for large $N$,

$$\ln(N!) \approx N\ln(N) - N \qquad [5]$$

Using **Factorial** as a subVI, write a program called **Stirling Test** that plots the percent deviation of Stirling's Approximation from the actual value of $\ln(N!)$ over a range of $N$-values. Use **Stirling Test** to determine the minimum value of $N$ so that the approximate $\ln(N!)$ deviates from the actual value by less than 1%.

## Problems

1.  I have agreed to hire you for a 30-day month. I pay you one penny on the first day, two pennies on the second day, and continue to double your daily pay on each subsequent day up (and including) to day 30. How many total dollars do you earn by the end of the month? Write a For Loop–based program with a shift register called **Month's Total Pay (Shift Register)** that determines the answer to this question. You may find the **Functions>>Programming>>Mathematics>>Elementary & Special Functions>>Exponential Functions** subpalette useful.

2.  In this problem, you will explore how LabVIEW handles an uninitialized shift register.

    (a)  Write the following VI called **Sum To Five**.

With **Sum To Five** open, run it three times in a row. What is the value of **Sum Total** after the first, second, and third run?

Now close **Sum To Five** (removing it from your computer's RAM), then open it again (return it to your computer's RAM). Run the VI three times in a row. What is the value of **Sum Total** after the first, second, and third run?

    (b)  Modify **Sum To Five** by deleting the **Numeric Constant** connected to the left shift register. The shift register will now be *uninitialized* when the program is run. After making this modification, save **Sum To Five**, close it, and then reopen it.

Sum Total

With **Sum To Five** open, run it three times in a row. What is the value of **Sum Total** after the first, second, and third run?

Now close Sum To Five (removing it from your computer's RAM), then open it again (reloaded into your computer's RAM). Run the VI three times in a row. What is the value of **Sum Total** after the first, second, and third run?

(c)  Based on your observations, what is the default value of an uninitialized shift register when the program is first loaded into RAM (i.e., before the program has been run the first time)?

(d)  Based on your observations, what is the value for an uninitialized shift register at the beginning of a program's second run? How about at the beginning of the program's third run? [Uninitialized shift registers are occasionally used in programs as a form of memory.]

3.  Let $x$ be a given positive number. Newton's iterative method for determining the square root $y = \sqrt{x}$ is as follows: As an initial guess for $y$, take $y_o = x/2$. Then, for the first iteration ($i = 0$) calculate the value for y to be $y_1 = \dfrac{1}{2}\left[y_0 + \dfrac{x}{y_0}\right]$. Iterate this process, where on the $i^{th}$ iteration $y_{i+1} = \dfrac{1}{2}\left[y_i + \dfrac{x}{y_i}\right]$, until the desired accuracy for $y$ is obtained. Write a VI called **Newton's Square Root**, which implements this method. Noting that $\sqrt{2} = 1.414213562\ ...$, how many iterations are required in order to obtain a value $y = \sqrt{2}$ that is accurate to the fourth decimal place (i.e., $\sqrt{2} = 1.4142$)?

4.  The icon **Reverse 1D Array**, found in **Functions>>Programming>>Array**, produces an output array whose elements are in reverse order to that of the array supplied at the icon's input. That is, if the array (0, 1, 2, 3) is input, then the array (3, 2, 1, 0) is output. Using any of the array-related icons available in **Functions>>Programming>>Array**, except for **Reverse 1D Array**, write a program called **Reverse Array Elements**, which accomplishes the same task as **Reverse 1D Array**. The front panel of **Reverse Array Elements** should contain an **Array Control** and an **Array Indicator** labeled **Input Array** and **Output Array**, respectively. Program **Input Array** with the array (0, 1, 2, 3), and then run you VI to demonstrate that it performs correctly.

5.  Write a program called **Running Average of Noisy Sine**, which generates and plots samples of a noisy sine wave along with an average of the four most recent samples of this waveform. To generate samples of a noisy sine wave, use the following code within a While Loop that iterates once every 50 ms until a Stop Button is clicked:

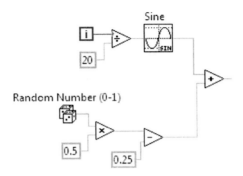

As each sample $y_i$ of the noisy sine wave is generated, plot it on a Waveform Chart, along with $\bar{y} = (y_i + y_{i-1} + y_{i-2} + y_{i-3})/4$, that is, the "running average" of the current sample and the three most recent previous samples. Refer to the Waveform Chart's Help Window to determine how to plot both $y_i$ and $\bar{y}$ on a single chart. Run **Running Average of Noisy Sine** and demonstrate that a running average effectively "filters out" random noise.

6.  Write a program called **Scale of A Major**, which plays one octave of the A major scale. In equal temperament tuning, the ratio of frequencies of neighboring notes is $2^{1/12}$. First, use a For Loop with shift register to generate an array of the frequencies $f_n$ (where $n = 0, 1, 2, \ldots, 12$) for the 13 notes within the octave from the notes A4 ($f_0 = 440$ Hz) to A5 ($f_{12} = 880$ Hz). Next, input this array to another For Loop and use it to play the eight notes of the A Major scale, which correspond to frequencies $f_0$, $f_2$, $f_4$, $f_5$, $f_7$, $f_9$, $f_{11}$, and $f_{12}$. Let each note sound for 500 ms. To play a note, use **Beep.vi**, which is found in **Functions>>Programming>>Graphics and Sound**, with its **use system alert?** input set to FALSE. The indices of the frequencies in the A Major scale can be stored either in a front-panel **Array Control** or in a block-diagram **Array Constant** (created by placing a **Numeric Control** within an **Array Constant** shell).

7.  Fourier analysis tells us that a square wave $y(x)$ with period and amplitude of one is given by the following sum of sine waves:

$$y(x) = \frac{4}{\pi} \sum_n \frac{\sin(2\pi n x)}{n} \qquad n = 1, 3, 5, 7, \ldots, \infty \qquad [6]$$

Write a VI called **Sum of Sines** that approximates the above sum by evaluating only its first $N$ terms (i.e., if $N = 3$, then only the $n = 1$, $n = 3$, and $n = 5$ terms are included in the sum). The front panel of your VI should appear as below with a **Numeric Control** called **Number of Terms** to input the desired value of $N$, and an **XY Graph** to plot $y$ versus $x$. To observe three high-resolution cycles of the square wave, evaluate the sum over the range from $x = 0$ to $x = 3$ with $\Delta x = 0.001$

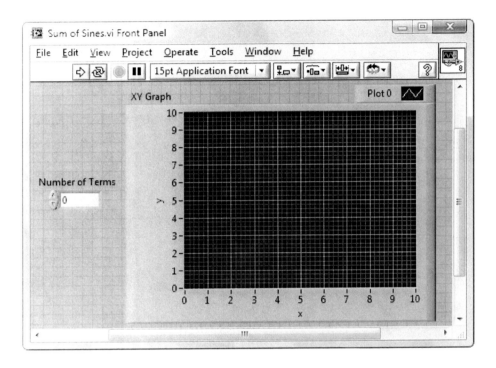

In designing your program, consider the following:

- All positive odd integers can be determined by $n = 2i + 1$, where $i = 0, 1, 2, 3, \ldots$

- If individual terms in a sum can be calculated one at a time, these terms can be summed using a For Loop with shift register. For the term associated with the odd integer $n$, form a 3001-element 1D array of values consisting of the quantity $\frac{4}{\pi}\sin(2\pi nx)$ evaluated over the range from $x = 0$ to $x = 3$ with $\Delta x = 0.001$. This 1D array can be summed with another similar-sized 1D array using the **Add** icon. Due to the polymorphic property of **Add**, if 1D arrays $y$ and $y'$ are input, each element of $y$ is added to the corresponding element of $y'$ and the resulting 1D array is output. That is, the $i^{th}$ element of the output array is equal to $y_i + y'_i$.

- To initialize an $N$-element array with each element equal to zero, use **Initialize Array** (found in **Functions>>Programming>>Array**) with **element** and **dimension size** wired to $0$ and $N$, respectively.

    Run **Sum of Sines** with a fairly small value of **Number of Terms**. As shown in the next illustration, near the transitions between $y = +1$ and $y = -1$, you will observe "overshoot" peaks (also known as ringing artifacts), an effect called the *Gibbs phenomenon*. As **Number of Terms** is increased, you should be able to minimize this effect. For what value of **Number of Terms** does the Gibbs phenomenon become negligible? (With our choice of $\Delta x$, the maximum allowed value for **Number of Terms** is 500. Why?)

8.  In the presence of air, a falling object's acceleration $a$ (= change in velocity per unit time) can be described by the following relation:

$$a = g - \alpha v^2 \qquad [7]$$

where $g = 9.8$ m/s$^2$ is the acceleration due to gravity in a vacuum, $v$ is the object's instantaneous speed in m/s, and $\alpha$ is a constant with units of m$^{-1}$. The velocity-squared term describes the effect of air resistance, and it becomes more and more significant as the object attains greater speeds. When the object achieves terminal velocity $v_T$, the object no longer accelerates because, at that speed, $g = \alpha v_T^2$.

Assume an elevated object is released from rest at time $t = 0$ and that the subsequent time is divided into $N$ small intervals, each of duration $\Delta t$. Then the object's fall can be sampled at $N$ times given by $t_n = n\Delta t$, where $n = 0, 1, 2, \ldots, N - 1$. Let $v_n$ be the object's instantaneous speed at time $t_n$. Then, if the object's speed $v_{n-1}$ at time $t_{n-1}$ is known and $\Delta t$ is small, Equation [7] can be used to determine $v_n$ as follows:

$$v_n = v_{n-1} + a\Delta t = v_{n-1} + \left(g - \alpha v_{n-1}^2\right)\Delta t \qquad [8]$$

(a)  Write a program called **Falling In Air**, which implements Equation [8] to determine a falling object's speeds $v_n$ at the $N$ times $t_n$ and then plots *Speed* versus *Time* on a Waveform Graph with calibrated $x$-axis. Take $\Delta = 0.001$s. The front panel of **Falling In Air** should appear as follows.

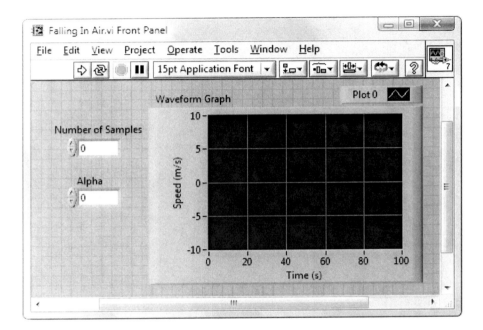

(b) For a 1.0 kg steel ball, a human skydiver, and a raindrop, the numerical value of $\alpha$ is about 0.00092, 0.0035, and 0.11, respectively. Run **Falling In Air** with each value of $\alpha$ and determine the resulting terminal velocity $v_r$ for each case. After being released, about how long does it take the object to attain $v_r$ in each case?

# Chapter 7

## The Case Structure

In a LabVIEW program, conditional branching is accomplished through the use of the *Case Structure*. This structure is analogous to an "if-then-else" statement in a text-based programming language. The Case Structure is found in **Functions>>Programming>>Structures** and is Boolean by default. That is, by wiring a TRUE or FALSE Boolean value to its *selector terminal* [?] , the structure will execute either the code within its TRUE window or its FALSE window, respectively. Using the ⭢, the selector terminal can be placed anywhere along the Case Structure's left border.

You may view only one of the Case Structure windows at a time. Above, the TRUE window is visible. To view the FALSE window, simply click the mouse cursor on the decrement (left) or increment (right) button in the *case selector label* ◄ True ▼► at the top of the structure. The FALSE window will then appear as shown here.

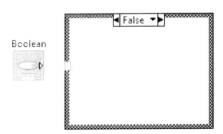

The Case Structure will automatically change its character from Boolean to numeric when you wire a numeric quantity to its selector terminal, such as the **I32** integer control shown next.

Initially only two Case windows (**0** and **1**) are available. Above, the case selector label ◄1 ▼► indicates that the **Case 1** window is currently visible. You can easily add another case by popping up anywhere on the Case Structure's border, and then selecting **Add Case After** in the pop-up menu.

You will then find that the structure now has three (**0, 1** and **2**) cases.

By repeating the above procedure, you may add as many (positive-integer) cases as you desire, unless you desire more than the maximally allowed 2,147,483,648($=2^{31}$)cases! If you wire a floating-point number to the selector terminal, LabVIEW will round that number to the nearest integer value. If the number supplied to the selector terminal is negative or if the number is larger than the highest-numbered case, the case designated (using the pop-up menu) as *Default* will be selected.

Finally, when wired to a selector terminal, an *Enumerated Type Control* can be used to self-document each of a Case Structure's cases, often making it the best choice for the case-selecting control. An Enumerated Type Control (called an *Enum*, for short) is a ring-style control, which associates a unique integer value with each item on a list of text descriptors. When an Enum is wired to the selector terminal of a Case Structure, its text descriptors (rather than the associated integers) appear in the case selector label. If you define these text descriptors wisely, they will intuitively describe the purpose of each Case diagram, obviating the need to document every case with, say, free labels. Such documentation is invaluable to other programmers (and your future self!) attempting to decipher your code.

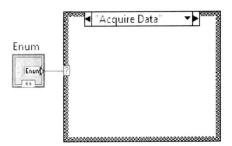

## 7.1 NUMERICAL INTEGRATION VIA SIMPSON'S RULE

In this chapter, we will write a VI that numerically integrates a discretely sampled data set using *Simpson's Rule*. Assume that a curve $y = f(x)$ has known values $y_1$, $y_2$ and $y_3$ at $x_1$, $x_2$, and $x_3$, respectively, where the succession of three $x$-values is equally spaced by a constant step $\Delta x$. In Simpson's Rule, one assumes that $f(x)$ in the region from $x_1$ to $x_3$ can be approximated by the quadratic $y = Ax^2 + Bx + C$, as shown next.

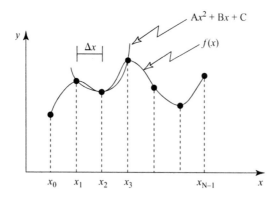

**Fig. 7.1**

Evaluating the posited quadratic at the three known data points yields the following three equations involving the three unknown constants $A$, $B$, and $C$:

$$y_1 = Ax_1^2 + Bx_1 + C$$
$$y_2 = Ax_2^2 + Bx_2 + C$$
$$y_3 = Ax_3^2 + Bx_3 + C$$

Using the fact that $x_2 = x_1 + \Delta x$ and $x_3 = x_1 + 2\,\Delta x$, linear algebra can be used to solve this set of equations for the three unknowns $A$, $B$, and $C$ (try it yourself), yielding

$$A = \frac{1}{2(\Delta x)^2}\left[y_1 - 2y_2 + y_3\right]$$

$$B = -\frac{1}{2(\Delta x)^2}\left[y_1(x_2 + x_3) - 2y_2(x_1 + x_3) + y_3(x_1 + x_2)\right] \qquad [1]$$

$$C = \frac{1}{2(\Delta x)^2}\left[y_1 x_2 x_3 - 2y_2 x_1 x_3 + y_3 x_1 x_2\right]$$

The area under the curve $f(x)$ can then be approximated as follows:

$$\int_{x_1}^{x_3} f(x)\,dx \approx \int_{x_1}^{x_3}\left(Ax^2 + Bx + C\right)dx$$
$$= \frac{1}{3}A\left(x_3^3 - x_1^3\right) + \frac{1}{2}B\left(x_3^2 - x_1^2\right) + C\left(x_3 - x_1\right) \qquad [2]$$

From Equation [1], we know the values of $A$, $B$, and $C$. Plugging these expressions into Equation [2], the right-hand side (RHS) of [2] becomes

$$\frac{y_1}{2(\Delta x)^2}\left\{\frac{1}{3}(x_3^3-x_1^3)-\frac{1}{2}(x_2+x_3)(x_3^2-x_1^2)+x_2 x_3(x_3-x_1)\right\}$$

$$-\frac{y_2}{(\Delta x)^2}\left\{\frac{1}{3}(x_3^3-x_1^3)-\frac{1}{2}(x_1+x_3)(x_3^2-x_1^2)+x_1 x_3(x_3-x_1)\right\}$$

$$+\frac{y_3}{2(\Delta x)^2}\left\{\frac{1}{3}(x_3^3-x_1^3)-\frac{1}{2}(x_1+x_2)(x_3^2-x_1^2)+x_1 x_2(x_3-x_1)\right\}$$

Then, using $x_2 = x_1 + \Delta x$ and $x_3 = x_1 + 2\Delta x$, this expression greatly simplifies after some (brutal) algebra, yielding the following famous result known as the *Simpson's "Three-Point" Rule*

$$\int_{x_1}^{x_3} f(x)dx \approx \left[\frac{1}{3}y_1+\frac{4}{3}y_2+\frac{1}{3}y_3\right]\Delta x \qquad [3]$$

This relation provides an improved numerical integration method over the "two-point" Trapezoidal Rule. Note that the above formula gives the integral over an interval of size $2\Delta x$, so the coefficients add up to 2. Also note that, based on our derivation, we expect Simpson's "Three-Point" Rule will give exact results when the function $f(x)$ is a polynomial of order 2 or less. Surprisingly, it can be shown that symmetries in the above equations lead to fortuitous cancellations, making Equation [3] exact even for polynomials of third order.

For a set of $N$ data points then, where

$$x_i = x_0 + i\,\Delta x \qquad i = 0, 1, 2, \ldots, N-1$$
$$y_i = f(x_i)$$

we can group the $N$ $x$-values into a succession of "three-point" units, so that

$$\int_{x_0}^{x_{N-1}} f(x)dx \approx \left[\frac{1}{3}y_0+\frac{4}{3}y_1+\frac{1}{3}y_2\right]\Delta x + \left[\frac{1}{3}y_2+\frac{4}{3}y_3+\frac{1}{3}y_4\right]\Delta x$$

$$+ \ldots + \left[\frac{1}{3}y_{N-3}+\frac{4}{3}y_{N-2}+\frac{1}{3}y_{N-1}\right]\Delta x$$

Let's call each term within square brackets a *"partial sum."* Then the above expression can be written as the following summation over partial sums

$$\int_{x_0}^{x_{N-1}} f(x)dx \approx \sum_i \left[\frac{1}{3}y_{2i}+\frac{4}{3}y_{2i+1}+\frac{1}{3}y_{2i+2}\right]\Delta x \quad i=0,1,2,\ldots,\frac{N-3}{2}\,(N\ \text{odd}) \qquad [4]$$

where the summation is over $(N-1)/2$ partial sums. We see then that this prescription for calculating the integral assumes that $N$ is an odd integer, where $N$ is the number of points in our data set. This, of course, is due to the fact that our $N$ data points can only be successfully grouped into a sequence of three-point units if $N$ is odd.

If given an even number of data points, one may calculate the integral numerically with the following approach: Use Simpson's Rule to find the contribution due to the first $N-1$ (an odd integer) points, then the Trapezoidal Rule to calculate the contribution due to the region between the two last points.

$$\int_{x_0}^{x_{N-1}} f(x)\,dx \approx \sum_i \left[\frac{1}{3}y_{2i} + \frac{4}{3}y_{2i+1} + \frac{1}{3}y_{2i+2}\right]\Delta x + \left[\frac{1}{2}y_{N-2} + \frac{1}{2}y_{N-1}\right]\Delta x$$

$$i = 0, 1, 2, \ldots, \frac{N-4}{2} \; (N \text{ even})$$

$\qquad\qquad\qquad\qquad\qquad\qquad\qquad\qquad\qquad\qquad\qquad\qquad$ [5]

The first term on the RHS of Equation [5] is the Simpson's Rule sum which contains $(N-2)/2$ partial sums, and the second term is the Trapezoidal Rule applied to the last two elements of the $y$ array.

In applying the above theory to integrate a discretely sampled set of $N$ data samples, three subtasks must be performed: (1) Given the $N$-element data array, determine whether $N$ is even or odd and then use the appropriate integral formula (Equation [4] or [5]) to find the number of partial sums to be calculated, (2) evaluate the Simpson's Rule term in the appropriate integral formula, and (3) calculate the Trapezoidal Rule term in Equation [5], if $N$ is even. Thus, to write a top-level program called **Simpson's Rule**, which carries out the above theory, we will first write a VI for each of the three identified subtasks. We will then use these three programs as subVIs in the top-level **Simpson's Rule** program.

## 7.2 PARITY DETERMINER USING A BOOLEAN CASE STRUCTURE

First, let's write a VI called **Parity Determiner**. Given an $N$-element **Array** as input, this program will determine whether $N$ is even or odd, and then calculate the number of partial sums in the Simpson's Rule term of the appropriate integral formula. From Equations [4] and [5], the number of partial sums is $(N-1)/2$ and $(N-2)/2$ when $N$ is odd and even, respectively.

Construct the following front panel with your own custom-designed icon. Create the **output cluster** by first placing a **Numeric Indicator** labeled **Number of Partial Sums** and then a Boolean indicator **Round LED** (found in **Controls>>Modern>>Boolean**) labeled **Odd?** within a **Cluster** shell. Use the cluster's pop-up menu option **Reorder Controls In Cluster...** to verify that **Number of Partial Sums** and **Odd?** is the cluster's index-zero and index-one element, respectively. Format the data type of the **Array** control and **Number of Partial Sums** indicator as **DBL** and **I32**, respectively. Use **File>>Save** to create a folder called **Chapter 7** with the **YourName** folder, then save your VI under the name **Parity Determiner** in **YourName\Chapter 7**.

Assign the connector terminal according to the following scheme.

Switch to the block diagram. Here, we will determine whether the number of elements in **Array** is odd or even using **Quotient & Remainder** (found in **Functions>>Programming>>Numeric**). The Help Window for this icon is shown next.

The **R** output of Quotient & Remainder, which corresponds to the modulo function *mod(x,y)* of other programming languages, calculates the remainder *R* when *x* is divided by $y$ using the following algorithm: $R(x,y) = x - y*floor(x/y)$, where *floor* truncates its argument to the next lowest integer. Thus $R(N,2)$ equals *0* and *1* for $N$ even and odd, respectively. Code the next diagram, where the Boolean output of **Equal?** (found in **Functions>>Programming>>Comparison**) will be TRUE (FALSE) when the number of elements in **Array** is odd (even). Wire this output to the second (index-one) input of a **Bundle** icon.

Now add a **Case Structure** (from **Functions>>Programming>>Structures**) to your block diagram and wire the **Equal?** output to its selector terminal.

When $N$ is odd, the number of partial sums to be calculated is $(N-1)/2$. Program this calculation into the Case Structure's TRUE window, and then wire its output to the first (index-zero) input of a **Bundle** icon. Then, wire Bundle's output to the **output cluster** terminal (if you get a broken wire, then the ordering of **Number of Partial Sums** and **Odd?** is incorrect within the front-panel cluster).

Note that the Case Structure's output tunnel is white and that the **Run** button is broken, indicating that the VI is not yet ready to run. Switch to the FALSE window and you'll find out why.

In a Case Structure, the data at all inputs (tunnels and the selector terminal) is available to all cases. Thus, the **Array Size** data tunnel, which was wired as an input in the TRUE window, will also make its data available in the FALSE window. However, a particular Case window is not required to use all available inputs. So even though nothing is connected yet, the input tunnel appears black, indicating a legal wiring configuration. Output tunnels obey a different rule: If any one Case window supplies data to an output tunnel, all other windows must do so too. At present, the FALSE window lacks this requisite output-tunnel wiring and this unlawful situation is signaled by the white appearance of the tunnel. The broken **Run** button generically signals this diagram error.

When $N$ is even, the number of partial sums to be calculated is $(N-2)/2$. Program this calculation into the Case Structure's FALSE window, as shown on the next page. As soon as you wire to the output tunnel, it will possess connections in all the Case windows (TRUE and FALSE). It will then turn black to denote it is now a legal output tunnel. Note the **Run** button is now whole.

As an alternate method of coding the FALSE window, you might explore the following method, if interested. First, select the FALSE window, pop up on the Case structure's border, and select **Delete This Case**. You will be automatically switched to the TRUE window, which you have previously coded. Then, since a very similar FALSE-window code is desired, pop up on the Case structure's border and select **Duplicate Case**. A clone diagram will appear in a newly created FALSE window, which you can then modify.

Return to the front panel and verify that the VI functions as expected by running it with **Array** programmed to have two, three, four, and then five elements (if you need to clear **Array** of all elements, pop up on its index display and select **Data Operations>>Reinitialize to Default Value**). Then save the final version of this VI in **YourName\Chapter 7** as you close it.

Parenthetically, as we found in Chapter 3, the Mathscript Node contains a conditional branching function which follows the if-then-else statement syntax. Above, we wrote Case Structure—based code to accomplish the following logic:

$$R = \text{mod }(N, 2)$$
$$if\ R = 1\ is\ TRUE\ then$$
$$\qquad Number\ of\ Partial\ Sums = (N-1)/2$$
$$else$$
$$\qquad Number\ of\ Partial\ Sums = (N-2)/2$$
$$end$$

Such conditional branching can alternately be written using a Mathscript Node in the following way. For the Mathscript Node's **R** output, **Choose Data Type>>Scalar>>Boolean** is selected.

## 7.3 SUMMATION OF PARTIAL SUMS USING A NUMERIC CASE STRUCTURE

Next, we will write a VI called **Sum of Partial Sums**, which calculates the Simpson's Rule term in Equations [4] and [5]. Construct the following front panel. Format the **Array** and **delta_x** controls as well as the **Simp Value** indicator as **DBL**, and the **Number of Partial Sums** control as **I32**. You may pop up on the DBL formatted objects and select **Display Format...**, and then uncheck **Hide trailing zeros** and make **Significant digits** something on the order of 6. Save this VI under the name **Sum of Partial Sums** in **YourName\ Chapter 7**.

Now begin coding the block diagram as shown next. Here is our programming strategy: Use a pair of nested For Loops to implement Simpson's Rule. The outer For Loop successively picks out three-point units from the original input array using the **Array Subset** icon. The inner For Loop is used to calculate the partial sum due to each three-point unit. The summation of all the partial sums is accumulated in the outer For Loop's shift register; the number of For Loop iterations required to complete this summation is supplied by the **Number of Partial Sums** control. Since the entire input array must be

passed to the **Array Subset** icon (as opposed to just one of its elements per loop iteration), you must pop up on the outer For Loop's tunnel and select **Disable Indexing**.

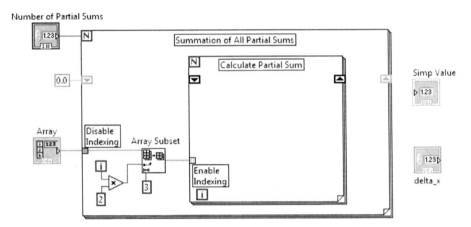

Now add the following code, which calculates each partial sum.

By enabling the indexing option on the inner For Loop, the three elements of the subarray created by **Array Subset** will be passed into the loop one at a time. There is no need
to wire the inner For Loop's count terminal; it will automatically be set to 3. Within the
loop, each of the three values is weighted by the proper factor consistent with Simpson's
Rule. The inner loop's shift register accumulates the current partial sum. By initializing
this register with the left terminal of the outer loop's register, the current partial sum is
added to the accumulation of all past partial sums.

In the above diagram, a numeric **Case Structure** is used to provide the correct 1–4–1
sequence of weighting factors dictated by Simpson's Three-Point Rule. Remembering

that the iteration terminal begins counting at *0*, the three Case windows should appear as illustrated.

Complete the diagram to include the multiplicative factor of $\Delta x$. Be sure to save your final work as you close this VI.

## 7.4 TRAPEZOIDAL RULE CONTRIBUTION USING BOOLEAN CASE STRUCTURE

Finally, we write a VI that calculates the second term on the RHS of Equation [5], that is, the Trapezoidal Rule contribution to the integral for an even-numbered data array. Build the following front panel and design an icon in the icon pane. The Boolean control is a **Round Button**. Save the VI under the name **Even Ends** in **YourName\Chapter 7**. Assign the connector terminal in a manner consistent with the Help Window shown.

Now write the following diagram. If the array possesses an even number $N$ of elements (**Odd?** is FALSE), the Case Structure's FALSE window determines the index of the second-to-last element ($i = N-2$), forms a subarray consisting solely of the original array's last two elements ($i = N-2$ and $i = N-1$), then numerically integrates this subarray using the **Trapezoidal Rule** VI.

Code the TRUE window as follows, reflecting the fact that for odd-numbered arrays, the Trapezoidal Rule contribution is zero.

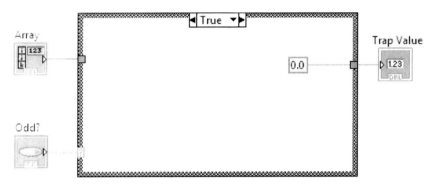

Be sure to save your final work on this VI as you close it.

## 7.5 NUMERICAL INTEGRATION USING SIMPSON'S RULE

Now we are ready to write the top-level **Simpson's Rule** VI. Assume that a set of $N$ equally spaced $(x, y)$ data samples exists, where the $x$-axis spacing between samples is $\Delta x$ and the variation of $y$ can be described by some function $f(x)$. With the $y$ array and $\Delta x$ as input, we wish to write a program that performs a Simpson's Rule integration of $f(x)$ between the extremal $x$-axis values of $x_0$ and $x_{N-1}$.

Construct the following front panel. Using the **Display Format...** option in its pop-up menu, set the **Value of Integral** indicator's **Significant digits** to *10* (or so). Save this VI under the name **Simpson's Rule** in **YourName\Chapter 7**.

Code the block diagram as shown below, and then save your work.

Let's see if the **Simpson's Rule** VI really works by supplying it with some known data from **Power Function Simulator**. Construct the following front panel and block diagram called **Simpson Test**. Save this VI in **YourName\Chapter 7**. To select the desired cluster item within **Unbundle By Name**'s output terminal, either click on the terminal using the 🖑, or pop up on the terminal and use the **Select Item** option.

Digitizing Parameters

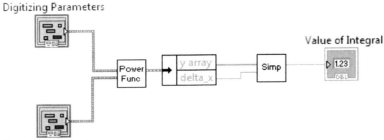

Function Parameters

With appropriate front panel settings, use **Simpson Test** to evaluate the integral $\int_0^1 5x^4 dx$ numerically (i.e., **Power Function Simulator** should be programmed to calculate $f(x) = 5x^4$). Try various values for **Number of Samples** and see how this affects the precision of **Value of Integral**.

## 7.6 COMPARISON OF THE TRAPEZOIDAL RULE AND SIMPSON'S RULE

Finally, let's compare the convergence properties of the Trapezoidal and Simpson's Rule methods and see if one proves to be the superior technique. To display the results of this study, place an **XY Graph** on a new front panel and save this VI under the name **Convergence Study (Trap vs. Simp)** in **YourName\Chapter 7**. Label the Graph's $x$- and $y$-axis **Number of Samples** and **Value of Integral**, respectively.

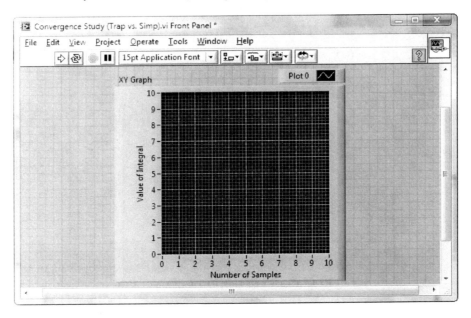

As shown next, start coding the block diagram, which will numerically integrate an
*N*-element array using both methods. Here, as *N* is varied from *10* to *200*, the integral
is calculated by the two methods, and the results are recorded in arrays at the For Loop
boundary. To plot all of the results on a single **XY Graph**, an *xy* cluster is formed for
the **Trapezoidal Rule** as well as the **Simpson's Rule** result. Then, a two-element array is built
with the Trapezoidal Rule and the Simpson's Rule *xy* cluster as its first (index 0) and
second (index 1) element, respectively. Such an *array of clusters* is the required input
for multiplots on an XY Graph.

To finish this diagram, we need to intercept the cluster emanating from the **Digitizing
Parameters** terminal and replace the front-panel **Number of Samples** value with the block-
diagram value for that quantity. Delete this cluster wire, then place a **Bundle By Name**
icon and wire it as shown. The value for **Number of Samples** that is passed to **Power Function
Simulator** is that created by the **Add** icon (rather than that passed from the front-panel
control).

Return to the front panel. Using the ↖, neaten the arrangement of objects. Then resize
the **Plot Legend** downward to accommodate information for two plots.

Using the **A** (or 🖑), highlight the text **Plot 0** in the Plot Legend, and then enter the text *Trap*. In a similar way, replace **Plot 1** by *Simp*.

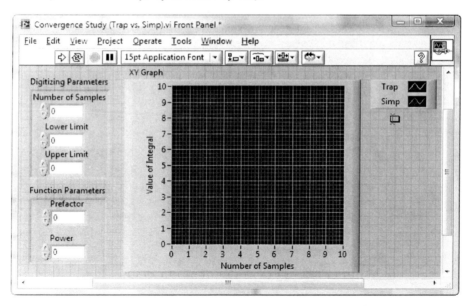

The plots can be distinguished from each other by choosing unique plot characteristics. For example, to chose a particular plot's color, simply pop up on its icon within the Plot

Legend, select **Color** and click on the hue you like (LabVIEW may have already chosen different colors for the two plots automatically).

Save your work. Then run **Convergence Study (Trap vs. Simp)** with **Power Function Simulator** programmed to calculate the function $f(x) = 5x^4$. Does one of the integration methods converge to an accurate value for the integral more quickly (that is, with a smaller **Number of Samples** value) than the other?

Now that you understand the inner workings of these two numerical integration methods, check out LabVIEW's built-in icon **Numeric Integration.vi** found in **Functions>>Mathematics>>Integration & Differentiation**.

## DO IT YOURSELF

Write a VI called **Five Blinking Lights** whose front panel has 5 **Round LED** Boolean indicators labeled *0* through *4*, as shown next. Using a **Case Structure**, develop a program that lights these LEDs one at a time in the order 0–1–2–3–4, and keeps repeating this sequence until the **Stop Button** is clicked. Make each LED stay lit for 0.2 second, so that an entire 0–1–2–3–4 sequence occurs once every second. Note that the value of a Boolean Constant can be changed using the 🖑.

## Problems

1. Construct a VI called **Seven-Segment Counter**. As shown below, use seven **Square LED** Boolean indicators, which have been appropriately resized and labeled **A** through **G**, to form a seven-segment display. Then, code the block diagram so that, when the Run button is pressed, this front-panel display counts from 0 to 9, with each digit illuminated for 1.0 second.

2. The Riemann Zeta function evaluated at the integers $n = 1, 2, 3, 4, \ldots$ is defined by

$$\zeta(n+1) = \frac{1}{n!} \int_0^\infty \frac{x^n}{e^x - 1} dx$$

269

This function appears in the theory of blackbody radiation, Bose-Einstein condensation, and the heat capacity of lattice vibrations. Taking $n = 2$ and using the fact that $2! = 2$,

$$\zeta(3) = \int_0^\infty \frac{1}{2} \frac{x^2}{e^x - 1} dx$$

Modify the Mathscript Node in **Power Function Simulator** so that it calculates the integrand of $\zeta(3)$, and then run **Simpson Test** with appropriate choices for **Number of Samples**, **Lower Limit**, and **Upper Limit**. Remember to use Mathscript element-wise operators. You'll encounter problems if you choose zero for **Lower Limit**. Find an acceptable solution to this problem. Also, you can't, of course, make **Upper Limit** equal to infinity. Because of the nature of the integrand, though, it's acceptable to take **Upper Limit** as something like *100*. Why? Optimize your choices to obtain as accurate a value as possible for $\zeta(3)$. This integral cannot be done analytically. However, high precision numerical calculations have determined that to a good approximation $\zeta(3) \approx 1.20205690031595942853997$. By comparison with this "correct" answer, how many decimal places of accuracy does **Simpson's Rule** deliver?

3.  As an alternative to implementing a loop's auto-indexing capability, an array can be built using the following subdiagram. Here, **Build Array** constructs an array by appending one element each iteration, and the ever-increasing array is stored from one iteration to the next in a shift register. The shift register is initialized with the **Empty Array Constant** shown, which can be created by popping up on the left shift register (after the code within the For Loop has been completed) and selecting **Create>>Constant**.

Using a modified form of the above diagram, write a program called **Multiples of 3, 4, or 5** that forms a single array containing all of the integers up to 30 that are evenly divisible by either 3, 4, or 5. You may find the **Compound Arithmetic** icon in **Functions>>Programming>>Boolean** useful.

4.  Write a Case Structure–based program called **Prime Detector**, which determines whether an input integer is a prime number. The front panel of this VI should appear as shown on the next page, where the integer to be tested is input at **Integer** and, if it is determined to be prime, the **Round LED** labeled **Prime?** is lit.

To construct your VI, consider the following properties of prime numbers. By definition, a prime number is evenly divisible by 1 and itself (i.e., only these two divisors result in a remainder of zero). Since even integers are evenly divisible by 2, even integers are not prime. An odd integer $N$ is not prime if it can be written as the product of two integers, where one integer is less than or equal to $\sqrt{N}$ and the other is greater than or equal to $\sqrt{N}$. Thus to test whether an odd integer is prime, it is sufficient to show that it is not evenly divisible by all of the odd integers less than or equal to $\sqrt{N}$, with the exception of 1. Note a sequence of odd integers, starting at 3, can be constructed from $N = 2i + 3$, where $i = 0, 1, 2, \ldots$

To test your VI, try inputting 104719 (prime) and 99763 (not prime). Which of the following integers are prime: 101467, 102703, 97861?

5.  Open **Simpson Test** and, using **File>>Save As...**, create a new VI called **Simpson Test (Built-In VI)**. On the block diagram, replace **Simpson's Rule** with **Numeric Integration.vi** (found in **Functions>>Mathematics>>Integration & Differentiation**) with its integration method input wired to **Simpson's Rule**. Then, with **Number of Samples** equal to *101*, run **Simpson Test (Built-In VI)** to evaluate the integral. Do you get the same result for **Value of Integral** as when running **Simpson's Rule** with the same front-panel inputs?

6.  Write a program called **Temperature Scale Converter**, which converts given temperatures between the Celsius and Fahrenheit scales continuously, until a **Stop Button** is clicked.

    (a) Place an **Enum** control on the front panel and program it with the following two items: *C to F* and *F to C*. On the block diagram, wire the Enum's terminal to the selector terminal of a Case Structure, then program the **C to F** and **F to C** case to execute the calculations $F = 1.8\,C + 32$ and $C = (F - 32)/1.8$, respectively. The completed front panel should appear as shown next. Run **Temperature Scale Converter** with several inputs and verify that it functions correctly, for example, $68°F \Leftrightarrow 20°C$, $100°C \Leftrightarrow 212°F$.

(b)   Add code to your VI so that the text above the **Input** control and **Output** indicator appropriately adapt to the **Conversion Mode** selection as the program runs. To accomplish this feat, note that the owned label of a control or an indicator cannot be changed during runtime, but its caption can be changed. Thus, in the front-panel pop-up menus of **Input** and **Output**, deactivate **Label** and activate **Caption**. Then, on the block diagram, pop up on the **Input** terminal and select **Create>>Property Node>>Caption>>Text**. Place the resulting **Property Node** within the **C to F** Case, pop up on it and select **Change To Write** to switch it from an indicator to a control. Pop up on the Property Node again and select **Create>>Constant**, then enter the text *Input deg C* into resulting **String Constant**. Repeat this procedure to make the caption for **Output** be *Output deg F*. Then, program the **F to C** Case so that the captions for **Input** and **Output** are *Input deg F* and *Output deg C*, respectively. Run your VI and verify that the captioning performs properly, for example, as shown below.

7.   Write the program shown below called **Guess Number**. This VI randomly generates an integer in the range from 1 to 10, and guides you in correctly guessing the integer. The front panel appears as shown next. Here, when the VI is running, you will type your

guessed integer into the **Guess** control and then pass this value to the block diagram by clicking on the **OK Button** labeled **Enter Guess**. You will receive feedback about your guess (e.g., too low, too high, correct) in the **Message** string indicator.

Code the block diagram shown below, which is based on a nested While Loop and Case Structure, a program architecture called a *state machine*. In a state machine, the While Loop executes continuously until its ⬤ is set to TRUE, and with each loop iteration, one of the Case Structure cases (termed a *state*) executes. The state that executes during a particular iteration performs some operation (e.g., check the equality of two quantities) and, in addition, selects which state will be executed during the next iteration.

When completed, run **Guess Number** and verify that it performs as expected. For this state machine, what quantity is stored in the upper shift register? How about the lower shift register? In each of the five states (*0, 1, 2, 3,* and *4*), what operation is carried out and what state is selected as the next state?

8. The response of a low-pass filter, that is, its gain *G* as a function of frequency *f*, is given by

$$G = \frac{1}{\sqrt{1 + \left(f/f_{3dB}\right)^2}}$$

where $f_{3dB}$ is the filter's 3dB-frequency. The following Mathscript-based subdiagram executes a linear plot of *G* versus *f* over the range from *f* =1 Hz to *f* = 10 kHz with $f_{3dB}$ = 2000 Hz. Remember .^ is the element-wise power operator.

The gain in decibels is defined as db = 20 $\log_{10}$ (G). A *Bode magnitude plot* graphs *Gain (in dB)* versus *f*, where the frequency-axis is logarithmic. By modifying the subdiagram given above, write a program called **Low-Pass Filter Plot**, which allows a user to graph the low-pass filter response on an XY Graph either as a linear plot or as a Bode magnitude plot. Let the user choose the type of plot via a Boolean switch labeled **Bode?**, which selects between two cases in a block-diagram **Case Structure**. For the logarithmic frequencies needed for the Bode magnitude plot, you can either "manually" take the base-10 logarithm of all the frequencies or (preferably) create a Property Node by popping up on the XY Graph terminal, selecting **Create>>Property Node>>X Scale>>Mapping Mode**, popping up on the Property Node and selecting **Change To Write**, and then wiring a constant *1* to it. For the linear case, you will then need a similar Property Node with a *0* wired to it. For extra flair, program the *y*-axis label as *Gain* or *Gain (dB)* for the Linear or Bode plot, respectively, using the Property Node **Y Scale>>Name Label>>Text**.

# Chapter 8 ■ ■ ■ ■ ■ ■ ■ ■ ■ ■ ■ ■ ■ ■ ■ ■ ■ ■ ■ ■

# The Sequence Structure

LabVIEW is a *dataflow* computer language. This approach to programming is contrary to the *control flow* mode in which most other programming languages operate. In control flow systems, the program elements execute one at a time in order that is coded explicitly within the program. The series of sentence-like statements in, for example, C and BASIC programs, which describe the sequential execution of Procedure A followed by Procedure B followed by Procedure C, are manifestations of the inherent control flow fashion in which these text-based languages are structured. In contrast, LabVIEW program elements abide by the principle of dataflow execution. Obeying a condition termed *data dependency*, a given object on the block diagram (called a *node*) will begin execution at the moment *all* of its input data become available. After completing its internal operations, this node will then present processed results at its output terminals. The interesting advantage of dataflow programming is that several nodes (for example, A, B, and C) can, through LabVIEW's multitasking ability, effectively execute in parallel. Such parallel execution will occur whenever the receipt of node A's inputs overlaps in time with the execution sequence of node B and/or C. This multitasking ability has the potential to enhance system throughput, especially in situations (not uncommon to data-acquisition operations) where a portion of a particular node's execution time involves waiting. In LabVIEW, useful parallel processes can be executed while one node waits for an event, whereas in a control flow system such waiting periods are simply dead time, putting the entire execution sequence on hold.

Despite the impression that you might glean from the previous paragraph, LabVIEW programs can, if desired, be written with a guaranteed one-step-at-time sequence of node execution. A proficient LabVIEW programmer may exploit the data dependency maxim in coding a block diagram where the output of node A is used as a source of input data to node B. Such a diagram will mimic control flow execution in that A must fully execute in order to enable the execution of B. The wiring configurations just described will many times provide an elegant solution to the need of ordered execution within a LabVIEW program. However, to create the correct configurations requires some skill on the programmer's part.

If an elegant data-dependent wiring configuration isn't forthcoming, LabVIEW provides the *Sequence Structure*, an explicit and foolproof way to obtain control flow within a program. The Sequence Structure comes in two styles—the *Flat Sequence Structure* and the *Stacked Sequence Structure*—both of which are found in **Functions >> Programming>>Structures**. In the developmental history of LabVIEW, the Flat form of the Sequence Structure was introduced later than the Stacked form, and most LabVIEW aficionados now agree that the Flat Sequence Structure produces the most readable code. Hence, we will only use the Flat Sequence Structure in the programs we develop in this chapter.

When it is initially placed on a block diagram, the **Flat Sequence Structure** looks like a single frame of movie film, as shown in the following illustration.

You can add a second frame, though, by popping up somewhere on the structure's border and selecting **Add Frame After**.

Then, the original and the newly created frame will appear side-by-side as below. The area enclosed by each frame can be resized using the ⬉.

By repeating this procedure, you can add any number of new frames. For example, below we have created a Flat Sequence Structure with three frames, and labeled the frames sequentially from left to right.

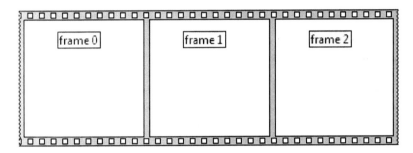

A Flat Sequence Structure executes sequentially from left to right. Hence, in a structure similar to that shown above, **frame 0** executes first, followed by **frame 1**, then **frame 2**, and so on until the last frame executes. Thus, by placing objects within different frames, you can use the Sequence Structure to control the order of execution among nodes that are not naturally coupled through data dependency. In addition, if a quantity produced in one frame is needed in a subsequent frame, that quantity can be passed between frames through a *tunnel*. For example, if one wanted to read the value of a front-panel Numeric Control, wait one-half second, then display the value on a front-panel Numeric Indicator, that three-step sequential process can be accomplished by the following code.

Finally, data can be input to any frame of the Flat Sequence Structure through a tunnel. Each frame can also emit output data through a tunnel, but, consistent with dataflow principles, this data will not be output from the structure until its last frame completes execution. Thus, the three-step sequential process described above can be alternately coded as shown next.

## 8.1 EVENT TIMER USING A SEQUENCE STRUCTURE

A timer, which measures the duration of a given event, provides an example of an instrument in which operations must proceed in a specific order—a START initiates the timing action, followed by a STOP that terminates it. In building a LabVIEW timer VI, then, it is perfectly natural to employ a Sequence Structure on the block diagram. Let's build such a timer and use it to measure something interesting. We will use our timer to compare the time it takes each of LabVIEW's two repetitive-operation structures, the For Loop and the While Loop, to build a given array of data. Our findings may surprise you!

Build the following front panel. Use **File>>Save** to create a folder called **Chapter 8** within the **YourName** folder, and then save this VI in **YourName\Chapter 8** under the name **Loop Timer (Sequence)**. Change the data-type representation of the **Number of Elements** control and the **For Loop Array** and **While Loop Array** indicators to **I32** and then resize these objects so that they may accommodate large integer values. Format the **For Loop Time (ms)** and **While Loop Time (ms)** indicators as **U32**.

Switch to the block diagram. The basis of our timer will be the **Tick Count (ms)** icon. This VI is found in **Functions>>Programming>>Timing** and its Help Window is reproduced next. **Tick Count (ms)** outputs an unsigned 32-bit (**U32**) integer that denotes the number of milliseconds that have elapsed since your computer was powered on.

Our strategy for measuring elapsed time will be this: Call **Tick Count (ms)** at the START of an event and call it at the event's STOP. Then by subtracting the two **Tick Count (ms)** output values, the event's elapsed time will be determined.

Place a **Flat Sequence Structure** on the block diagram. Inside the Sequence Structure's frame, put a **Tick Count (ms)** icon and wire its **millisecond timer value** output to a tunnel on the frame's border. This wire will contain your computer's internal clock value at the START of the For Loop, which will be contained in the next frame.

Create a second frame by selecting **Add Frame After** from the Sequence Structure's pop-up menu. Inside this frame, code a For Loop which creates an *N*-element array, where each array element has a numerical value equal to the value of its index. Output this array to the **For Loop Array** indicator on the front panel so that you may view its elements. *N* is determined by the **Number of Elements** control which is wired from outside the Sequence Structure, through a tunnel, to the For Loop's count terminal.

Create a third frame. Place a **Tick Count (ms)** icon there to ascertain your computer's clock value at the For Loop STOP. Then use **Subtract** to find the difference in milliseconds between the For Loop STOP time and START time (obtained in the "tunneled" wire from the initial frame), and output the result to the front panel's **For Loop Time (ms)** indicator. Finally, the **Tick Count (ms)** output in this frame will also be used as the While Loop START value. Wire the output of this icon to a tunnel, so that its value will be available for use in later frames.

Create a fourth frame and enclose within it the While Loop–based code shown next. As we learned in Chapter 2, by wiring $N - 1$ to the lower terminal of the **Equal?** icon, the While Loop will create an $N$-element array. Remember to select **Enable Indexing** at the While Loop tunnel; the While Loop default is **Disable Indexing**.

Finally, create a fifth frame. Here, determine the While Loop STOP through the use of a **Tick Count (ms)**, then (using the "tunneled" wire from the third frame) subtract the While Loop START from it. Output this value for the elapsed While Loop execution time to the front panel's **While Loop Time (ms)** indicator.

The VI is now complete, so return to the front panel and save your work. We will use **Loop Timer (Sequence)** to place the For Loop and While Loop in head-to-head competition to see if either provides superior performance in the task of assembling an *N*-element array.

Prior to running this program, you will need to input an appropriate value for *N* in the **Number of Elements** control. The proper choice for *N* will depend on the speed and memory capabilities of your computer. You should make *N* large enough so that it takes each of the two loops at least a sizable number of (at least 10) milliseconds to complete the array-building process. For times less than 10 milliseconds, the accuracy of **Tick Count (ms)** becomes an issue, as discussed below. Since the data type for **Number of Elements** is **I32**, the largest allowed value for *N* is $2^{31} - 1 = 2,147,483,647$. However, if you make *N* too large, you will find that **Loop Timer (Sequence)** produces a runtime error indicating that the program has overwhelmed your computer's free memory.

To get started, choose a value for *N* in the range of *100,000* to *1,000,000*. Then, through an iterative process, determine an appropriate *N* value for your system.

Once you've established a good value for *N*, get a fresh start by closing **Loop Timer (Sequence)**, then reopening it. Also, close as many other application programs as possible, so that your processor can devote the bulk of its resources to running LabVIEW. Input your *N* value, and then run the VI. Note the resultant values of **For Loop Time (ms)** and **While Loop Time (ms)**. You should find that the For Loop produces the array several times faster than the While Loop does. You may wish to peruse the array indicators to verify that the exact same array is being produced by the For Loop and While Loop methods.

Without closing the VI, try rerunning it a second, third, fourth, and more times and noting **For Loop Time (ms)** and **While Loop Time (ms)** for each run. You should find that the second run produces output times noticeably shorter than the initial program execution (that is, the first run after opening the VI). Subsequent runs then reproduce nearly the same times, with only a small fluctuation on the order of milliseconds.

Why does the While Loop take significantly more time to produce an array than the For Loop? In comparing the array-producing code for each loop, note that the While Loop requires an **Equal?** icon to determine if further iterations are required, but the For Loop has no such requirement. Perhaps then it is the extra execution times required for the **Equal?** operation during each iteration that accounts for the slower While Loop speed. To test this idea, make the loops equivalent (in regard to the **Equal?** icon) by adding an **Equal?** within the second frame's For Loop as shown below.

Run the VI. How do **For Loop Time (ms)** and **While Loop Time (ms)** compare once this addition is made to the VI? Does the extra time required for executing the **Equal?** icon significantly slow down the For Loop–based array-building code?

To understand the true reason that a While Loop takes more time to produce an array than a For Loop, you must first understand how LabVIEW uses memory. When LabVIEW launches, a single block of memory (either physical RAM or virtual memory) is allocated for all of the program's editing, compiling, and execution operations. During runtime, a memory manager allocates memory for tasks as needed, with the constraint that arrays and strings must be stored in contiguous blocks of memory. If the memory manager is unable to find a block of unused memory that is large enough for a particular string or array, a dialog box appears to indicate that LabVIEW was not able to allocate the required memory.

Now let's consider what happens when using the auto-indexing feature of a looping structure to build a data array. In the case of a For Loop, LabVIEW can predetermine the size of the array to be built based on the value wired to the Loop's count terminal. Thus the memory manager is only called once as the loop initiates execution and at that time allocates the appropriate block of memory necessary to hold the array.

With a While Loop, however, the final array size cannot be known in advance. A new element is appended to the existing array with each loop iteration and this process continues until a TRUE Boolean value at the conditional terminal causes the loop to complete its execution. Because the array is constantly increasing in size as the loop iterates, the memory manager must be called continually in order to find an appropriately sized chunk of RAM to hold the ever-growing data array. These repeated calls to the manager take time, especially if memory starts to become scarce. In a memory-tight situation, the manager may take extra time as it tries to shuffle around other blocks until a suitable space opens up.

Thankfully, because of some built-in LabVIEW intelligence, a While Loop's array-building capability is not as handicapped by the memory manager as it might appear from the above description. To avoid calling the manager with each iteration, the While Loop auto-indexing feature instructs the manager to allot enough new memory to store not just a single new array element, but rather a large number of additional array elements each time it is called. Through this trick, the number of memory-manager calls necessary in building a given-sized array becomes rather small. Of course, when the loop completes its execution, there most likely will be some unused memory associated with the array because of the overgenerous manner in which the manager delivered memory on its last call. Thus, when the loop terminates, LabVIEW simply directs the manager to dissociate this excess memory from the now-complete array. The result of this shrewd use of the memory manager is that the While Loop's and For Loop's array-building capabilities are not widely divergent (say, by powers of ten) in their performance, as you discovered through **Loop Timer (Sequence)**.

Above, we noted that the **Loop Timer (Sequence)** VI runs slower during its first run than it does during its subsequent executions. This observation can be traced to use of the memory manger. During the program's first run, the manager works out the memory allotments peculiar to the particular need of that VI. During subsequent runs, many of the first-run memory assignments are simply reused, diminishing the time-consuming use of the manager.

Finally, what causes the small fluctuations in execution times during the second, third, fourth,...runs? Assuming that your processor is not being asked to multitask between many open applications, these fluctuations are caused by the intrinsic accuracy of LabVIEW's built-in timing functions. These icons mark time by counting interrupts to your system's CPU which occur every 1 ms on a Windows system. It is this 1-ms resolution in the **Tick Count (ms)** icon that accounts for the observed timing fluctuations. The lesson to be learned here is that LabVIEW's timing VIs provide a simple and effective method of measuring the duration of an event with 1 ms resolution. In situations that demand submillisecond timing (such as is common in acquiring a sequence of analog-to-digital conversions), however, these icons are totally inadequate. For these high-accuracy timing applications, the submicrosecond-resolution clock on a National Instrument's data acquisition (DAQ) device can be used.

If interested, you might try exploring the performance of alternate array-building diagrams. These diagrams purposely avoid use of a loop's auto-indexing feature in an effort to make explicit what this feature does automatically. For example, try the following For Loop–based method, which initializes an appropriate-sized array (with each element defined as zero) prior to the loop structure, then avoids further calls to the memory manager through the use of **Replace Array Subset** within the loop itself. Use Help Windows to understand the functioning of the unfamiliar icons.

A While Loop–based diagram is shown next, which initializes the shift register with an empty array, then appends elements to it with each loop iteration. Why is this diagram such a poor performer? Although not time-efficient, in nondemanding applications (e.g., involving a small-sized array), this diagram is commonly used to perform "real-time" graphing (with a calibrated *x*-axis) inside a While Loop by feeding the **Build Array** output to the terminal of a **Waveform Graph**.

Another While Loop–based diagram initializes a larger-than-needed array (a 2,000,000-element array is shown; your choice of $N$ may require something different), sequentially places data values in the first $N$ elements of this array through the use of **Replace Array Subset**, then lops off the excess remainder of the array using **Array Subset**. Why does this diagram execute quickly?

## 8.2 EVENT TIMER USING DATA DEPENDENCY

A skillful programmer can exploit LabVIEW's data-dependent mode of execution to force two or more nodes to execute sequentially. Thus it is almost always possible to program a diagram without the use of a Sequence Structure.

In some cases, it is natural for node A's output to be used as the input of node B. Then these "chained together" nodes will execute in the order A-B. LabVIEW's File I/O icons provide such an example. Let A and B be the two file-related icons **Open/Create/Replace File** and **Write To Text File**. Typical of all File I/O VIs, each of these icons has an input and output called **file (path** or **refnum)** and **refnum out**, respectively. If A is (somehow) provided an input **file path**, when it completes its execution, **refnum out** is output. By wiring this output to B's **file refnum** input, one guarantees that A executes fully before B. Look back at your work in Chapter 5. We implemented this scheme in the **Spreadsheet Storage (OpenWriteClose)**. Additionally, you will find that LabVIEW icons with an associated purpose (e.g., file I/O, data acquisition) have **error in** input and **error out** output terminals that can be used to chain together several related nodes to ensure sequential execution.

Many times, however, the output of A does not constitute a natural input for B. In such situations, ordered execution can still be obtained by an alternate method called *artificial data dependency*. In this scheme, it is the mere arrival of data, with no regard to its actual value, that triggers the execution of a node. An example of artificial data dependency will be constructed on the following pages.

Let's try to write a timer diagram, whose ordered execution is controlled by artificial data dependency. First, open **Loop Timer (Sequence)** and use **Save As...** to create a new VI called **Loop Time (Data Dependency)** in **YourName\Chapter 8**. Keep the front panel unchanged as shown.

Switch to the block diagram. Eliminate the Flat Sequence Structure by popping up on its border and selecting **Remove Sequence**. Code the For Loop shown at the left in the next diagram, which creates an *N*-element array with the numerical value of each element equal to its index (the right portion of this diagram has not been changed yet). Wire the output of a **Tick Count (ms)** icon to the border of the For Loop. **Tick Count (ms)**'s output then is an input to the For Loop, meaning that the loop cannot initiate execution until the Tick Count (ms) icon produces a value. This clock value will be used as the For Loop START. The START value is never used within the For Loop, so it is simply the arrival of Tick Count's output that triggers the loop execution—an example of artificial data dependency.

After the For Loop ceases execution, we want to obtain the For Loop STOP value using a **Tick Count (ms)** icon, then calculate the elapsed execution time in milliseconds. The trick for assuring this sequence of operations is to enclose the "STOP-value" code within a single-frame Flat Sequence Structure as shown below. The code within this Sequence Structure will execute upon receipt of a value at the structure's input tunnel. Wire the Start value, through the For Loop, to an output tunnel (disable the loop's auto-indexing feature). Then wire this For Loop output over to the Flat Sequence Structure input. The completion of the For Loop will then trigger the execution of the Sequence Structure's code, as desired.

Also, when coding the Sequence Structure above, the enclosed **Tick Count (ms)**'s output is wired to an output tunnel on the structure's border. We will next turn our attention to a While Loop, which creates an *N*-element array. This tunnel outputs the While Loop START value, and the arrival of this value at the While Loop will trigger the loop's execution—another example of artificial data dependency.

Code the array-building While Loop, whose execution is triggered by arrival of the While Loop START value, as shown next. For array building, you will have to enable auto-indexing at the loop border.

Complete the diagram as follows to calculate the time it takes for While Loop–based array building.

Return to the front panel and save your work. Run the VI with your optimized choice for **Number of Elements**. Check to see that the expected *N*-element array is built by both the For Loop and the While Loop. If your diagram is correct, the performance of **Loop Time (Data Dependency)** should be equivalent to that of **Loop Timer (Sequence)**.

## 8.3 HIGHLIGHT EXECUTION

To cement your understanding of the manner in which this VI executes, let's implement one of LabVIEW's handiest (and coolest!) debugging tools, called *Highlight Execution*. Input something small, such as *10*, for **Number of Elements** on the front panel, and then switch to the block diagram. In the toolbar, enable **Highlight Execution** by clicking on the button containing the light bulb.

Then click on the **Run** button. The VI will execute in slow-motion animation, with the passage of data marked by bubbles moving along the wires and important data values given in automatic pop-up probes. Because the diagram executes in slow motion, the values for **For Loop Time (ms)** and **While Loop Time (ms)** obtained under Highlight Execution will be meaningless. However, this mode of execution provides a beautiful visual demonstration of the concept of artificial data dependency.

Use **Highlight Execution** liberally in your future work. This LabVIEW feature is an invaluable debugging tool for troubleshooting VIs that are causing you problems.

## DO IT YOURSELF

Write a VI called **Reaction Time** that measures a user's reaction time, defined as the elapsed time from when **Square LED** is lit until the user clicks the **Stop Button**. The front panel for this program includes a **Stop Button** and **Square LED** Boolean indicator, suitably enlarged, as well as a **Numeric Indicator** labeled **Reaction Time (ms)**, as shown below.

**Reaction Time** should execute in the following sequence of steps: (1) The user clicks on the **Run** button, and then quickly moves the mouse cursor over to the **Stop Button**; when the **Run** button is pressed, **Square LED** and **Reaction Time (ms)** are initialized as *unlit* and *0*, respectively; (2) at a random time (within the range of *3* to *8* seconds) after the **Run** button has been clicked, **Square LED** becomes lit; (3) on the **Reaction Time (ms)** indicator, the value of the elapsed time in milliseconds since **Square LED** was lit is displayed continuously, then the displayed value is halted at the moment that the user clicks on the **Stop Button**; and (4) **Square LED** becomes unlit.

A few helpful tips follow.

1. A random number in the range from *0* up to *1* can be obtained using the **Random Number (0–1)** icon, which is found in **Functions>>Programming>>Numeric**.
2. Think carefully about the appropriate choice for the **Mechanical Action** option in the Stop Button's pop-up window.
3. Properties of a front-panel object (e.g., its color, visibility, position) can be controlled from the block diagram using a **Property Node**. For example, to create a **Property Node** that lights **Square LED**, pop up on its block-diagram icon terminal and select

Create>>Property Node>>Value. The resultant **Property Node** can then be placed anywhere on the block diagram (i.e., removed from the **Square LED**'s icon terminal) and will appear as shown at the left in the next illustration. The outward-directed arrow indicates that this icon is presently configured as an indicator, which reads **Square LED's** current **Value** (TRUE or FALSE). To change this icon into a control, pop up on it and select **Change To Write**. Its arrow now is directed inward, indicating the **Property Node** is a control, as shown next at the right. By wiring a TRUE Boolean constant to this icon, **Square LED**'s **Value** will be TRUE, causing the front-panel indicator to be lit. Conversely, wiring a FALSE Boolean constant to the icon will make **Square LED** unlit. You are free to place an unlimited number of Property Nodes on the block diagram for any given front-panel object.

## Problems

1. Write a VI called **Five Blinking Lights (Sequence Structure)**, which sequentially illuminates five LEDs on its front panel. Place five **Round LED**s on the front panel and label them *0* through *4*. Using a Sequence Structure, develop a program that lights these LEDs one at a time in the order 0-1-2-3-4. Make each LED remain lit for 0.2 second and have the 0-1-2-3-4 lighting pattern repeat continuously until a front-panel **Stop Button** is clicked. To make multiple copies of an LED's block-diagram terminal, pop up on the terminal and select **Create>>Property Node>>Value** as described in the **Do It Yourself** project.

2. To demonstrate that a nested While Loop and Case Structure can accomplish the same function as a Sequence Structure, write a program called **Five Blinking Lights (State Machine)**. On the block diagram, use a nested While Loop and Case Structure to create code that repeatedly blinks five front-panel **Round LED**s in the order 0-1-2-3-4 until a front-panel **Stop Button** is clicked, as outlined in Problem 1. You should find that this block-diagram architecture (which is called a *state machine*) provides a more elegant solution to the Five Blinking LED problem than a Sequence Structure (e.g., no Property Nodes are required). You may find the following icon useful: **Quotient & Remainder** in **Functions>>Programming>>Numeric**.

3. Write a program called **Magic Stop Button**, whose front panel initially appears blank, as shown in the next illustration.

As shown below, when this VI is run, a blinking Stop Button appears on the front panel until the user clicks on it. The Stop Button then disappears and the program stops. To write this program, first place a Stop Button on the front panel, pop up on it, and select **Visible>>Caption**, and then make the caption read *Press this button to stop the VI*. Then pop up on the Stop Button and select **Advanced>>Hide Control**, and then secure this choice by selecting **Edit>>Make Current Values Default** and saving the VI.

Code the block diagram so that when the program is run, first the Stop Button becomes visible, then it blinks until it is clicked, and finally the Stop Button is made not visible. The features of the Stop Button (e.g., its visibility, blinking) can be controlled by creating appropriate Property Nodes.

4. The subdiagram shown below generates and plots 100 data samples on a Waveform Chart.

(a) Write a VI called **Two Charts (Simultaneous)**, which performs two such 100-sample plots simultaneously. That is, place two Waveform Charts on the front panel, labeled **Chart 1** and **Chart 2,** and program the block diagram so that the 100-sample plots on **Chart 1** and **Chart 2** are performed over the same time interval.

(b) Write a VI called **Two Charts (Data Dependency)**, which first completes the 100-sample plot on **Chart 1**, and then performs the 100-sample plot on **Chart 2**. Use artificial data dependency on the block diagram to accomplish this feat (i.e., do not use a Sequence Structure).

(c) Finally, add code to **Two Charts (Data Dependency)** so that the two Waveform Charts are cleared at the end of a run. To accomplish this task, place a single-frame Sequence Structure on the block diagram and use artificial data dependency to assure that this block-diagram object is the last item to execute before the VI completes a run. Within the Sequence Structure, place two plot-clearing Property Nodes, one associated with each Chart. The appropriate Property Node is made by popping up on the Chart's terminal and selecting **Create>>Property Node>>History Data**. After changing this Property Node to a control by selecting **Change to Write** in its pop-up menu, **Create>>Constant** will produce the needed input, called an **Empty Array,** as shown below.

5. Write a program named **Parallel While Loops with Reset** in which two While Loops execute in parallel on the block diagram, simultaneously producing independent plots of random numbers (in the range 0 to 1) on two front-panel Waveform Charts, until stopped by the click of a single front-panel Stop Button. Label one of the Waveform Charts **Chart 1**, the other **Chart 2**.

(a) First, try coding **Parallel While Loops with Reset** using the following block diagram to accomplish the intended goal (i.e., halt simultaneously executing While Loops with a single Stop Button). Run this program and show that it does not produce two simultaneous plots. Describe briefly what this diagram does do instead and explain why.

(b)    For the correct diagram, one While Loop is stopped by wiring directly to the Stop Button's terminal and the other by wiring to a **Value** Property Node associated with this terminal. To produce the needed Property Node, pop up on the Stop Button's terminal and select **Create>>Property Node>>Value**. When wiring this Property Node, you will get a broken wire until the Stop Button's **Mechanical Action** is changed to something other than a "latch" mode (e.g., select **Mechanical Action>>Switch When Pressed**.) Complete the diagram for **Parallel While Loops with Reset** and verify that it runs as intended.

(c)    When associated with a **Value** Property Node, the Stop Button must be in a "change value" (rather than a "latch") mode. Thus, when the Stop Button is pressed in order to stop the VI, the button will be left in its TRUE value as the VI completes execution, which is inconvenient for subsequent uses on this program. To remedy this problem, output a quantity from each While Loop and use artificial data dependency to reset the value of the Stop Button back to FALSE before the VI completes its execution. Run this final version of **Parallel While Loops with Reset** and verify that it executes as intended.

6.    Code **Real-Time Waveform Graph**, whose block diagram is shown below. This diagram builds an array of data within a While Loop, and this ever-growing array is plotted on a Waveform Graph during each iteration.

On the front panel, turn off autoscaling on the Waveform Graph's *x*-axis and manually scale this axis to run from *0* to something large (e.g., *20000*). Run **Real-Time Waveform Graph** and, based on your observations, identify the circumstances under which (e.g., after how many iterations) the design of this VI leads to poor performance. Explain why this diagram is a poor performer when run for many iterations.

7. Write a VI called **For Loop Time (Icon vs. Mathscript)**, which compares the required array-building time *T* of a LabVIEW For Loop icon and a Mathscript Node. To build the arrays, use the diagrams shown below.

Icon Array                    Mathscript Array

For various values of *N*, what is *T* for each diagram? Which diagram is the best performer?

8. The **error in** and **error out** terminals present on many LabVIEW icons can be used to sequence the execution of a collection of these icons via data degeneracy. As an example of this technique, use two **Elapsed Time** and one **Prompt User for Input** Express VIs (found in the subpalettes of **Functions>>Express**) to construct a program named **Time To Press OK**, which times how long it takes a user to click on a dialog-box **OK** button. Configure the Express VIs as shown on the next page, then wire them together so that data dependency due to the error terminals causes them to execute sequentially from left to right. The **Present (s)** terminal of Elapsed Time outputs the (universal) time in seconds at which this icon executes, so the difference of this value from the two Elapsed Time icons gives the time elapsed from when the VI was started until the user clicks on the dialog-box OK button. Add code that calculates and displays this elapsed time (in seconds) on a front-panel numeric indicator named **Time To Press OK (s)**.

# Chapter 9

# Analysis VIs: Curve Fitting

**BEFORE YOU COME TO LAB**

In the coming lab exercises, you will be using a *thermistor* as a temperature sensor. A thermistor has the useful characteristic that its resistance versus temperature $(R-T)$ curve can be fit to a well-known analytic function, facilitating its use as a calibrated thermometer. In Table 9.1, the $R-T$ data for a particular thermistor (Epcos Model #B57863S0103F040) is given. If you will be using a different thermistor in the **Do It Yourself** project at the end of this chapter, please use the $R-T$ data for the thermistor that will be used in your particular experimental set-up.

**Table 9.1** $R-T$ Data for Epcos Model #B57863S0103F040 Thermistor

| TEMPERATURE (°C) | RESISTANCE (KΩ) |
|---|---|
| −50.00 | 670.1 |
| −45.00 | 471.7 |
| −40.00 | 336.5 |
| −35.00 | 242.6 |
| −30.00 | 177.0 |
| −25.00 | 130.4 |
| −20.00 | 97.07 |
| −15.00 | 72.93 |
| −10.00 | 55.33 |
| −5.00 | 42.32 |
| 0.00 | 32.65 |

*(Continued)*

**Table 9.1** (*Continued*)

| TEMPERATURE (°C) | RESISTANCE (KΩ) |
|---|---|
| 5.00 | 25.39 |
| 10.00 | 19.90 |
| 15.00 | 15.71 |
| 20.00 | 12.49 |
| 25.00 | 10.00 |
| 30.00 | 8.057 |
| 35.00 | 6.531 |
| 40.00 | 5.327 |
| 45.00 | 4.369 |
| 50.00 | 3.603 |
| 55.00 | 2.986 |
| 60.00 | 2.488 |
| 65.00 | 2.083 |
| 70.00 | 1.752 |
| 75.00 | 1.481 |
| 80.00 | 1.258 |
| 85.00 | 1.072 |
| 90.00 | 0.9177 |
| 95.00 | 0.7885 |
| 100.00 | 0.6800 |
| 105.00 | 0.5886 |
| 110.00 | 0.5112 |

In this chapter, you will learn how to calibrate a thermistor by fitting its $R-T$ data to a formula known as the *Steinhart-Hart Equation*. This accomplishment is contingent on the ability to read the calibration data into your fitting program. One method we will explore for inputting data into a VI is through reading a disk file that contains the relevant information.

It would be helpful if, prior to coming to lab, you create and save a spreadsheet-formatted file containing a thermistor's $R-T$ data (either using the provided table or data for your own device). In Chapter 5 you learned that in the spreadsheet format, tabs separate columns, and end of line (EOL) characters separate rows.

You can use either a spreadsheet data analysis program or a word processor to create the desired file and store it under the name **Thermistor R-T Data.txt**. For example, I used Microsoft Word to construct the ASCII spreadsheet file shown in the following illustration of the thermistor resistance (in kΩ) versus temperature (in °C) data. When saving the file, it is important to use the **Save** or **Save As...** command, then select **Plain Text (\*.txt)**, as opposed to **Word Document (\*.doc)**, in the **Save as type:** box. This procedure creates an ASCII text file devoid of any Word formatting characters. Take care not to inadvertently add an extra EOL character at the end of the file.

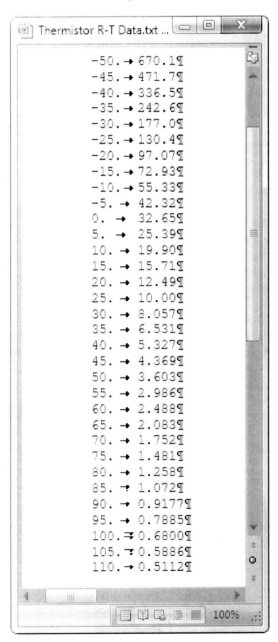

Alternately, if you use a spreadsheet-based program to create the file, you will want to save it as a **Text (Tab delimited)** file. This option may be listed under a **Save, Save As...** or **Export** menu item, depending on the program you use. Make sure that you save this file in the correct manner to avoid including a bunch of program-dependent formatting characters in the file. Perhaps the best way to check that you've produced the desired file is to open it using a word processor to see if it appears as above.

Bring your spreadsheet file of $R-T$ data with you to lab. You'll need it during your work in this chapter.

## 9.1 TEMPERATURE MEASUREMENT USING THERMISTORS

Among the many electronic temperature sensors available—including platinum resistance thermometers, thermocouples, and semiconductor devices such as the Analog Devices AD590—thermistors are used in applications that require high sensitivity, small size, ruggedness, fast response time, and low cost. When utilized with the proper support circuitry, a thermistor can measure temperatures to an accuracy of 0.01°C over a fairly wide range (about 100°C). The only price to be paid for the thermistor's high sensitivity is that its resistance is a nonlinear function of temperature. In the past, the nonlinearity of thermistors was a distinct disadvantage, causing users to employ resistor-based linearizing networks or less-than-accurate resistance-to-temperature conversion methods such as the Beta formula (see below). Fortunately, the widespread availability of microprocessors has enabled the use of accurate (and somewhat complex) temperature-calibration models, greatly simplifying the use of thermistors in accurate temperature measurement.

The active layer of most thermistors is composed of a semiconducting metal-oxide alloy. As is characteristic of a semiconductor, when the thermistor's temperature is elevated, loosely bound (valence) electrons within its active layer are thermally released from their respective binding sites to become conduction electrons. By this mechanism the thermistor's resistance decreases with temperature and results in the utility of this device as a temperature sensor. Since the bound electronic state is of lower energy, an activation energy barrier $E$ separates the conducting state from the insulating state. Thus at a given temperature $T$ (in Kelvin), we expect that the probability of such activation, and therefore the density $n$ of conduction electrons, is proportional to a Boltzmann factor. That is,

$$n = n_0 \exp\left[-\frac{E}{kT}\right]$$  [1]

where $n_0$ is the density of conduction electrons at very high temperatures and $k$ is Boltzmann's constant. It is easily shown that these conduction electrons imbue the material with a resistivity $\rho$ given by

$$\rho = \frac{m}{ne^2\tau}$$  [2]

where $e$ and $m$ are the charge and mass of an electron, respectively. The collision time $\tau$ is the average time traveled by a conduction electron until it is scattered by either a lattice vibration, impurity, or some other scattering mechanism peculiar to the material.

Thus, assuming that $e$, $m$, and $\tau$ are independent of temperature, we expect that the temperature dependence of $\rho$ is entirely due to the Boltzmann factor related to $n$. Putting [1] into [2], we predict that

$$\rho = \rho_0 \exp\left[\frac{E}{kT}\right] \qquad [3]$$

where $\rho_0$ is the temperature-independent constant given by

$$\rho_0 = \frac{m}{n_0 e^2 \tau} \qquad [4]$$

The macroscopic resistance $R$ of a sample is determined by

$$R = \rho \frac{L}{A} \qquad [5]$$

where $L$ and $A$ are the sample's length and cross-sectional area, respectively.

Because a solid sample's geometric shape is little changed with temperature, we expect that the temperature dependence of $R$ will simply mimic that of $\rho$. Thus

$$R = R_0 \exp\left[\frac{E}{kT}\right] \qquad [6]$$

where $R_0$ is the (assumed) temperature-independent constant

$$R_0 = \frac{mL}{A n_0 e^2 \tau} \qquad [7]$$

If all of our logic is correct, we predict that the temperature dependence of a thermistor's resistance should be described by Equation [6]. By taking the logarithm of both sides, this relation can be rewritten as the following linear relation

$$\frac{1}{T} = -\frac{k}{E} \ln R_0 + \frac{k}{E} \ln R \qquad [8]$$

or

$$\frac{1}{T} = A + B \ln R \qquad [9]$$

where $A$ and $B$ are constants defined to be $A \equiv -k \ln R_0 / E$ and $B \equiv k/E$. Equation [9] is the so-called *Beta formula* ($B$ is sometimes written as $\beta$) and it predicts that when a thermistor's temperature-dependent resistance data is plotted as $1/T$ (*y*-axis) versus $\ln R$ (*x*-axis), a straight line will result with the *y*-intercept and slope equal to $A$ and $B$, respectively.

In spite of this expectation, it is found experimentally that a real thermistor's $R-T$ data exhibits some nonlinearity when plotted as $1/T$ versus $\ln R$. Rather than being constant, the slope $B$ decreases with decreasing temperature. Over a narrow temperature range (say, of 10°C), $B$ varies so slightly that the Beta formula can predict the thermistor's behavior with a temperature uncertainty of about 0.01°C. However, for the much larger temperature spans over which a thermistor is useful, the Beta formula inadequately models the data.

Something then must be missing in our theory. In reviewing the above derivation, one might suspect it is our assumption of a temperature-independent collision time. It is easy to imagine that as temperature increases, the conduction electrons move faster and/or lattice vibrations become more pronounced, and consequently shorten the scattering time $\tau$. While the exponential Boltzmann factor activation of conduction electron density would still dominate the resistance's temperature dependence, the more mild temperature-dependent scattering time can possibly account for the observed slight nonlinearity of a $1/T$ versus $\ln R$ plot.

How then can we more accurately model the $R-T$ data? Unfortunately, the theory for electrical conduction in metal oxide thermistors is still incomplete and cannot offer us a more refined relation to replace Equation [9]. (In fact, the semiconductor energy band model advocated above is not accepted by all researchers. Some investigators, instead, hold that electric conduction in these materials is due to "hopping" of charge carriers from one ionic site to the next.) In light of this theoretical vacuum, the most recent literature on thermistors accounts for the nonlinearity of the $1/T$ versus $\ln R$ curve by using the standard curve-fitting technique of considering $1/T$ to be a polynomial of $\ln R$. That is, the inverse temperature is written as the following $n^{th}$ order polynomial:

$$\frac{1}{T} = A + B \ln R + C \left(\ln R\right)^2 + \dots + Q \left(\ln R\right)^n. \qquad [10]$$

From this point of view, Equation [9] is such a polynomial that has been truncated at the first-order term. One then is led to the conclusion that by retaining higher-order terms, the equation will remain valid over a wider range of temperatures.

Starting with this idea, researchers have found that the 100°C temperature span over which a typical thermistor is active can be accurately modeled by a third-order polynomial:

$$\frac{1}{T} = A + B \ln R + C \left(\ln R\right)^2 + D \left(\ln R\right)^3 \qquad [11]$$

This relation can be used to convert resistance to temperature with a precision equal to that of the original $R-T$ data (typically 0.005 to 0.01°C).

In 1968, the two oceanographers Steinhart and Hart discovered that, when modeling up to 200°C spans contained within the temperature range of $-70$°C to 135°C (they were especially interested in the oceanographic range of $-2$°C to $+30$°C), the second-order term in the above relation could be neglected without significant loss of accuracy.

Then Equation [11] reduces to

$$\frac{1}{T} = A + B \ln R + D \left( \ln R \right)^3 \qquad [12]$$

Equation [12] is called the *Steinhart-Hart Equation* and is widely used in thermistor calibration. For a typical thermistor, the constants $A$, $B$, and $D$ have order of magnitude values of $10^{-3}$, $10^{-4}$, and $10^{-7}$, respectively, when resistance is measured in ohms. The temperature must, of course, be on the Kelvin scale.

## 9.2 THE LINEAR LEAST-SQUARES METHOD

The process of scientific inquiry many times proceeds as follows: Researchers perform an experiment to investigate a particular physical phenomenon and to obtain a set of $N$ data samples $(x_j, y_j)$. After shutting off their instruments, the investigators apply appropriate curve-fitting techniques to determine the functional relationship $y = f(x)$ manifest in their data. Once armed with this $f(x)$, the scientific community can then judge the veracity of theoretical models posited to explain the phenomenon under investigation by testing each model's capacity for correctly predicting the experimentally observed $f(x)$. In this section, we will explore a curve-fitting method that allows one to extract an accurate $y = f(x)$ from a given data set $(x_j, y_j)$, a skill that plays a crucial role in the above-described process. We will put this skill to immediate use in calibrating a temperature-sensing thermistor.

Quite often it is convenient to model a functional relationship $y = f(x)$ as the linear combination of a set of basis functions $b_k$. The general form of this kind of model is

$$y = f \left( x \right) = \sum_{k=0}^{M-1} a_k b_k \left( x \right) = a_0 b_0 + a_1 b_1 + \ldots + a_{M-1} b_{M-1}. \qquad [13]$$

where the basis functions $b_k(x)$ are known functions of $x$ and the $a_k$ are the linear expansion coefficients. For example, if $b_k = x^k$, then $f(x) = a_0 + a_1 x + a_2 x^2 + \ldots + a_{M-1} x^{M-1}$ is a polynomial of order $M-1$. Alternately, if $b_k = \cos(kx)$, then $f(x)$ is a function described by a Fourier series. Note that the functions $b_k(x)$ can be nonlinear functions of $x$. However, the coefficients $a_k$ appear in a linear fashion in Equation [13].

Once the decision is made to model $f(x)$ as a linear combination of a set of basis functions, one needs some method to determine the appropriate choice of the $a_k$ in Equation [13], so that a given data set is accurately described. The theory of data analysis provides just such a method. Because all data-taking processes are subject to random errors, the repeated acquisition of a particular experimental quantity produces an array of values that is distributed as a Gaussian ("bell curve") distribution. By carefully considering the ramification of this fact, it can be shown that Equation [13] will best describe a given $N$ data samples $(x_j, y_j)$ when the $a_k$ are chosen such that they minimize an error function $e$ defined as

$$e \equiv \sum_{j=0}^{N-1} \left[ y_j - f(x_j) \right]^2 = \sum_{j=0}^{N-1} \left[ y_j - \sum_{k=0}^{M-1} a_k \, b_k(x_j) \right]^2 \qquad [14]$$

Note that $e$ is a measure of the deviation between the experimentally determined $y_j$ and the $y$ obtained from the fitting function $y = f(x_j)$, which depends on the values chosen for the coefficients $a_k$. Various algorithms have been proposed to identify the $a_k$ that minimizes $e$ for a given data set, all of which are classified under the rubric of *linear least-squares* methods. In LabVIEW, the available methods are the Singular Value Decomposition (SVD), Givens, Householder, LU (Lower Triangular-Upper Triangular) Decomposition, and Cholesky algorithms. The interested reader may refer to a data analysis text such as *Numerical Recipes* (Cambridge University Press) for a description of the details of these methods.

LabVIEW provides built-in VIs that will implement a linear least-squares fit of your data to the functional form of Equation [13]. For applications that require high performance (e.g., fastest possible execution time, smallest possible files) or detailed control over the fitting process (e.g., choice of least-squares algorithm), the low-level curve-fitting icons found in **Functions>>Mathematics>>Fitting** should be employed. In less demanding situations, however, the high-level *Curve Fitting Express VI* will provide the same results as code written with low-level icons, with much more programming ease.

In the following pages, you will write a program which fits a thermistor's $R - T$ data to the Steinhart-Hart Equation. For this task, you will employ the Curve Fitting Express VI to carry out the linear least-squares fitting algorithm.

## 9.3 INPUTTING DATA TO A VI USING A FRONT-PANEL CONTROL

To write your curve-fitting VI, you must first understand how to input the given array of $R - T$ data into a LabVIEW program. The most straightforward method for inputting an array of values is from a front-panel Array Control. Let's write a VI that works this way. Construct the front panel shown here. Place two Array Controls (select **Array** shell first, then place **Numeric Control** inside) and label one **T (Celsius)**, the other **R (kilohm)**. Next, add an **XY Graph** with its *x*- and *y*-axis labeled **Temperature (Celsius)** and **Resistance (kilohm)**, respectively, so that the thermistor data may be viewed graphically. Using **File>>Save,** create a folder named **Chapter 9** within the **YourName** folder, then store this VI as **R-T Data Input (Array Control)** in **YourName\Chapter 9**.

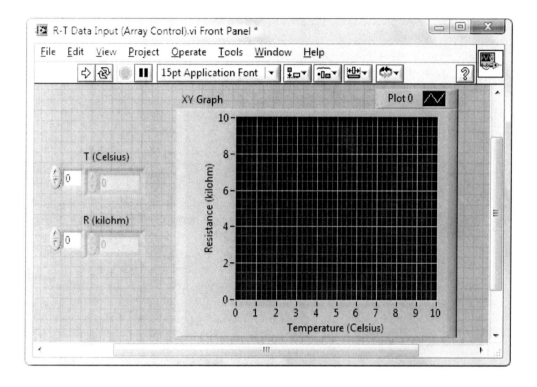

Switch to the block diagram. Write the following code that receives the input arrays, bundles them into a cluster, and then passes this cluster to the **XY Graph** for plotting.

The **Bundle** icon in this diagram is creating a cluster consisting of two elements. The input attached to Bundle's top and bottom terminal becomes the index 0 and index 1 element of the cluster, respectively. When this cluster is passed to the **XY Graph** icon terminal, the index 0 and index 1 elements are plotted as the *x*- and *y*-axis variable, respectively. Thus, the above diagram yields a resistance (*y*-axis) versus temperature (*x*-axis) plot.

Return to the front panel. OK, here would be the hard part, where we manually type all of the temperature values from the thermistor calibration data table into the **T (Celsius)** control, and then type all of the resistance values from the data table into the **R (kilohm)** control. Demonstrate for yourself how this procedure works by just entering the first five data values from the given thermistor calibration table.

| Temperature: | −50.00, | −45.00, | −40.00, | −35.00, | −30.00 |
|---|---|---|---|---|---|
| Resistance: | 670.1, | 471.7, | 336.5, | 242.6, | 177.0 |

Remember that the small left-hand box portion of the Array Control (called the *index display*) indicates the index of the element being displayed, and the larger right-hand box (the *element display*) is the actual numerical value of that element. You increment or decrement the index display using the 🖑. Also use this tool to highlight the interior of the element display, then type in the numerical value that you desire.

Now we want to store the information in these two arrays permanently. That is, we wish for the arrays to be initialized with these values each time the VI is opened. To accomplish this feat, select **Edit>>Make Current Values Default**. Save your work.

Run the VI, and you should see a graph of the thermistor's resistance as a function of temperature over the range from −50°C to −30°C. Try closing, then opening the VI. Repeat this process a few times. You should find that the two front-panel arrays are initialized with the thermistor data each time the program is loaded.

For future reference, you can also input data into a program by placing an array directly on the block diagram using the **Array Constant** icon found in **Functions>>Programming>>Array**. The **Array Constant**, just like the front-panel **Array** shell, comes equipped with an index display as well as an element display receptacle that you stuff with a constant of desired data type (numeric, Boolean, string, or cluster). In this approach, you first position an **Array Constant** on your block diagram, then place a **Numeric Constant** from **Functions>>Programming>>Numeric** (with Representation **DBL**) within its element display receptacle. You then program the Array Constant with the relevant sequence of data values. The resulting block diagram would appear as below.

It is considered good LabVIEW style to group related programming objects in a cluster. This approach leads to well-organized front panels, reduces the quantity of block-diagram wires (termed *wire clutter*), and, coupled with the use of **Unbundle By Name**, produces a self-documented VI. Let's re-form **R-T Data Input (Array Control)** in this manner. Place a **Cluster** shell (found in **Controls>>Modern>>Array, Matrix & Cluster**) on a blank region of your VI's front panel and label it **R-T Data**. Using the 🔧, first drag **T (Celsius)**, then **R (kilohm)** into the **Cluster** shell. The cluster can then be automatically resized by popping up on its border and selecting **AutoSizing>>Size to Fit**.

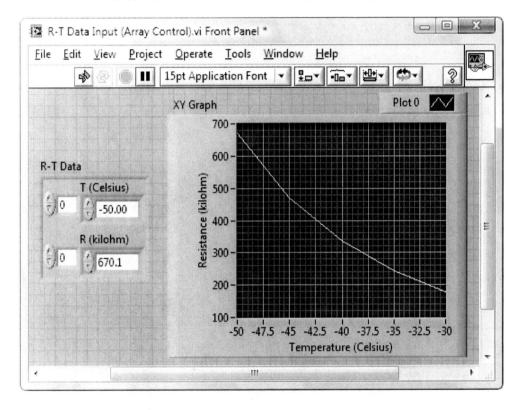

Switch to the block diagram. Remove the broken wires as well as the **Bundle** icon. Then, using the 🪡, connect the **R-T- Data** and **XY Graph** icon terminals with a cluster wire.

We're done. That's our program. Return to the front panel and save your work. Run the VI to verify that it is working properly.

There is one subtle point that is important to understand so that you obtain a valid resistance ($y$-axis) versus temperature ($x$-axis) plot. Namely, the elements in a cluster control are ordered. This ordering is not based on the position of the elements within the cluster shell, but initially is determined by the order in which the elements were placed in the Cluster shell during programming. The first element placed in the Cluster shell is indexed zero, the second element placed there is indexed one, and so on. Given a two-element cluster as input, an XY Graph associates the index 0 and index 1 element with the $x$- and $y$-axis, respectively. Thus, in our present program, we want the **T (Celsius)** and **R (kilohm)** arrays within the Cluster shell to be indexed 0 and 1, respectively. To verify this ordering, pop up on the border of the Cluster shell and select **Reorder Controls In Cluster...** The front panel will change its appearance as shown next.

At the bottom right corner of each cluster element, its current cluster index is given in the white box. The mouse cursor has morphed into the cluster order cursor, which can be used to change the indexing of the cluster elements by clicking on their associated black boxes. Once the new index assignments have been made, they can be secured by clicking on the **Confirm** button ☑ or reverted to the original settings with the **Cancel** button ☒.

Using **Reorder Controls In Cluster...** in the Cluster shell's pop-up menu, first order the cluster so that your VI will plot temperature (*y*-axis) versus resistance (*x*-axis). Click on the **Confirm** button, and then run the VI to view the $T-R$ plot. Next, by changing the cluster ordering, create an $R-T$ plot, that is, graph resistance (*y*-axis) versus temperature (*x*-axis). Save your work as you close this VI.

## 9.4 INPUTTING DATA TO A VI BY READING FROM A DISK FILE

An alternate method for inputting data into a VI is by reading it in from a previously created disk file. Let's write a LabVIEW program that can receive and plot the thermistor spreadsheet data that you generated prior to coming to lab. If you haven't created this file yet, now is the time to do it.

Construct this front panel, which simply contains an **XY Graph** with *x*- and *y*-axis labeled **Temperature (Celsius)** and **Resistance (kilohm)**, respectively. Design an icon for this VI, then save it in **YourName\Chapter 9** under the name **R-T Data Input (Spreadsheet)**.

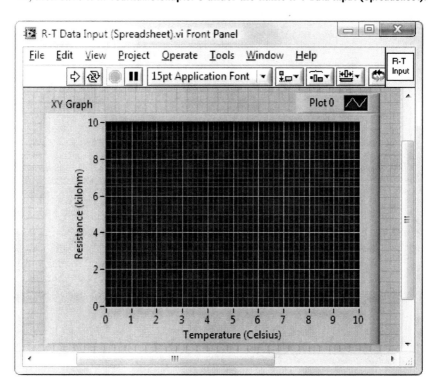

Switch to the block diagram. An ASCII spreadsheet file can be read into your VI using the **Read From Spreadsheet File.vi** icon found in **Functions>>Programming>>File I/O**. The *Simple Help Window* for this icon (when configured for double-precision floating-point numbers as explained below) is shown in the following illustration. As with all of the more sophisticated LabVIEW VIs, **Read From Spreadsheet File.vi** can be used for a lot of specialized purposes and therefore has many input and output connections. All of these available connections are described on the *Detailed Help* window, which can be accessed by clicking on the *Simple/Detailed Help button* 🔳 in Help Window's bottom left-hand corner. However, for most applications (this time, ours included), you need only connect wires to the input and outputs displayed in the Simple Help. The inputs and outputs that appear in boldface (**Read From Spreadsheet File.vi** has none of these) are those which must be wired, or else the icon will not execute. All other terminals denote optional features available for your use, if desired. Default values for inputs are shown in parentheses.

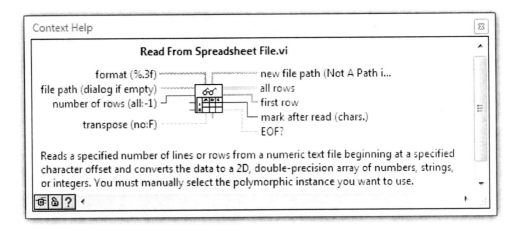

Place **Read From Spreadsheet File.vi** on your diagram. This icon is termed polymorphic, meaning that it can be configured to read spreadsheet files consisting of double-precision floating-point numbers, integers, or strings. This format choice is made using the icon's *Polymorphic VI Selector*. Operate the Polymorphic VI Selector either by clicking on it with the 🖐 as shown below at the left, or by popping up on it and choosing **Select Type**. Since our data file consists of floating-point numbers, use one of these methods to program the icon for its **Double** mode.

Use **Create>>Constant** to wire a **Path Constant** containing your spreadsheet-formatted $R-T$ data file's path to the **file path** input. If you leave this input unwired, a dialog box will appear when the program is run, which will prompt you for the desired file.

Also, increase the precision of numerical values from the insufficient default value of *3* decimal places to *4* (or more, if needed) by wiring a **String Constant** containing *%.4f* to the **format** input. Finally, since we desire to read the entire file, which is the default value for the **number of rows** input, leave this input unwired.

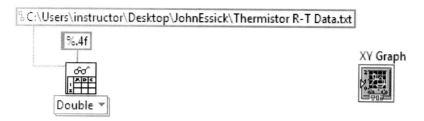

## 9.5 SLICING UP A MULTIDIMENSIONAL ARRAY

**Read From Spreadsheet File.vi** will present your data at its **all rows** output in the form of a 2D array. This array's two columns then will have to be separated ("sliced off") into two 1D arrays to facilitate an XY plot. Separating out the columns is done using the **Index Array** icon (found in **Functions>>Programming>>Array**). Place this icon on your diagram. Note that it originates with just one **index** input.

Next, wire Read From Spreadsheet File.vi's **all rows** output to the **n-dimension array** input of Index Array. The 2D array wire will appear and **Index Array** responds by automatically producing a second **index** input.

The top and bottom index inputs, which appear as small filled and hollowed-out black boxes, are the row and column indices, respectively. First, let's retrieve the temperature values from the 2D array. We know that the temperature values are all contained in the index-zero (first) column. So wire a **Numeric Constant** containing a zero (*0*) to the column (bottom) index of **Read From Spreadsheet File.vi**.

Note that once it is wired to the **Numeric Constant**, the column index input becomes a filled black box, which indicates that one particular column is selected. Now if we wanted to pick out, for example, the eighth temperature value in this column, which has the row index of seven, we would wire a **Numeric Constant** containing *7* to the top index input. **Index Array** would then output this single array element.

However, instead of a single temperature value, we want the entire column of temperature values. To instruct **Index Array** to "slice off" the entire column, we leave the row index input unwired. When unwired, row indexing is disabled (i.e., no particular row is selected), and the icon will output all rows in the selected column in the form of a 1D array. Note that the unwired row index input appears as a hollowed-out box, indicating that indexing at this input has been toggled off.

Repeat the above procedure to slice off the resistance data, which is in the index-one (second) column.

The **Index Array** icons below will each output a 1D array. Bundle these two 1D arrays together to form an *xy* cluster and wire this cluster to the XY Graph's terminal.

Return to the front panel, save your work, and then run your VI. You should be presented with a beautiful graph of your thermistor's resistance (*y*-axis) temperature (*x*-axis). Note the exponential-looking decay of resistance with increasing temperature. This is a signature of the thermally activated process taking place within the material of which the thermistor is composed.

If you encounter an error when your VI tries to open the data file, carefully check that the file's path within the block diagram's **Path Constant** is perfectly correct. Further check to make sure your data file is correctly constructed—such an error may be easily detected by running your program with strategically placed Probes and/or using Highlight Execution. An extraneous EOL character at the end of the data file is a common problem.

In the following exercise, we will use **R-T Data Input (Spreadsheet)** as a subVI within a program that fits $R-T$ data to the Steinhart-Hart equation. We will need access to the given temperature and resistance data in two different guises: Celsius-kilohm as well as Kelvin-ohm.

To provide the data in the two required forms, add two indicator clusters to your VI as follows. First, pop up on the cluster wire emanating from the **Bundle** icon and create an indicator cluster called **Celsius-kilohm** using **Create>>Indicator**.

Switch to the front panel. Position the **Celsius-kilohm** cluster indicator appropriately. Make the label of each array within this cluster visible by popping up on its index display and selecting **Visible Items>>Label**. Then, label the top and bottom array within it as **T (Celsius)** and **R (kilohm)**, respectively.

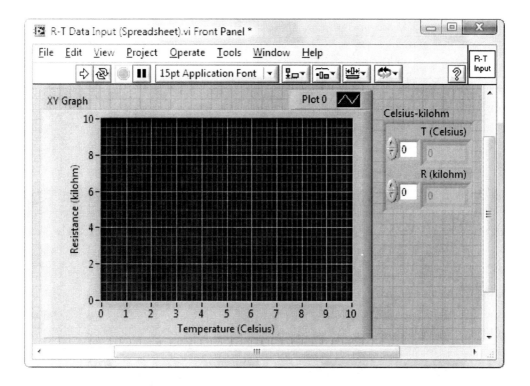

Switch back to the block diagram and complete it as shown next. This code converts the array of temperatures given in Celsius to the Kelvin scale (by adding the scalar constant *273.15* to each element of the Celsius temperature array). Also, the resistances given in kilohms are converted to ohms (by multiplying each element of the kilohm resistance array by the scalar constant *1000*). If you use **Create>>Constant** to produce the two constants (*273.15* and *1000*), create these constants first (i.e., before connecting the array wires to the **Add** and **Multiply** icons), or else an Array Constant will be created, rather than a scalar constant. After you connect the two inputs to **Bundle**, the **Kelvin-ohm** indicator cluster can be produced using **Create>>Indicator**.

Switch to the front panel. Position the **Kelvin-ohm** indicator cluster appropriately, then name the top and bottom array as **T (Kelvin)** and **R (ohm)**, respectively.

Finally, assign the connectors consistent with the Help Window shown next.

Run the VI and make sure that the correct temperature and resistance values appear in each cluster. For example, make sure that there is not an extra zero as the last element in an array (if an extra zero is present, you need to correct your **Thermistor R-T Data** file). Then, save your work as you close this VI.

## 9.6 CURVE FITTING USING THE LINEAR LEAST-SQUARES METHOD

Now that you know how to input your thermistor data into a program, let's write a VI that uses the linear least-squares method to fit these data to the Steinhart-Hart Equation (Equation [12]). At first glance, the Steinhart-Hart Equation does not appear to follow the form assumed by the linear least-squares theory (Equation [13]). However, if we define $y \equiv 1/T$ and $x \equiv \ln(R)$, the Steinhart-Hart Equation becomes

$$y = A + Bx + Dx^3 \qquad [15]$$

which is of the form given in Equation [13] with the basis functions $b_0 = 1$, $b_1 = x$, and $b_2 = x^3$ and the coefficients $a_0 = A$, $a_1 = B$, and $a_2 = D$. Equation [15] is called the *linearized Steinhart-Hart Equation*. The strategy for our curve-fitting VI is this: Input the thermistor's $R-T$ data, recast these data as $x = \ln R$ and $y = 1/T$, and then use the **Curve Fitting** Express VI to perform a linear least-squares fit to obtain the values of $A$, $B$, and $D$.

The **Curve Fitting** Express VI is found in **Functions>>Express>>Signal Analysis**. The Help Window for this icon (when configured for linear least-squares method) is shown in the next illustration.

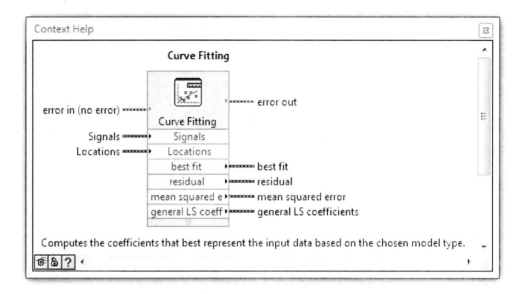

Here, it is assumed an experimentalist has acquired $N$ data samples $(x_j, y_j)$. These data are input to the Express VI in the form of an $x$ and $y$ $N$-element array at its **Locations** and **Signals** input, respectively. After programming it with $M$ basis functions $b_k(x)$, the VI will find the particular linear combination of the given $b_k(x)$ that best describes the data. That is, the data input to this icon is fit to Equation [13] and the optimized values of the coefficients $a_k$ are output at the **general LS coefficients** terminal. Besides the optimized values for $a_k$, this icon outputs the array of "best fit" $y$-values where

$$y_j^{\text{best fit}} \equiv f(x_j) = \sum_{k=0}^{M-1} a_k b_k(x_j) \qquad [16]$$

Also output are the $N$-element residual array $R$, where this array's $j^{\text{th}}$ element is defined as

$$R_j \equiv y_j - y_j^{\text{best fit}}, \qquad [17]$$

and the *mean square error* (mse), which is defined as

$$\text{mse} \equiv \frac{1}{N} \sum_{j=0}^{N-1} R_j^{\,2} \qquad [18]$$

The residual array as well as the mse can be taken as a measure of the error in your fitted curve.

Build the front panel given below. To compare the resulting best fit $y$-values with the original data, include an **XY Graph** with two differentiated plots called **Data** and **Best Fit** (for example, select different colors or point styles and disable interpolation for one curve in the **Plot Legend**). Also, display the determined values of $a_k$ in an Array Indicator labeled **Best-Fit Coefficients**. Pop up on this indicator and format it to display numerics in scientific notation (using **Display Format...**). Save this VI as **R-T Fit and Plot** in **YourName\Chapter 9**.

Switch to the block diagram and place the **Curve Fitting** Express VI on it. When first placed on the block diagram, the Express VI will open the dialog window shown next, where you will configure its operation. From among the several possible fitting options, select **General least squares linear**. The box labeled **Models** will become activated, and it is here that you define the basis functions to be used in the linear least-squares fitting operation. We will be fitting our data to the linearized Steinhart-Hart Equation, whose basis functions are $b_0 = 1$, $b_1 = x$, and $b_2 = x^3$, so enter $1$, $x$, and $x^{\wedge}3$ on the first, second, and third line of the **Models** box, respectively.

Click on the **OK** button. The dialog window will close and the **Curve Fitting** icon will appear on your block diagram with its input and output terminals expanded (except for **error in** and **error out**).

Place **R-T Data Input (Spreadsheet)** on the block diagram and then complete the code shown in the next illustration, which delivers the $y = 1/T$ and $x = \ln(R)$ array at Curve Fitting's **Signals** and **Locations** input, respectively. **Natural Logarithm** is found in **Functions>> Mathematics>>Elementary & Special Functions>>Exponential Functions**.

Express VIs accept and return information in a format called the *Dynamic Data Type (DDT)*, which is rendered as a dark blue banded wire. The DDT bundles related information about the data it carries. For example, when carrying an array of numeric values, the DDT also includes the data's label name and the date and time the data was acquired. When you wire **Reciprocal's** output to Curve Fitting's **Signals** input, the **Convert to Dynamic Data** icon will be inserted into the wire automatically. This icon converts the 1D floating-point numeric array produced by **Reciprocal** to the dynamic data type used by Express VIs such as Curve Fitting. Similarly, the **Convert to Dynamic Data** icon will be inserted automatically when wiring **Natural Logarithm's** output to Curve Fitting's **Locations** input. If you ever need to place **Convert to Dynamic Data** on a block diagram manually, it is found in **Functions>>Express>>Signal Manipulation**.

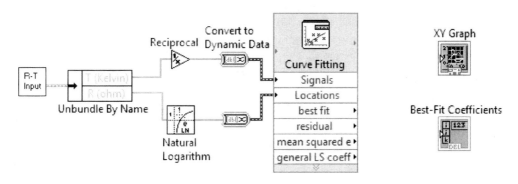

Finally, add the following code which plots the original data and the best-fit results for comparison. Here, you will have to place the **Convert from Dynamic Data** icons (found in **Functions>>Express>>Signal Manipulation**) on the diagram manually. When **Convert from Dynamic Data** is first placed on the block diagram, its dialog window will open; select **Conversion>>1D array of scalars-automatic** and **Scalar data type>>Floating point numbers**, then press the **OK** button. Curve Fitting's **best fit** output contains the array of $y_j^{\text{bestfit}}$ values. Since $y = 1/T$, where $T$ is in Kelvin, this array must be converted to Celsius temperatures prior to the comparison plot.

325

Return to the front panel and run your VI. Upon completion, the **Best-Fit Coefficients** array indicator will display the Steinhart-Hart's *A*, *B*, and *D* coefficients at its *0*, *1* and *2* indices, respectively.

For convenient reading, you can make all three array elements visible at once, if you like, by the following procedure. First, place the ⬚ in the corner of the **Best-Fit Coefficients** indicator on the front panel so that it morphs into a *grid cursor* ⬚.

Then drag the grid cursor to the right, until it has created two additional indicators.

When you release the mouse button, indicators for array elements *0*, *1* and *2* will then be visible. The index display provides the index for the element displayed in the leftmost indicator.

Record the resulting values of *A*, *B*, and *D* for your thermistor for future use. Do your *A*, *B*, and *D* have the order of magnitude values of $10^{-3}$, $10^{-4}$, and $10^{-7}$, respectively, which are typical of most thermistors? Does the **XY Graph** indicate that the fitted Steinhart-Hart Equation accurately describes the temperature-dependent resistance of your thermistor?

## 9.7 RESIDUAL PLOT

Is there a quantitative method for judging how well the Steinhart-Hart Equation describes the thermistor's $R-T$ data? Yes, there is, and the Curve Fitting Express VI provides us with such a method. After it is has determined the array of best-fit *y*-values, Curve Fitting calculates the residual defined as $R_j \equiv y_j - y_j^{\text{bestfit}}$ for each index *j*. We see that the residual is simply the deviation of the best-fit value from the actual *y*-value data input at **Signals** and so provides a quantitative measure of the "goodness" of the fitting procedure. All of the residual values $R_j$ from $j = 0$ to $j = N-1$ are output as an *N*-element array (contained within a dynamic data type wire) at the Curve Fitting's **residual** output.

In general, for a quantitative display of the deviation of Curve Fitting's best-fit values from the input experimental data, one can simply plot the **residual** array output versus some appropriate quantity (such as the *x*-array input at **Positions**). However, in our present problem, we linearized the Steinhart-Hart Equation via $y = 1/T$, and so the residual calculated by Curve Fitting is the deviation of the inverse Kelvin temperature. That is, the **residual** output *R* is given by

$$R \equiv y - y^{\text{best fit}} = \frac{1}{T} - \frac{1}{T^{\text{best fit}}} \qquad [19]$$

where *T* is the experimental Kelvin temperature and $T^{\text{bestfit}}$ is Curve Fitting's best-fit Kelvin temperature. Defining $\Delta y \equiv y - y^{\text{bestfit}}$ and $\Delta(1/T) \equiv 1/T - 1/T^{\text{bestfit}}$, we see that

$$R \equiv \Delta y = \Delta\left(\frac{1}{T}\right) \qquad [20]$$

To obtain a more intuitive measure, let's define the residual of Kelvin temperature as the "temperature residual" $R^T \equiv T - T^{\text{bestfit}} = \Delta T$. Then, if $y = 1/T$, we know that for small $\Delta T$ (in comparison to $T$ itself),

$$\Delta y \approx \frac{dy}{dT}\Delta T = \frac{d}{dt}\left(\frac{1}{T}\right)\Delta T = \left(-\frac{1}{T^2}\right)\Delta T \qquad [21]$$

Since $R = \Delta y$ and $R^T = \Delta T$, Equation [21] shows that the temperature residual can be determined from Curve Fitting's residual output using the relation

$$R^T \approx -T^2 R \qquad [22]$$

Since $R^T$ measures a difference of Kelvin temperatures, its units can be in either Kelvin or Celsius degrees because the degree interval is of the same size on both of these temperature scales.

Place a second **XY Graph** on the front panel of **R-T Fit and Plot** and label its $x$- and $y$-axis **Temperature (Celsius)** and **Temperature Residual (Celsius)**, respectively.

Switch to the block diagram and add the code shown below, which calculates the temperature residual and sends it to the front panel for display.

Return to the front panel and save your work. Then, run your VI. Is there a tempera-ture range within which the Steinhart-Hart consistently overestimates the temperature value? Is there a range within which the temperature is consistently underestimated? If you need your thermistor to be calibrated with an accuracy of $\pm 0.05°C$ (i.e., the Steinhart-Hart values falls within $\pm 0.05°C$ of the actual temperature), within what tem-perature range can you operate your thermistor?

When curve-fitting, one expects that a better fit can be achieved if the number of basis functions used in the fitting algorithm in increased. Thus, we would expect that a third-order polynomial (Equation [11]) should fit our thermistor data better than the Steinhart-Hart Equation (Equation [12]), which deletes the second-order term.

Let's see if this conjecture is true. Switch to the block diagram and double-click on the **Curve Fitting** Express VI. When the dialog window opens, select **General least squares linear**, then in the **Models** box, program the VI to perform a fit using the four basis functions $b_0 = 1$, $b_1 = x$, $b_2 = x^2$, and $b_3 = x^3$. Then click on the **OK** button.

With this modification in place, return to the front panel. Expand **Best-Fit Coefficients** to include four indicators, and then run the VI. How has the accuracy (as determined from the Temperature Residual plot) of your fit improved? Is your result consistent with Steinhart-Hart's claim that the second-order term can be deleted from the fitting equa-tion without significant loss of accuracy? Remember that this claim comes to us from an era when each additional term would have to have been calculated manually or with additional analog circuitry, and so any labor-saving tip would have been welcome.

Note that, by adding the $(\ln R)^2$ term, the new coefficients for the 1, $\ln R$ and $(\ln R)^3$ terms become different from those derived in the Steinhart-Hart case. This phenomenon results from the fact that the functions $x^k$ that we are using as a basis are not *orthogonal*. Take a linear algebra course for more details.

Finally, try fitting the thermistor data to the Beta formula (Equation [9]). In this first-order polynomial fit, only two basis functions are required: $b_0 = 1$, $b_1 = x$. Run **R-T**

**Fit and Plot**, and then comment on the accuracy achieved by the Beta formula in comparison to the Steinhart-Hart Equation.

## DO IT YOURSELF

Build a computer-based digital thermometer that uses the thermistor that you have calibrated in this chapter as its temperature sensor.
    Construct this instrument in the following steps:

1.    Build the circuit shown in Figure 9.1, which produces a constant current of 0.1 mA through
    the thermistor. You may wish to include the high input-impedance unity-gain op-amp
    buffer to isolate the thermistor circuit from the voltage-sensing circuitry. If available, use an
    ammeter to determine the current through your thermistor precisely. It probably will differ
    slightly from 0.1 mA due to resistor tolerances. With a voltmeter, check that the voltage
    difference between $V_{out}$ and GND is within the expected range (e.g., for a thermistor whose
    resistance is about 10 kΩ near room temperature, $V_{out} = IR \approx (0.1\,mA)(10\,k\Omega) = 1\,V$.

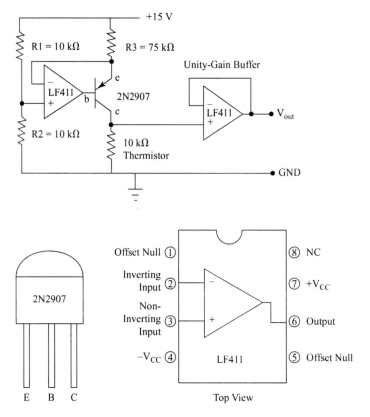

**Fig. 9.1    Hardware circuit for Do It Yourself project. Constant current of 0.1 mA flows
through thermistor.**

2. Connect the positive and negative pins of a DAQ device's differential analog input channel to $V_{out}$ and GND, respectively.

3. Write a VI called **Digital Thermometer**, whose front panel has a **Numeric Indicator** labeled **Temperature (deg C)** and a **Stop Button** as shown next. Save this VI in the **YourName\ Chapter 9** folder.

On the block diagram, place a **DAQ Assistant** icon. When the **Create New Express Task...** dialog window appears, configure it to perform an Analog Input operation that outputs temperature from a current-excited thermistor (**Acquire Signals>>Analog Input>> Temperature>>Iex Thermistor**) on the particular AI channel that you have connect to the circuit (e.g., **ai0**). When the **DAQ Assistant** dialog window then appears, configure it with the Steinhart-Hart values appropriate for your thermistor. The three Steinhart-Hart coefficients, which we called $A$, $B$, and $D$ in this chapter, are respectively labeled $A$, $B$, and $C$ in the DAQ Assistant dialog window. Also choose **Selected Units>>deg C, Iex Source>>External, Iex Value>>100 u** (which means 100 $\mu A$ = 0.1 mA; use your measured value for the current, if available), **Configuration>>4-Wire**, and **Acquisition Mode>>1 Sample (On Demand)**. Then close the dialog window (by clicking the **OK** button). When the DAQ Assistant icon appears on your block diagram, enclose it within a **While Loop** and complete the required extra code, so that temperature readings are produced and displayed on a front-panel **Temperature (deg C)** indicator once every, say, 0.5 or 1 second, and this process repeats continually until a front-panel **Stop Button** is clicked.

Once completed, run **Digital Thermometer** and use it to measure the room's temperature and the temperature of your skin.

## Problems

1. According to the theoretical considerations discussed at the beginning of this chapter, the value of $B$ determined from fitting a thermistor's $R-T$ data to the Steinhart-Hart Equation is related to the activation energy barrier $E$ for electronic conduction in the thermistor. Use the value you obtained for $B$ to obtain $E$ in the units of electronvolt (eV) for your thermistor. The value for Boltzmann's constant is $k = 8.6 \times 10^{-5}$ eV/K.

2.  A Type T thermocouple is constructed by joining the ends of a copper and a constantan wire. This junction between the two dissimilar metals produces a temperature-dependent voltage, which can be used as a temperature sensor. The Type T thermocouple's voltage as a function of temperature is given in the following table.

| TEMPERATURE $T$ (°C) | VOLTAGE $V$ (MICROVOLT) |
| --- | --- |
| 0 | 0.000 |
| 50 | 2036 |
| 100 | 4279 |
| 150 | 6704 |
| 200 | 9288 |
| 250 | 12013 |
| 300 | 14862 |
| 350 | 17819 |
| 400 | 20872 |

The calibration equation for this temperature sensor is commonly obtained by fitting this *Temperature vs. Voltage* data to the following equation

$$T = a_1 V + a_2 V^2 + a_3 V^3 + a_4 V^4 + a_5 V^5 + a_6 V^6 \qquad [23]$$

Write a program called **Thermocouple Fit** that uses the **Curve Fitting** Express VI to determine the coefficients $a_1$ through $a_6$ in Equation [23]. Consider entering the given $T-V$ data into a spreadsheet, and then using **Spreadsheet Read** (the **Do It Yourself** project in Chapter 5, or a slightly modified version of it) to input the data to your program. Alternately, you can use Array Controls to input the data. (You might be interested in comparing your values for $a_1$ through $a_6$ to those used in the **Volts to Temperature.vi** subVI within **Convert Thermocouple Reading.vi**, which is found in **Programming>>Numeric>>Scaling**. These values of $a_1$ through $a_6$ are derived from a more complete $T-V$ data set than used in this problem.)

3.  The following subdiagram implements **Wait Until Next ms Multiple**, which is found in **Functions>>Programming>>Timing**, to produce a $y$-array of millisecond timer values and a corresponding $x$-array of integers that index the $y$-values.

(a) Refer to **Wait Until Next ms Multiple**'s Help Window and then explain why the $(x,y)$ data this subdiagram produces should obey the equation of a straight line $y = mx + b$, where $m$ and $b$ are the line's slope and $y$-intercept, respectively. What is the expected value for $m$ of this line?

(b) Place the above subdiagram on the block diagram of a VI called **Wait Until Next ms Fit**. On this same block diagram, place a **Curve Fitting** Express VI and program it to perform a linear fit on the given $(x, y)$ data and output the resulting value for the slope to the front panel. When you run your completed VI, do you obtain the expected value for $m$?

(c) If you observe any (slight) deviation of the observed $m$ from its expected value, offer an explanation for this deviation. Is there a way that you could test whether your explanation is valid?

4. A resonant circuit composed of an inductor, resistor, and capacitor is powered by an AC input voltage whose amplitude $V_{in}$ is constant, but whose frequency $f$ can be varied. An experimentalist measures the circuit's AC output voltage amplitude $V_{out}$ at several frequencies $f$, and then calculates the circuit's gain at each frequency, where gain is defined as the ratio of the output to input voltage amplitudes, that is, $G \equiv V_{out}/V_{in}$. The results from this experiment are shown in the following data table.

| FREQUENCY $f$ (Hz) | GAIN $G$ |
|---|---|
| 1500 | 0.070 |
| 1600 | 0.084 |
| 1700 | 0.102 |
| 1800 | 0.128 |
| 1900 | 0.168 |

(*Continued*)

| | |
|---|---|
| 2000 | 0.235 |
| 2100 | 0.374 |
| 2200 | 0.740 |
| 2300 | 0.861 |
| 2400 | 0.442 |
| 2500 | 0.281 |
| 2600 | 0.206 |
| 2700 | 0.164 |
| 2800 | 0.136 |
| 2900 | 0.117 |
| 3000 | 0.103 |

A theoretical analysis of this resonant circuit predicts that the frequency-dependent gain $G(f)$ is given by

$$G = \frac{1}{\sqrt{1 + \left[ \frac{1}{\Delta f}\left( f - \frac{f_o^2}{f} \right) \right]^2}}$$

[24]

where the constants $f_o$ and $\Delta f$ are the resonant frequency and full-width at half-power maximum, respectively (there is one frequency below $f_o$, and, one frequency above $f_o$, where $G = 1/\sqrt{2}$; the full-width is defined as the difference between these frequencies).

Write a VI called **Resonant Circuit** that fits the given experimental $G-f$ data to the theoretically predicted relation for $G(f)$ (Equation [24]). On the VI's front panel, compare the data and the best-fit curves by plotting both on the same **XY Graph** and also display the best-fit values for constants $f_o$ and $\Delta f$ in an Array Indicator.

A few suggestions for your work:

- Consider entering the given $G-f$ data into a spreadsheet, and then using **Spreadsheet Read** (the **Do It Yourself** project in Chapter 5, or a slightly modified version of it) to input the data to your program. Alternately, you can use Array Controls to input the data.

- Since the parameters $f_o$ and $\Delta f$ appear in Equation [24] in a nonlinear fashion, use the **Model Type>>Non-linear** option of the Curve Fitting Express VI to perform the fitting procedure. For the **Model Parameters**, let $f_o = a$ and $\Delta f = b$. You will have to supply **Starting Values** for $a$ and $b$. For the nonlinear fitting algorithm implemented by Curve Fitting to function properly, these initial guesses for $a$ and $b$ need to be somewhat close to the final best-fit values.

- After you run your program once with the dialog window for **Curve Fitting** closed, the **Signals** and **Locations** arrays will be loaded into this Express VI. You can then open **Curve Fitting** (by double-clicking on its icon) and run it interactively. Thus, you can iteratively run **Curve Fitting** with a succession of choices for **Starting Values**, until you determine a choice that yields a successful fit.

5.  The Steinhart-Hart Equation can be written as, $T = \dfrac{1}{A + Bx + Dx^3}$ , where $x = \ln R$. In this form, the constants $A$, $B$, and $D$ appear in a nonlinear manner. Starting with **R-T Fit and Plot**, use **File>>Save As…** to create a new VI named **R-T Fit and Plot (Nonlinear)**. Modify the block diagram so that **Curve Fitting** fits the input thermistor data to the above form of the Steinhart-Hart Equation (you will have to program **Curve Fitting** to perform a nonlinear fit; see suggestions given in Problem 4). Verify that you obtain (nearly) the same values for $A$, $B$, and $D$ as well as the temperature residual $R^T \equiv T - T^{bestfit}$ as you obtained from the linear least-squares method implemented in this chapter.

6.  Write a program called **Noisy Sine Fit**, which fits a sine function to a noisy sine wave input.

    (a)  First, place a **Simulate Signal** Express VI, found in **Functions>>Express>> Input**, on the block diagram. When its dialog window opens, program this Express VI to produce 100 samples of a sine wave with **Uniform White Noise** added at a sampling frequency of 10000 Hz. Close the window and configure the **Simulate Signal** icon so that you have front-panel control of the sine-wave frequency and amplitude as well as the noise amplitude shown below.

    Place a Waveform Graph on the front panel and plot the noisy sine-wave output from **Simulate Signal**. Run the VI with **Amplitude, Frequency,** and **Noise Amplitude** set to *1, 250,* and *0.6,* respectively, to verify that your program functions correctly.

    (b)  Place an Array Indicator on the front panel and label it **Best-Fit Amplitude and Frequency**. Then on the block diagram, place a **Curve Fitting** icon and program it to perform a nonlinear fit to the function $a^*\sin(2^*3.14^*b^*x)$. You will also need to supply initial guesses for the parameters $a$ and $b$. After closing the dialog window, connect the output of Simulate Signal to the **Signals** input of Curve Fitting. The resulting dynamic data type wire will automatically create the needed independent variable $x$ for Curve Fitting (i.e., there is no need to explicitly wire anything to the **Locations** input). Wire the nonlinear coefficients output to the front-panel array indicator. Finally, if interested, simply join a wire from Curve

Fitting's **best fit** output to the noisy sine-wave wire emanating from **Simulate Signal**. A **Merge Signals** Express VI will automatically appear, causing both curves to be plotted on the Waveform Graph.

(c)  Run your completed VI with various choices for **Amplitude, Frequency,** and **Noise Amplitude**. To achieve a good fit of the sine function to the noisy sine wave, you will most likely have to experiment with the choice of initial guesses to the fitting parameters. For which parameter must the initial guess be most carefully chosen?

7.  Beginning at time $t = 0$, a capacitor of unknown value $C$ is discharged through a resistor $R = 47\,k\Omega$. In order to determine $C$, as the capacitor is discharging, the voltage difference $V$ between its plates is measured at 5 millisecond intervals, resulting in the following data set.

| $t$ (s) | 0.005 | 0.010 | 0.015 | 0.020 | 0.025 | 0.030 |
|---------|-------|-------|-------|-------|-------|-------|
| $V$ (Volt) | 8.7 | 6.3 | 4.5 | 3.3 | 2.4 | 1.7 |

Theoretically, the relation between $V$ and $t$ is given by

$$V = V_0 e^{-t/RC} \qquad [25]$$

where $V_0$ is the voltage across the capacitor at $t = 0$.

(a)  By taking the logarithm of both sides, rewrite Equation [25] in a linearized form, that is, in the form $y = mx + b$. Then write a program called **Discharging Capacitor (Linear)**, which determines $C$ (in $\mu F$) and also $V_0$ from the given $V$ versus $t$ data. Build this VI so that the data set is input to its block diagram, where the data are used to construct appropriate $x$ and $y$ arrays, which are then input to a **Curve Fitting** icon. Program **Curve Fitting** to fit the input arrays to the equation of a straight line, then use the resulting best-fit parameters $m$ and $b$ to calculate $C$ and $V_0$, and then display these values on the front panel.

(b)  Write another program called **Discharging Capacitor (Nonlinear)**, which also determines $C$ and also $V_0$ from the given $V$ versus $t$ data. Build this VI so that the data set is input to its block diagram, where a **Curve Fitting** icon is programmed to perform a nonlinear fit (see suggestions in Problem 4) of the input arrays to the equation $y = ae^{-x/b}$. Use the resulting best-fit parameters $a$ and $b$ to calculate $C$ (in $\mu F$) and $V_0$, and then display these values on the front panel.

(c)  Run **Discharging Capacitor (Linear)** and **Discharging Capacitor (Nonlinear)**. What values do you obtain for $C$ and $V_0$ from each VI? Do the two approaches yield the same result?

8.  Write a Mathscript-based program called **Discharging Capacitor (Mathscript)** that analyzes the discharging capacitor experiment described in Problem 7. In particular, use the Mathscript command $[p, s] = polyfit(x, y, n)$ to perform a fit of the *Voltage* versus *Time* data to a linearized form of Equation [25]. A description of this command can be found by typing *help polyfit* in the Mathscript Interactive Window. Construct this VI in the following steps:

(a) Enter the given $V$ versus $t$ data as row vectors in a Mathscript Node. The $V$ row vector is created with the command $V = [8.7 \quad 6.3 \quad 4.5 \quad 3.3 \quad 2.4 \quad 1.7]$. Then plot $V$ versus $t$ on a front-panel XY Graph with the data points as solid dots without interpolation.

(b) By taking the logarithm of both its sides and defining $y = \ln(V)$ and $x = t$, Equation [25] can be linearized in the form of the first-order polynomial $y = a_0 + a_1 x$, where $a_0$ and $a_1$ are constants. Add the required code within your Mathscript Node to form the $y$ array (use *help basic* to find the Mathscript command for the natural logarithm) and then use $[p, s] = polyfit(x, y, 1)$ to determine the constants $a_1$ and $a_0$. These constants will be given as the first and second elements of the $p$ row vector, that is, $p(1) = a_1$ and $p(2) = a_0$ (unlike LabVIEW arrays, Mathscript indexing begins with *1*, rather than *0*). Using these best-fit $a_1$ and $a_0$, calculate $C$ (in $\mu F$) and $V_0$ and display these values in front-panel indicators.

(c) Finally, generate an array of time values called *tfit* from $t = 0$ to $0.03$ in $0.001$ increments. Then create another array called *Vfit*, which is produced by evaluating the best-fit form of Equation [25], that is, $Vfit = V_0 \exp(-tfit/RC)$, at all of the *tfit* values. Plot *Vfit* versus *tfit* as an interpolated line on the same XY Graph used to plot the $V$ versus $t$ experimental data.

(d) Run **Discharging Capacitor (Mathscript)**. What values are obtained for $C$ and $V_0$? Does the best-fit curve accurately describe the $V$ versus $t$ data?

9. Curve fitting of data must be guided by a theory describing the phenomenon under investigation. Consider the following hypothetical situation: An experiment acquires seven samples of quantity $y$ as a function of quantity $x$ with the results given in the table below. A theory of the phenomenon being studied suggests that the measured quantities should be related by a second-order polynomial $y(x) = a_0 + a_1 x + a_2 x^2$, where $a_0 = 6.0$, $a_1 = 4.0$, and $a_2 = 2.0$ in appropriate units.

| $x$ | 1.0 | 2.0 | 3.0 | 4.0 | 5.0 | 6.0 | 7.0 |
|---|---|---|---|---|---|---|---|
| $y$ | 10 | 24 | 34 | 54 | 75 | 101 | 133 |

(a) Write a program called **Polynomial Fit** that plots the given data on an XY Graph and also inputs these data to a **Curve Fitting** Express VI. Program the Express VI to fit the data to a second-order polynomial and display the resulting **polynomial coefficients** $a_0$, $a_1$, and $a_2$ in an indicator on the front panel. Also, on the block diagram, include the code shown below to generate a well-resolved representation of the best-fit polynomial $y_{fit}$ and plot this curve, along with the given data, on the front-panel XY Graph (the icon **Reverse 1D Array** is needed because the function *polyval* requires the polynomial coefficients to be in reverse order of that provided by **Curve Fitting**). How do your fitted values of $a_0$, $a_1$, and $a_2$ compare with the theoretically expected values? Why do you suppose that the experimentally determined values for these coefficients don't perfectly match the theoretical values? Characterize qualitatively (in a sentence or two) how well the best-fit polynomial describes the measured data points.

(b)  One might assume that a better fit to the data can be obtained by using a higher-order polynomial. To explore this assumption, reprogram **Curve Fitting** so that it fits the given data to a fourth-order polynomial. From the XY Graph, how does $y_{fit}$ appear to compare with the experimental data? What values do you obtain for $a_0$, $a_1$, $a_2$, $a_3$, and $a_4$? How do the first three of these values compare to the values expected from theory? Speculate on why this fit to a higher-order polynomial is (choose one) better or worse than the fit obtained in part (a).

(c)  Interestingly, a fitted polynomial of order $N-1$ will perfectly pass through $N$ given data samples. Demonstrate this fact using your program. Would this fitted curve be useful in predicting the value of $y(x)$ in a region between the data samples?

# Chapter 10- ■ ■ ■ ■ ■ ■ ■ ■ ■ ■ ■ ■ ■ ■ ■ ■ ■ ■ ■ ■ ■

# Analysis VIs: Fast Fourier Transform

## 10.1 THE FOURIER TRANSFORM

Imagine you are investigating a physical phenomenon by measuring the analog quantity $x$ that varies continuously with time $t$, that is, $x = x(t)$. Based on our discussion at the beginning of the previous chapter, we expect that $x(t)$ can be modeled as the linear combination of a set of $M$ basis functions $b(t)$ such that

$$x(t) = \sum_{k=0}^{M-1} a_k \, b_k(t) \tag{1}$$

The Fourier transform is an example of just such a model. The seminal work of nineteenth-century mathematician Baron Jean Baptiste Joseph Fourier, with subsequent elaboration by others, showed that the infinite set of complex exponentials $\exp(i2\pi ft)$, where $f$ is all the frequencies from $-\infty$ to $+\infty$ and $i = \sqrt{-1}$ , is a complete basis set with which one can form linear combinations to model any arbitrary function. Since an infinite number of exponentials with infinitesimally close frequency spacing is required to model an arbitrary function, the summation in Equation [1] becomes an integral, yielding

$$x(t) = \int_{-\infty}^{+\infty} X(f) e^{i2\pi ft} df \tag{2}$$

where the "*Fourier component*" $X(f)$ is the amplitude (I'll be much more precise in my use of the word *amplitude* shortly) of the exponential with frequency $f$. Generally, $X(f)$ is a complex-valued function, carrying both *magnitude* and *phase* information, a point we will carefully consider in this chapter.

Conversely, the Fourier components $X(f)$ may be obtained by the following equation

$$X(f) = \int_{-\infty}^{+\infty} x(t) e^{-i2\pi ft} dt \tag{3}$$

Equations [3] and [2] are called the *Fourier transform* and the *inverse Fourier transform*, respectively.

## 10.2 DISCRETE SAMPLING AND THE NYQUIST FREQUENCY

In a real experiment, one cannot sample the quantity $x$ continuously, but must instead settle for discretely sampling it $N$ times. If the samples are taken at equally spaced times, where the time difference between adjacent samples is $\Delta t$, then the sampling frequency $f_s$ is given by

$$f_s = \frac{1}{\Delta t} \qquad [4]$$

and the $N$ consecutive sampling times are

$$t_j = j\,\Delta t \qquad j = 0, 1, 2, \ldots, N-1 \qquad [5]$$

Given such a sampling scheme, there is a maximum-frequency sine wave that we can expect to detect. This limitation is simply due to the fact that to observe a sine signal we must at the very least, for example, first sample its positive peak, then its negative trough at the next sample, followed by its positive peak at the third sample, and so on. Thus, the minimal sampling of a sine wave is two sample points per cycle. In an experiment then, where your instruments are acquiring data every $\Delta t$ seconds (at a sampling rate $f_s = 1/\Delta t$), the maximum frequency sinusoidal signal you can detect has a period $T = 2\,\Delta t$. This detection limit is called the *Nyquist frequency* $f_{nyquist}$ and is given by

$$f_{nyquist} = \frac{1}{2\Delta t} = \frac{f_s}{2} \qquad [6]$$

Additionally, by sampling a waveform $x(t)$ at $N$ equally spaced times $t_j$, one can only expect to determine the Fourier components $X(f)$ of $N$ equally spaced frequencies $f_k$ in the Fourier transform. This statement follows from the tacit assumption that $x(t)$ is periodic such that $x(t) = x(t + N\,\Delta t)$, a criterion that forces the waveform $x(t)$ to be composed only of component frequencies that fit an integer number of cycles into the sequence of $N$ samples. The constant (zero-frequency) function obviously meets this criterion. The next lowest acceptable frequency has a period of $N\,\Delta t$ and thus a frequency of $f_1 = 1/N\,\Delta t\;(=f_s/N)$. Within the acceptable range of $-f_{nyquist}$ to $+f_{nyquist}$, the rest of the frequencies $f_k$ are the integer multiples of $f_1$. Thus, defining the spacing between adjacent frequencies to be $\Delta f = f_s/N$, the $N$ equally spaced frequencies $f_k$ determined by discretely sampling data are

$$f_k = k\left(\frac{f_s}{N}\right) = k\Delta f \qquad k = -\frac{N}{2}+1, \ldots, 0, \ldots, +\frac{N}{2}. \qquad [7]$$

Although Equation [7] appears deficient in its neglect of the frequency $-f_{nyquist}$ (given by $k = -N/2$), the two extreme frequencies $\pm f_{nyquist} = f_s/2$ actually describe the same basis function due to the periodic nature of $x(t)$. Thus the range of $k$-values in the above relation rightfully runs from $-N/2 + 1$ to $+N/2$.

## 10.3 THE DISCRETE FOURIER TRANSFORM

With the above-described constraints due to discrete sampling, we then approximate the Fourier transform integral (Equation [3]) by the following sum:

$$X(f_k) = \int_{-\infty}^{+\infty} x(t) e^{-i2\pi f_k t} dt \approx \sum_{j=0}^{N-1} x(t_j) e^{-i2\pi f_k t_j} \Delta t \qquad [8]$$

Note, using Equation [4], [5], and [7],

$$f_k t_j = (k\Delta f)(j\Delta t) = k j \frac{f_s}{N} \Delta t = \frac{kj}{N} \qquad [9]$$

Putting [9] into [8], and realizing that $\Delta t$ is a constant, we find

$$X(f_k) \approx \Delta t \sum_{j=0}^{N-1} x(t_j) e^{-i2\pi jk/N} \equiv \Delta t X_k \qquad [10]$$

where we have defined the *discrete Fourier transform* to be

$$X_k \equiv \sum_{j=0}^{N-1} x(t_j) e^{-i2\pi jk/N} \qquad k = -\frac{N}{2}+1, ..., 0, ..., +\frac{N}{2} \qquad [11]$$

Finally, we approximate the inverse Fourier transform integral (Equation [2]) by the following sum:

$$x(t_j) = \int_{-\infty}^{+\infty} X(f) e^{i2\pi f t_j} df \approx \sum_{k=0}^{N-1} X(f_k) e^{i2\pi f_k t_j} \Delta f \qquad [12]$$

Putting [10] into [12],

$$x_j \approx \sum_{k=0}^{N-1} X_k \, \Delta t \, e^{i2\pi f_k t_j} \, \Delta f \qquad [13]$$

Then using Equation [9] and also noting that $(\Delta t)(\Delta f) = (1/f_s)(f_s/N) = 1/N$, Equation [13] becomes the *discrete inverse Fourier transform*

$$x_j \approx \sum_{k=0}^{N-1} \frac{X_k}{N} e^{i2\pi f_k t_k} = \sum_{k=0}^{N-1} \frac{X_k}{N} e^{i2\pi k j/N} \equiv \sum_{k=0}^{N-1} A_k e^{i2\pi k j/N} \qquad [14]$$

Equations [11] and [14] provide us with a method of performing spectral analysis on an experimental signal $x$. Given that $x$ has been sampled discretely at $N$ evenly spaced times $t_j$, we first calculate the $N$ values of $X_k$ using the definition of the discrete Fourier transform (Equation [11]). Then, (in the spirit of Fourier) viewing the signal $x$ as really the composite of oscillations at $N$ different frequencies $f_k$, Equation [14] tells us that the amplitude of oscillation $A_k$ at each $f_k$ is given by

$$A_k = \frac{X_k}{N} \qquad [15]$$

Since $A_k$ is derived from the complex-valued $X_k$, in general, $A_k$ is a complex-valued function. Let's then name $A_k$ the *complex-amplitude*. The plot of complex-amplitude $A_k$ versus frequency $f_k$ is called the *frequency spectrum* of the sampled signal $x_j$.

Up to now we have taken $k$ to be all of the integers within the range of $-N/2+1$ to $+N/2$. However it is easy to show that Equation [11] is periodic in $k$, with a period of $N$:

$$X_{k+N} = \sum_{j=0}^{N-1} x\left(t_j\right) e^{-i2\pi j(k+N)/N} = \sum_{j=0}^{N-1} x\left(t_j\right) e^{-i2\pi j k/N} e^{-i2\pi j} = X_k \qquad [16]$$

where we have used the fact that $\exp(-i2\pi j) = 1$ because $j$ is an integer. Based on Equation [16], the negative $k$-values in the range of $-N/2+1$ to $-1$ are equivalent to the $k$-value range of $+N/2+1$ to $+N-1$. Because the array indices in computer programs generally are positive integers, algorithms to evaluate Equation [11] typically let $k$ vary from 0 to $N-1$ (which covers one complete period). This is the convention that LabVIEW's Fourier Transform–based VIs follow. Thus, in the $N$-element array of discrete Fourier transform values $X_k$ that such a VI outputs, the elements with indices from 0 to $N/2$ contain the sequence of $X_0$ to $X_{N/2}$ (affiliated with positive frequencies in the range of $0 \leq f \leq +f_{nyquist}$) and elements with indices from $N/2+1$ to $N-1$ contain the sequence of $X_{-N/2+1}$ to $X_{-1}$ (affiliated with negative frequencies in the range of $-f_{nyquist} < f < 0$).

## 10.4 FAST FOURIER TRANSFORM (FFT)

Since the mid-1960s, a clever algorithm to calculate the discrete Fourier transform (Equation [11]) in an efficient manner has been widely used. This algorithm, known as the *Fast Fourier Transform (FFT)*, exploits inherent symmetries in the calculation of Equation [11] that arise due to the periodic way in which the original data were taken ($N$ evenly spaced samples taken with time spacing $\Delta t$). Where the "brute-force" evaluation of Equation [11] would require on the order of $N^2$ complex multiplication operations, the FFT manages this task with only $N \log_2 N$ such operations. The savings in time are immense, especially when $N$ is large. As an example, consider a digital oscilloscope configured to produce a data set of 1024 points (a typical mode of operation). To analyze the frequency spectrum of this set, the FFT algorithm will be 100 times faster than the straightforward evaluation of Equation [11]. For a very large data set of $10^6$ points, the FFT is 50,000 times faster!

In addition to the assumption of evenly spaced data points, there is one extra restriction to the use of FFTs. In deriving a FFT algorithm, one must assume that the size of the data set is a power of two (i.e., $N = 2^m$ where $m = 1, 2, 3, \ldots$) to obtain maximum calculational efficiency. If $N$ is not equal to a power of two, algorithms (called *Discrete Fourier Transforms* or *DFTs*) exist to evaluate the discrete Fourier transform, but with a speed much slower than that of the FFT.

## 10.5 FREQUENCY CALCULATOR VI

The LabVIEW FFT icon that you will implement observes the following convention: The discrete Fourier transform values $X_k$ are output as an $N$-element array, whose elements are indexed 0 to $N-1$, which are associated with $N$ evenly spaced frequencies (where $\Delta f = f_s/N$) in the range of $-f_{nyquist} < f \leq +f_{nyquist}$. The array elements with indices from 0 to $N/2$ contain the sequence of $X_0$ to $X_{N/2}$ associated to the positive frequencies from 0 to $+f_{nyquist}$, and the elements with indices from $N/2 + 1$ to $N-1$ contain the sequence of $X_{-N/2+1}$ to $X_{-1}$ associated with the negative frequencies from $-f_{nyquist} + \Delta f$ to $-\Delta f$.

Let's first write a program, to be used later as a subVI, which calculates the frequency associated with each array element of the FFT output. Start by creating the following front panel. Use **Select a Control...** in the Controls Palette to obtain the **Digitizing Parameters** control cluster in **YourName\Controls** (if you haven't created this cluster yet, see section 3.11 "**Control and Indicator Clusters**"). Assign the connector terminals consistent with the Help Window shown and design an icon. Use **File>>Save** to create a folder called **Chapter 10** in the **YourName** folder, and then save this VI under the name **Frequency Calculator** in **YourName\Chapter 10**.

Now write a block diagram that creates an $N$-element array of frequencies $f_k$, indexed from 0 to $N-1$. We desire the frequency at index $k$ to be the $f_k$ associated with the $X_k$ at that same index in FFT icon's output array. The separation between adjacent frequencies $f_k$ is $\Delta f = f_s/N$, where $f_s$ is the sampling frequency. From the above paragraph's discussion, we see that the $f_k$ can be calculated as follows:

$$ f_k = \begin{cases} k\,\Delta f & k = 0, 1, 2, \ldots, \dfrac{N}{2} \\[4mm] (k-N)\Delta f & k = \dfrac{N}{2}+1, \ldots, N-1 \end{cases} $$

[17]

Program this formula using a **Case Structure** as shown in the following illustration.

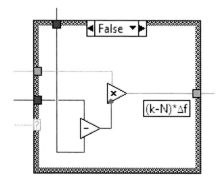

Alternately, Equation [17] can be coded using a **Mathscript Node** as shown below. Here, the command *cat(2,f1,f2)* concatenates the two row vectors *f1* and *f2* (type *help cat* in the Mathscript Interactive Window's Command Window for a detailed explanation of this command). For the **f** output, remember to pop up and select **Choose Data Type>>1D-Array>DBL 1D**.

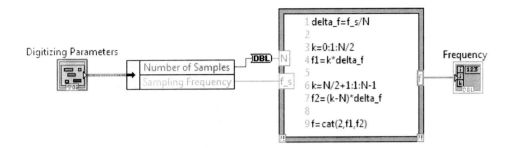

Return to the front panel. Input the values *1024* and *2000* to the **Number of Samples** and **Sampling Frequency** controls, respectively. Run the VI. If it functions properly, the positive frequencies of 0 to +1000, in increments of $2000/1024 \cong 1.95$, will appear in the **Frequency** array at the indices 0 through 512. The negative frequencies of (approximately) $-998.05$ to $-1.95$ will reside at the array elements with indices from 513 through 1023.

## 10.6 FFT OF SINUSOIDS

To get a handle on how FFT signal processing works, let's synthesize some data consisting of sinusoidal waves with known complex-amplitudes $A_k$ and frequencies $f_k$, feed this data to the FFT algorithm, divide the resultant $X_k$ values by $N$ to produce the spectrum of $A_k$ (see Equations [14] and [15]), and determine if this spectrum matches the known input. We will use our old friend **Waveform Simulator**, which you created in Chapter 3 and saved in **YourName\Chapter 3**, to synthesize an $N$-element array of data and our new friend **Frequency Calculator** to produce the array of $f_k$ values.

Construct the front panel shown next, saving it under the name **FFT of Sinusoids** in **YourName\Chapter 10**. You can obtain the control clusters in **YourName\Controls** using **Controls>>Select a Control...** or else through the **Create>>Control** pop-menu option once you place **Waveform Simulator** on the block diagram. We will display the complex-amplitudes $A_k$ on the **XY Graph**, so label its *x*- and *y*-axis **Frequency** and **Complex-Amplitude**, respectively. Since the $A_k$ are complex numbers, we will display the real and imaginary parts of the $A_k$ via two plots. Using the **Plot Legend**, set up these two plots, and label them **Re(A)** and **Im(A)**. Distinguish these plots from each other by color.

The discrete Fourier transform values $X_k$ output by the FFT algorithm are complex numbers. LabVIEW includes a complex number data representation, so $X_k$ is output from the **FFT.vi** icon as this data type. The Help Window for **FFT.vi** (found in **Functions>>Digital Signal Processing>>Transforms**) is shown next. In our program, the data array we will input at **x** is purely real. Since the FFT icon is polymorphic, it will adapt itself automatically to receive this purely real input. Alternately, if we input a complex-valued array at **x**, the icon would again automatically adapt. The complex-valued $X_k$ are output at **FFT {x}**.

Now write the block diagram as follows. The **Complex To Re/Im** icon is found in **Functions>>Programming>>Numeric>>Complex.**

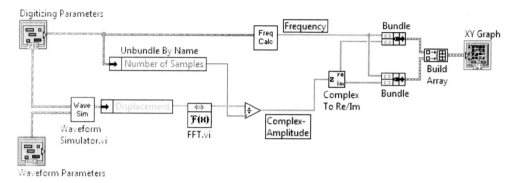

Return to the front panel of **FFT of Sinusoids** and save your work.

## 10.7 APPLYING THE FFT TO VARIOUS SINUSOIDAL INPUTS

Using the **Waveform Parameters** control, program your VI to produce a *Cosine* function with amplitude of *4.0* and frequency of *250* Hz. Note: I will use the word *amplitude* to mean the peak-height (relative to the zero level) of a sinusoidal function. For the **Digitizing Parameters**, choose *1024* and *2000* for **Number of Samples** and **Sampling Frequency**, respectively. Remember that the FFT algorithm requires the number of samples in your input data to be a power of two, so we have chosen $N = 2^{10} = 1024$. Run your program. The VI will find the real and imaginary parts of the complex-amplitude $A_k$ for positive and negative frequencies $f_k$ up to the Nyquist frequency $\pm f_{nyquist} = \pm f_s/2$. For our choice of $f_s = 2000$ Hz, the Nyquist frequency is 1000 Hz, so the 250 Hz cosine oscillation frequency that you have input will fall within the FFT's range of detection.

If all works well, the VI should output the Fourier transform of your cosine input in the following form: The real part of the complex-amplitude $A$ is equal to zero at all frequencies $f$, except $A = +2$ at $f = \pm 250$ Hz. The imaginary part of $A$ is zero at all frequencies $f$.

Why is this output the correct representation for the cosine input? Shouldn't the amplitude $A$ be 4, not 2? To understand your result, plug the output back into Equation [14]. Then

$$x_j \approx \sum_{k=0}^{N-1} A_k e^{i2\pi f_k t_j} \quad = (+2)e^{i2\pi(250)t_j} + (+2)e^{i2\pi(-250)t_j}$$

$$= 4\left[\frac{e^{i2\pi(250)t_j} + e^{i2\pi(-250)t_j}}{2}\right]$$

$$= 4\cos\left[2\pi(250)t_j\right]$$

where we have used the identity $\cos x = (e^{ix} + e^{-ix})/2$ for the last step. So we see that the output complex-amplitudes are indeed the correct representation for an input cosine function of amplitude 4.0.

Try a sine wave input and see what you get. Using **Waveform Parameters**, program **FFT of Sinusoids** to produce a *Sine* function with amplitude of *4.0* and frequency of *250* Hz, while keeping *1024* and *2000* for **Number of Samples** and **Sampling Frequency**, respectively. Now run the VI. You should find that the imaginary part of the complex-amplitude is $A = -2$ at $f = +250$ Hz and $A = +2$ at $f = -250$ Hz, and is zero elsewhere. The real part of $A$ is zero at all frequencies. To see that this is the correct representation of your input, we again start with Equation [14] and note the following:

$$x_j \approx \sum_{k=0}^{N-1} A_k e^{i2\pi f_k t_j} = (-2i) e^{i2\pi(250) t_j} + (+2i) e^{i2\pi(-250) t_j}$$

$$= 4 \left[ \frac{e^{i2\pi(250) t_j} - e^{i2\pi(-250) t_j}}{2i} \right]$$

$$= 4\sin \left[ 2\pi (250) t_j \right]$$

where we have used the facts that $i = -1/i$ and $\sin x = (e^{ix} - e^{-ix})/2i$.

There are two frequencies that behave differently from the rest, zero-frequency and the Nyquist frequency. First, program **FFT of Sinusoids** with the constant *DC Level* (also called *zero-frequency*) function

$$x_j = 4.0$$

then run the VI with **Number of Samples** and **Sampling Frequency** equal to *1024* and *2000*, respectively. Note that, rather than finding half of the amplitude at a positive and negative frequency as in the previously studied non-zero frequency cases, the full amplitude (called the *DC component*) appears at the single frequency $f = 0$.

Now, program **FFT of Sinusoids** with the following function which oscillates at the Nyquist frequency

$$x_j = 4.0 \cos \left[ 2\pi (1000) t_j \right]$$

then run the VI again with **Number of Samples** and **Sampling Frequency** equal to *1024* and *2000*, respectively. If a mysterious diagonal line appears in the resulting FFT plot, remove it by popping-up on the **Plot Legend** and deselecting the "connect-the-dots" mode of plotting in the **Interpolation** option. Similar to the zero-frequency case, you'll discover that the full oscillatory amplitude appears at the single frequency $f_{nyquist}$.

Here is a synopsis of our last observation: The cosine function, which is used to generate the data set, takes on its peak value when the 2000 Hz sampling process begins at time $t = 0$. The 1000 Hz Nyquist-frequency waveform then is sampled twice each cycle, producing a data set that has the following sequence–peak value, trough value, peak value, trough value.... and so on. We find that by applying the FFT algorithm of this data set, the correct value for the amplitude of oscillation (i.e., 4.0) is determined at $f_{nyquist}$.

Unfortunately, by exploring the Nyquist-frequency situation a little further, a troublesome problem emerges. Try programming **FFT of Sinusoids** to produce a Nyquist-frequency oscillation, but this time one that follows the sine function:

$$x_j = 4.0 \sin [2\pi(1000)t_j]$$

then run the VI with **Number of Samples** and **Sampling Frequency** equal to *1024* and *2000*, respectively. You'll find that the FFT detects no amplitude of oscillation at $f_{nyquist}$ in this case. Can you explain why the 1000 Hz sinusoidal oscillation is invisible in this data set?

If interested, you can try including a phase constant $\delta$ (in radians) in the sine function's argument to produce a data set that neither takes on its peak value or zero-value (like the cosine and sine function, respectively) at time $t = 0$. To create the desired function, open **Waveform Simulator** by double-clicking on its icon on **FFT of Sinusoids'** block diagram. Then program Waveform Simulator's **User-Defined** function to be

$$x_j = 4.0 \sin [2\pi(1000)t_j + \delta]$$

Run **FFT of Sinusoids** with **Shape** selected as *User-Defined* in the **Waveform Parameters** control cluster. Does the FFT yield the correct amplitude of oscillation (i.e., 4.0) for any non-zero value of $\delta$?

To summarize our Nyquist-frequency findings: In taking the FFT of a discretely sampled data set, the resultant value for the amplitude at $f_{nyquist}$ is only accurate when the Nyquist-frequency oscillation is "in phase" with the sampling process. Since one cannot guarantee that this will be the case in an actual experimental situation, the Fourier amplitude at $f_{nyquist}$ should always be viewed with suspicion. Thus, in our work below, we will ignore the Fourier component derived at the Nyquist frequency.

Next, program the **User-Defined** function on **Waveform Simulator's** block diagram to be a function that includes two simultaneous sinusoidal oscillations such as

$$x_j = 8.0 + 4.0 \sin[2\pi(250)t_j] + 6.0 \cos[2\pi(500)t_j]$$

and then run **FFT of Sinusoids**. Do you understand the output?

Finally, what happens when the input contains a sine and cosine oscillation, both at the same frequency? Run **FFT of Sinusoids** with the **User-Defined** function

$$x_j = 4.0 \sin[2\pi(250)t_j] + 6.0 \cos[2\pi(250)t_j]$$

You may have to change, for example, the **Point Style** of each plot to view the output accuracy. Do you understand the real and imaginary plots?

## 10.8 MAGNITUDE OF THE COMPLEX-AMPLITUDE

To this point, we have been representing a complex number as the sum of a real and an imaginary part, that is, $z = x + iy$. However, as you know, the complex number $z$ may also be represented as a *magnitude r* times a *phase factor $e^{i\phi}$*, where $r = \sqrt{x^2 + y^2}$ and $\tan\phi = y/x$. Let's say that the oscillation of a system at frequency $f$ is, as in the above example, the composite of a sine and cosine function with (real) amplitudes $B$ and $C$, respectively:

$$x = B \sin (2\pi ft) + C \cos(2\pi ft) \tag{18}$$

Then, defining $\theta = 2\pi ft$, and using the complex exponential representation of the sine and cosine functions,

$$x = B\frac{e^{i\theta} - e^{-i\theta}}{2i} + C\frac{e^{i\theta} + e^{-i\theta}}{2}$$

$$= \frac{1}{2}(C - iB)e^{i\theta} + \frac{1}{2}(C + iB)e^{-i\theta}$$

From Figure 10.1, we see that $(C + iB) = re^{+i\phi}$ and $(C - iB) = re^{-i\phi}$, where $r = \sqrt{B^2 + C^2}$ and $\phi = \tan^{-1}(B/C)$.

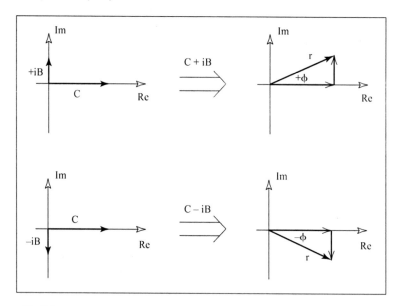

**Fig 10.1**

Thus,

$$x = \frac{1}{2}\left(re^{-i\phi}\right)e^{i\theta} + \frac{1}{2}\left(re^{+i\phi}\right)e^{-i\theta}$$

$$= r\frac{e^{i(\theta-\phi)} + e^{-i(\theta-\phi)}}{2}$$

$$= r\cos\left(\theta-\phi\right) \tag{19}$$

Then, equating [18] with [19] and remembering $\theta = 2\pi ft$, $r = \sqrt{B^2 + C^2}$ and $\phi = \tan^{-1}(B/C)$, we find that

$$x = B\sin\left(2\pi ft\right) + C\cos\left(2\pi ft\right) = \sqrt{B^2 + C^2}\cos\left(2\pi ft - \phi\right) \tag{20}$$

Equation [20] tells us that a sine and cosine oscillation at the frequency $f$ combine to produce a single phase-shifted cosine oscillation at frequency $f$ with a net amplitude equal to the vectorial sum of the sine and cosine amplitudes.

Let's then write a VI called **FFT (Magnitude Only)** that, given an input data set $x_j$, finds the net amplitude of oscillation at the frequencies $f_k$, but ignores the phase information (that is, neglects whether this oscillation is in the form of a pure sine wave or a pure cosine wave or a composite of sine plus cosine). With **FFT of Sinusoids** open, select **File>>Save As…**, and create a new VI named **FFT (Magnitude Only)** in **YourName\ Chapter 10**. Use the **Plot Legend** to request only a single plot—labeled **Mag(A)**—on the **XY Graph**. Relabel the XY Graph's $y$-axis as **Magnitude of Complex-Amplitude**.

Now modify the block diagram so that, using the complex-amplitude's real and imaginary parts, its magnitude is calculated and displayed. Replace the **Complex To Re/Im** icon with **Complex To Polar** (found in **Functions>>Programming>>Numeric>> Complex**). The simple way to make this swap is by popping up on **Complex To Re/Im**, then selecting the **Replace** menu item. It will be obvious what to do from that point. Complete the modification necessary to produce the block diagram shown here.

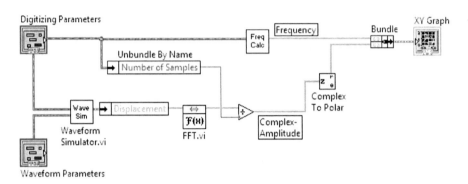

Program the **User Defined** function of **Waveform Simulator** to calculate the sum of a sine and cosine function, both having frequency $f = 250$ Hz, with amplitudes 3.0 and 4.0 respectively:

$$x = 3.0 \sin[2\pi(250)t] + 4.0 \cos[2\pi(250)t]$$

From Equation [20], the above waveform should be equivalent to

$$x = \sqrt{3^2 + 4^2}\cos\left[2\pi(250)t - \tan^{-1}(3/4)\right]$$

Run **FFT (Magnitude Only)** with **Number of Samples** and **Sampling Frequency** equal to *1024* and *2000*, respectively. Is the output representative of a single 250 Hz cosine curve with net amplitude $\sqrt{3^2 + 4^2} = 5$?

The Fourier transform method represents a function as the linear combinations of the complex exponential basis set. Because $\cos(2\pi ft) = (e^{i2\pi ft} + e^{-i2\pi ft})/2$, a cosine function's amplitude is equally divided between the complex-amplitudes of the positive and negative frequency basis functions $e^{\pm i2\pi ft}$. Two exceptions to this "equal division of amplitude" occur, of course, for the basis functions at zero frequency (i.e., $e^{i2\pi(0)t} = 1$) and at the Nyquist frequency $f_{nyquist}$. When displaying the spectrum of a given input data set, this symmetry of complex-amplitudes in frequency-space is commonly exploited by simply doubling the magnitude of the positive frequency's complex amplitude and only plotting the positive-frequency axis. Such a plot displays the net amplitude of sinusoidal oscillation at each frequency $|f_k|$ directly, while phase information is completely neglected.

Modify **FFT (Magnitude Only)** to plot the input data's spectrum in the manner described above. Let's ignore the Fourier component at $f_{nyquist}$ because, as we saw previously, its value is sometimes suspect. Since the $y$-axis values are now equivalent to the net amplitude of sinusoidal oscillation at each (positive) frequency $f_k$, relabel the XY Graph's $y$-axis and Plot Legend as **Amplitude**. The following block diagram accomplishes the desired features.

Run the VI and verify that it performs correctly. Once satisfied, save your work.

## 10.9 LEAKAGE AND WINDOWING

I have good news and I have bad news.

## 10.10 OBSERVING LEAKAGE

First, the bad news. Reprogram **FFT (Magnitude Only)** with the function

$$x = 4.0 \cos[2\pi(250)t]$$

and then run the VI using the values *1024* and *2000* for **Number of Samples** and **Sampling Frequency**, respectively. Its spectrum will appear as shown.

Using the **Plot Legend**, make the individual plotted points visible by changing the **Point Style**. Then magnify the spectral peak at $f = 250$ Hz by changing the x-axis endpoints. An easy way to zoom in on a peak is to use the Graph's *Zoom Tool*. To access the Zoom Tool, first pop up on the Graph's interior and select **Visible Items>>Graph Palette**. Using the ☝, click on the Zoom Tool ⊕, which looks like a magnifying glass, and select the **X-Axis Zoom** option.

The mouse cursor will become a small magnifying glass when positioned over the plot. Click and drag it to select the region which you would like to zoom in on.

When you release the mouse button, the *x*-axis will be rescaled in the desired manner.

If you want to return to the original plot, select **Zoom to Fit** from the Zoom Tool menu, which will cause all of the data to be displayed again by autoscaling both axes.

The spectrum looks perfect, doesn't it? The amplitude is equal to zero everywhere, except at the frequency of 250 Hz, where it equals +4.0, just as we expected.

So, what's the problem? Try repeating the above procedure after programming **FFT (Magnitude Only)** with the following function

$$x = 4.0\cos[2\pi(249)t]$$

Surprisingly, the resulting spectrum will no longer appear as a "delta-function" spike with a height of 4.0, but instead will be a broadened peak of maximum height less than 4.0, as shown.

Zoom in on the peak to get a closer look.

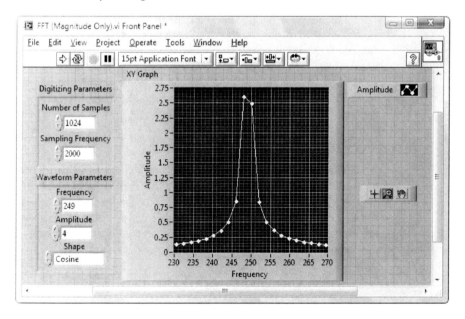

At face value, this spectrum seems to be telling us that the input data is oscillating at several different frequencies in the approximate range of 230 to 270 Hz. However, we know that this is simply untrue. We input a pure sine wave with the exact frequency of 249 Hz.

We are then faced with a mystery. Why does our FFT-based VI work perfectly for an input frequency of 250 Hz, but fails when that frequency is changed to 249 Hz? Well, here's the resolution to this paradox: Above, I've been sneaky and always, with the exception of $f = 249$ Hz, asked you generate input data using sine and cosine functions that oscillate at one of the $N$ discrete frequencies $f_k$. For example, via Equation [7], when acquiring 1024 equally spaced data samples at a sampling rate of 2000 Hz, the FFT algorithm produces complex-amplitudes $A_k$ at the $N$ frequencies $f_k$ given by

$$f_k = k \left( \frac{2000\,\text{Hz}}{1024} \right) \qquad k = -511, \ldots, 0, \ldots, +512$$

You can verify that 250 Hz and 500 Hz are both one of the $f_k$. In particular, they are $f_{128}$ and $f_{256}$, respectively.

A way of summarizing our above observations is this: If you input a sinusoidal oscillation at exactly one of the frequencies $f_k$, for example $f_{128} = 250.00$ or $f_{256} = 500.00$ Hz, the FFT algorithm will perfectly produce the frequency spectrum of that data (that is, a delta-function spike of the correct height at the correct frequency). However, if instead one inputs a sinusoidal oscillation at a frequency $f$ not equal to one of the $f_k$, such as 249 Hz, which falls between $f_{127} = 248.05$ Hz and $f_{128} = 250.00$ Hz, the resulting spectrum is imprecise. In fact, it is as if its spectral amplitude, which rightfully should be a spike at frequency $f$, has diffused from this central point and distributed itself among the neighboring $f_k$. This smearing of spectral information—termed *leakage*—is an artifact of the finite number $N$ of data samples contained in our discretely sampled data set. You can prove this for yourself by rerunning **FFT (Magnitude Only)** programmed to calculate a 249.0 Hz cosine wave, first making **Number of Samples** equal to *512*, then *1024*, then *2048*, followed by *4096*, and so on. You will find that as $N$ increases, the spectrum becomes much more delta-function–like (it will be helpful here to pop up on the Graph and deselect **X Scale>>AutoScale X**).

## 10.11 ANALYTIC DESCRIPTION OF LEAKAGE

It's not too hard to derive analytic expressions that explain these observations. Consider an oscillatory waveform with frequency $f$ that is described by the complex exponential $x(t) = A\exp(i2\pi ft)$, where $A$ is a constant. Once the features of this waveform's "single-spike" Fourier transform are understood, the properties of a sinusoid's "dual-spike" spectrum will be obvious. Applying the discrete Fourier transform (Equation [11]) to the complex-exponential waveform, the values of $X_k$ are

$$X_k = \sum_{j=0}^{N-1} A\, e^{i2\pi f(j\Delta t)} e^{-2\pi j k/N}$$

so

$$X_k = A \sum_{j=0}^{N-1} \left[ e^{i2\pi(f\Delta t - k/N)} \right]^j \qquad [21]$$

This series is the well-known finite geometric series, whose form and summation value are given by

$$\sum_{j=0}^{N-1} x^j = \frac{1-x^N}{1-x} \qquad [22]$$

After using [22] to evaluate the sum in [21], a few lines of algebra and trigonometric relations yield the following expression for the magnitude of the discrete Fourier transform values $X_k$

$$|X_k| = \left| A \frac{\sin\left[\pi N\left(f\Delta t - k/N\right)\right]}{\sin\left[\pi\left(f\Delta t - k/N\right)\right]} \right| \qquad [23]$$

This equation then describes the quantity determined by **FFT (Magnitude Only)**.

Let's investigate the meaning of Equation [23]. First, consider the case when the input frequency $f$ exactly equals one of the frequencies $f_k$, say $f_{k'}$. Setting $f = f_{k'} = k'(f_s/N) = k'(1/N\Delta t)$, Equation [23] becomes

$$|X_k| = \left| A \frac{\sin\left[\pi\left(k' - k\right)\right]}{\sin\left[\pi(k' - k)/N\right]} \right| \qquad [24]$$

For $k \neq k'$, the difference $(k' - k)$ will be an integer less than $N$, making the numerator and denominator of [24] zero and non-zero, respectively. Thus, $|X_k| = 0$ for $k \neq k'$. However, for $k = k'$, both numerator and denominator of [24] are zero and l'Hopital's rule gives $|X_k| = AN$. So we expect the resulting frequency spectrum to be a delta function with a single spike at frequency $f_{k'}$ of height $|A_{k'}| = |X_{k'}|/N = A$. This prediction is consistent with what we observed when we input the oscillations at $f = f_{128} = 250$ Hz and $f = f_{256} = 500$ Hz inputs to the FFT algorithm.

Second, consider the case when the input frequency $f$ is not equal to one of the frequencies $f_k$. We can simplify the appearance of Equation [23] by noting from Equations [4] and [7]

$$f\Delta t - \frac{k}{N} = \frac{f}{f_s} - \frac{f_k}{f_s} = \frac{f - f_k}{f_s}$$

Then Equation [23] becomes

$$|X_k| = \left| A \frac{\sin\left[\pi N\left(\dfrac{f - f_k}{f_s}\right)\right]}{\sin\left[\pi\left(\dfrac{f - f_k}{f_s}\right)\right]} \right|$$

and, since $\Delta f = f_s/N$,

$$|X_k| = \left| A \frac{\sin\left[\pi\left(\dfrac{f-f_k}{\Delta f}\right)\right]}{\sin\left[\pi\left(\dfrac{f-f_k}{f_s}\right)\right]} \right| \qquad [25]$$

Inspecting Equation [25], we see that, under our assumption that $f$ does not equal one of the $f_k$, none of the $|X_k|$ will be zero. However, the frequencies $f_k$ that fall closest to $f$ will make the denominator of Equation [25] the smallest. Thus the $|X_k|$ associated with the $f_k$ neighboring $f$ will take on the largest values. To better visualize the meaning of Equation [25], I use it to plot the magnitude of the complex-amplitude $|A_k| = |X_k|/N$ versus $f_k$ with the following familiar choice of parameters: $f_s = 2000$ Hz, $N = 1024$, $A = 4.0$, and $f = 249$ Hz. Then $\Delta f = f_s/N = 2000/1024$ and $f_k = k\Delta f$, where $k = -511$, ...,0,..., +512. This plot will predict the output of **FFT (Magnitude Only)** when a 249 Hz cosine function of amplitude 4 is input.

Zooming in on the peak to get a closer look, we see that Equation [25] provides a perfect theoretical prediction of the spectral leakage we observed previously with a 249 Hz input to **FFT (Magnitude Only)**.

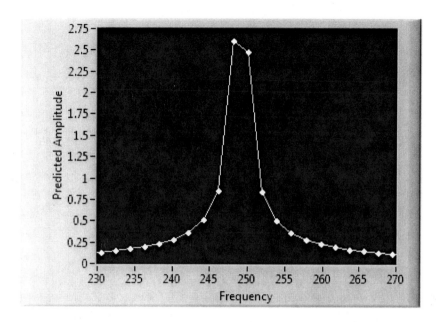

## 10.12 DESCRIPTION OF LEAKAGE USING THE CONVOLUTION THEOREM

The *Convolution Theorem*, a powerful result from higher mathematics, provides another vantage point from which to understand the problem of leakage. From this point of view, the situation appears as follows: When we acquire a finite number $N$ of discretely sampled points for FFT spectral evaluation, we are in effect observing an infinite set of data $d_j (j = -\infty, \ldots, -1, 0, +1, \ldots, +\infty)$ through a rectangular viewing window in time. Mathematically, we can define the rectangular window function $w(t_j)$ to be zero at all times $t_j = j\Delta t$, except during the "data-viewing" time interval from $j=0$ to $j=N-1$, when it is equal to one. Then our finite set of $N$ sampled data points $x_j$ is given by the product $x_j = d_j w_j$. This idea is illustrated in Figure 10.2.

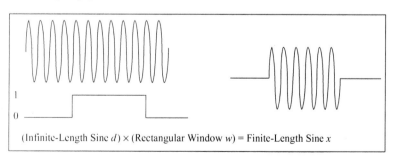

(Infinite-Length Sine $d$) × (Rectangular Window $w$) = Finite-Length Sine $x$

Fig 10.2

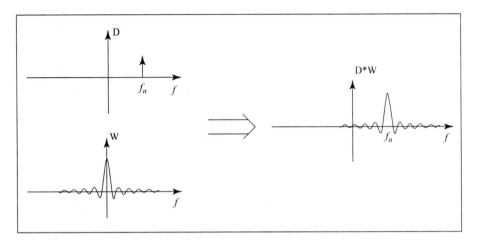

**Fig 10.3**

Let's say that the Fourier transforms of $d_j$ and $w_j$ are $D_k$ and $W_k$, respectively. The question relevant to our finite-length data set then becomes "What happens when one takes the Fourier transform of the product $d_j w_j (= x_j)$?" Well, according to the famous Convolution Theorem, the Fourier transform of the product of two functions $d_j w_j$ is equal to the *convolution* of the two Fourier transforms $D_k$ and $W_k$. The convolution for continuous functions, denoted by $D*W$, is defined by

$$D(f)*W(f)=\int_{-\infty}^{+\infty}D(\phi)W(f-\phi)d\phi \qquad [26]$$

In the discrete case, this definition becomes

$$(D*W)_k = \sum_{m=-N/2+1}^{N/2} D_m W_{k-m} \qquad [27]$$

The convolution of two functions, while complicated in general to determine, is simple to ascertain in the following important case. Let $D_m$ be a unit-amplitude delta function located at the frequency $f_n$, that is, $D_m = 0$ for all $m$, except $D_{m=n} = 1$. Then, from [27], we find that $(D*W)_k = W_{k-n}$, meaning that the convolution is just the function $W$ displaced so that it is now centered at $f_n$ rather than $f = 0$. This idea is illustrated in Figure 10.3 using continuous functions.

This example provides new insight into the leakage phenomenon. Consider the case of a finite-length complex-exponential input of the form $x(t_j) = A\exp(i2\pi f t_j)$, which can be described as the product of an infinite-length complex-exponential $d(t_j)$ and a rectangular window function $w(t_j)$. The Fourier transform $A$ of the infinite complex-exponential is, of course, a delta function of height $A$ located at frequency $+f$, while the discrete Fourier transform of the rectangular window (apart from a unity-amplitude phase factor) is easily shown to be

$$\left(W_{rectangle}\right)_k = \frac{\sin\left[\dfrac{\pi f_k}{\Delta f}\right]}{\sin\left[\dfrac{\pi f_k}{f_s}\right]} \qquad [28]$$

A plot of Equation [28] (treating frequency as a continuous variable) with $f_s = 2000$ Hz and $N = 1024$ is shown in the following diagram. Note the substantial amplitudes of $W_{rectangle}$ at high frequencies, which we know qualitatively result from the sharp turn-on and turn-off at the edges of the rectangular window (recall from the Fourier analysis of a square wave, its sharp edges are produced by the presence of high-frequency harmonics). The values of $W_{rectangle}$ at the frequencies $f_k$, that is, $(W_{rectangle})_k$, are given by the dots.

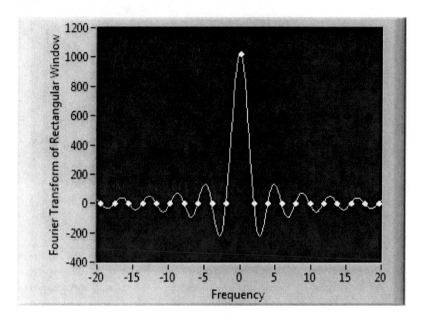

The magnitude of the finite-length complex-exponential's Fourier transform $X$ is formed from the convolution of $W_{rectangle}$ with the height-$A$ delta function at $+f$. This convolution is simply $W_{rectangle}$, shifted along the frequency axis so that it is centered at frequency $+f$, and then multiplied by $A$. Thus,

$$\left|X_k\right| = \left|(D*W)_k\right| = \left|A\,\frac{\sin\left[\dfrac{\pi\left(f_k - f\right)}{\Delta f}\right]}{\sin\left[\dfrac{\pi\left(f_k - f\right)}{f_s}\right]}\right| \qquad [29]$$

363

Note that Equation [29] is equivalent to the "leakage description" given by Equation [25].

Next, I plot Equation [29] for $f_s = 2000$ Hz and $N = 1024$ with $f$ equal to one of the $f_k$, namely, $f = f_{128} = 250$ Hz. I take $A = 1$ for simplicity. Note that all of the values of $|X_k|$, which are given by the dots, fall on the zero-crossing of Equation [29], except $|X_{128}|$ at $f_{128} = 250$. Thus the discrete Fourier spectrum will be a delta function in this case.

Changing $f$ to the non-$f_k$ value of 249 Hz, the following plot of Equation [29] is obtained. Note that the $|X_k|$, again indicated by the dots, now fall at non-zero locations on the curve defined by Equation [29], resulting in the spectral leakage phenomenon.

The Convolution Theorem then gives us the following insight: It is the wide-ranging frequency content of the rectangular window function (as given in Equation [28] and shown in the previous plot) that produces leakage of spectral amplitude to frequencies $f_k$ far removed from the input frequency $f$.

## 10.13 WINDOWING

Now the good news. Thanks to the power of your high-speed computer, you have the option of windowing your finite-length data set with some function other than a rectangle. Once the rectangular-windowed data set $(x_{rectangle})_j$ has been collected into your computer, all you have to do is construct a more desirable window $w_j$ in software, then form the product $x_j = w_j(x_{rectangle})_j$. Leakage can simply be suppressed by choosing a window whose Fourier transform has minimal high-frequency components. Qualitatively, we know that the leakage-producing high-frequency components result from the discontinuous jumps at the edges of the rectangular function, so the "desirable" software window functions should be constructed to have a much smoother turn-on and turn-off. The *Hanning* window, which is defined to be

$$\left( w_{hanning} \right)_j = 0.5 \left[ 1 - \cos\left( \frac{2\pi j}{N} \right) \right] \qquad j = 0, 1, \ldots, N-1$$

is a popular choice for such a software window. The Hanning and Rectangle windows are plotted here for comparison.

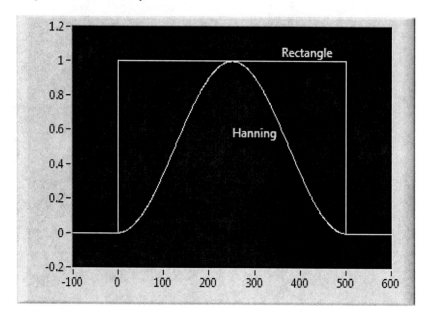

Let's try windowing our data set in **FFT (Magnitude Only)** and see how this improves things. To window the **Waveform Simulator**–produced data set, you will use **Scaled Time Domain Window.vi**, found in **Functions>>Signal Processing>>Windows**. Its Help Window is shown next.

For this icon, you provide the "to-be-windowed" data array at the **x** input and select the desired window at **window** using an Enumerated Type ("Enum") control. The resultant array (equal to the product of the data and the window), is output at **Windowed x**. The **window properties** cluster is a two-element bundle of the selected window's *equivalent noise bandwidth (ENBW)* and *coherent gain*. These window properties are important for use in some calculations, as will be shown in a few minutes.

When using **Scaled Time Domain Window.vi**, the user must choose one window, from among the approximately 20 available, to be applied to the input data array. Each window has its own properties, such as central-lobe width and roll-off rate, which are designed to optimize its use in particular applications. Generally, the Hanning window is a satisfactory choice in most situations. This window suppresses spectral leakage, allowing frequencies within the input signal to be well resolved. Table 10.1 lists some of the commonly used windows, along with the applications for which they are most suitable.

**Table 10.1**  Commonly Used Windows

| WINDOW | OPTIMAL APPLICATION |
|--------|---------------------|
| Hanning | General-purpose applications |
| Hamming | Resolution of closely spaced sine waves |
| Flat Top | Accurate amplitude measurement of isolated frequency |
| Kaiser* | Resolution of two closely spaced frequencies with widely differing amplitudes |
| Rectangle | Resolution of two closely spaced frequencies with almost equal amplitudes |

\* This filter's central-lobe width controlled by value at **window parameter.**

Switch to the block diagram of **FFT (Magnitude Only)**. Include **Scaled Time Domain Window.vi** as shown, so that the data set obtained from **Waveform Simulator** is windowed prior to being passed to the **FFT.vi** icon. Additionally, pop up on Scaled Time Domain Window.vi's **window** input and select **Create>>Control** to create the front-panel Enum control that facilitates window selection.

Switch to your front panel, find the Enum control labeled **window** that now appears there, then position it aesthetically, then save your work.

One of the benefits derived from having implemented **Create>>Control** to produce the front-panel **window** control is that, while creating it, LabVIEW automatically loads this Enum with the sequence of text messages that label the available options at Scaled Time Domain Window.vi's **window** input. Using the 🖑, review the sequence of selections offered by the **window** control.

Let's explore the positive effects of windowing. Let **Number of Samples** and **Sampling Frequency** be *1024* and *2000*, respectively, and program your VI to produce the waveform $4.0\sin[2\pi(249\ \text{Hz})t]$. Zoom in on the expected peak by scaling the axes appropriately, say, *230* to *270* for the *x*-axis and *0* to *4* for the *y*-axis, then pop up on the Graph, and use its pop-up menu to turn off the *x*- and *y*-axis autoscaling options. Run the VI without the benefit of windowing by selecting *Rectangle* in the **window** control.

Now select the *Hanning* window and rerun the VI.

Note the dramatic decrease in leakage through the use of a Hanning window. See, I told you I had good news!

To demonstrate that certain windows perform better in a particular application, consider the challenge of resolving two sine waves of nearly the same frequency but widely differing amplitudes, a problem that might be encountered in, say, speech recognition research. Use the **User-Defined** option of **Waveform Simulator** to input the following waveform to **FFT (Magnitude Only)**:

$$x = 4.0 \sin[2\pi(249)t] + 0.1 \sin[2\pi(253)t]$$

With this input waveform (you will need to adjust the Graph's y-axis scaling appropriately), run **FFT (Magnitude Only)** several times, each time with a different choice for **window**. For the kaiser window, you must supply a value (try *3*) at the **window parameter** input of Scaled Time Domain window.vi. Which windows perform best for this particular application? For which windows does the smaller 253 Hz peak get obscured by the 249 Hz peak (due to large spectral leakage)?

Change the input waveform to two sine waves of nearly the same frequency with equal amplitudes:

$$x = 4.0 \sin[2\pi(249)t] + 4.0 \sin[2\pi(253)t]$$

Which window best resolves these peaks now (e.g., shows the deepest dip between the peaks)?

Finally, the *Flat Top* window is optimized to produce a peak with an amplitude that closely matches that of the input waveform. Input $4.0 \sin[2\pi(249 \text{ Hz})t]$ to **FFT (Magnitude Only)** with various choices for **window**. Is the amplitude of the spectral peak produced from this input closest to the value of 4.0 when using the Flat Top window (in comparison to when other windows are used)?

## 10.14 ESTIMATING FREQUENCY AND AMPLITUDE

I have even more good news. Define the *peak power* at frequency $f_k$ to be $P_k \equiv A_k^2$. Then the sinusoidal input frequency $f$ that generated the finite-width spectral peaks observed above can be estimated by a weighted sum, where each observed $f_k$ is weighted by the peak power at that frequency. That is,

$$f \approx \frac{\sum_k f_k P_k}{\sum_k P_k} = \frac{\sum_k f_k A_k^2}{\sum_k A_k^2} \qquad [30]$$

where the sum is over the $k$-values that span the peak.

Also the peak power $P \equiv A^2$ of the sinusoid with amplitude $A$ can be estimated by

$$P \approx \frac{\sum_k P_k}{\text{ENBW}} = \frac{\sum_k A_k^2}{\text{ENBW}} \qquad [31]$$

where *ENBW* is the effective noise bandwidth of the window used in the analysis process that produced the peak. Once $P$ is determined, then the sinusoidal amplitude is $A = \sqrt{P}$.

Write a VI called **Estimated Frequency and Amplitude** that implements Equations [30] and [31]. Store this program in **YourName\Chapter 10.** The front panel should look as shown next. The **Positive Frequency** and **Amplitude** arrays need to be the index 0 and index 1 elements, respectively, of the **Input Arrays** control cluster in order to be consistent with how these arrays are bundled together on the **FTT (Magnitude Only)** block diagram. Pop up on this cluster and use **Reorder Controls In Cluster...** to make sure the two arrays are indexed properly.

Assign the connector terminals consistent with the following Help Window.

Now code the following block diagram. Because it is easy to do, we will perform the sums in Equations [30] and [31] over all $k$, but it is only necessary to include $k$-values for which the $A_k$ are significantly non-zero, that is, $k$ in the neighborhood of the peak.

Save this VI as you close it.

Return to the block diagram of **FTT (Magnitude Only)** and incorporate **Estimated Frequency and Amplitude** as shown below. Use **Unbundle By Name** to obtain the selected window's ENBW (labeled **eq noise BW**) from the **window properties** cluster that is output from Scaled Time Domain Window.vi. Use **Create>>Indicator** to create the **Estimated Values** cluster.

Switch to the front panel and position **Estimated Values** nicely. Within this cluster, pop up on both the **Frequency** and **Amplitude** indicators, and using the **Display Format...** option, deselect **Hide trailing zeros**. Save your work.

With the values for **Digitizing Parameters** and **Waveform Parameters** shown above, run the VI using various windows. I think you'll be very pleased with the estimated values of **Frequency** and **Amplitude!** Can you make any qualitative conclusions about the strengths and weaknesses of the different windows?

## 10.15 ALIASING

Finally, let's use **FFT (Magnitude Only)** to demonstrate aliasing, the phenomenon in which an incoming signal is digitally sampled too infrequently, resulting in a digitized waveform whose frequency is much lower than that of the actual signal. In Chapter 4, we found that when sampling a signal of frequency $f$ at sampling frequency $f_s$, the digitized waveform will also have frequency $f$ only if $f \leq f_{nyquist}$, where $f_{nyquist} = f_s/2$ is the Nyquist frequency. If, instead, the frequency of the incoming signal is $f > f_{nyquist}$, then the digitized waveform will have frequency $f_{alias}(\neq f)$ given by

$$f_{alias} = |f - nf_s| \qquad n = 1, 2, 3, \ldots \qquad [32]$$

where $f_{alias}$ is in the range $0 \leq f_{alias} \leq f_{nyquist}$.

Program **FFT (Magnitude Only)** so that the input waveform is a sine wave of amplitude 4.0. Choose your favorite window, say, the general-purpose *Hanning*. In **Digitizing Parameters**, set **Number of Samples** and **Sampling Frequency** to be *1024* and *2000*, respectively. Then, we expect aliasing to occur when the frequency of our input sine wave exceeds the Nyquist frequency $f_{nyquist} = f_s/2 = (2000 \text{ Hz})/2 = 1000 \text{ Hz}$. Run the VI several times with the following sequence of input sine-wave frequencies: *1200 Hz, 1500 Hz, 1800 Hz, 2000 Hz, 2800 Hz, 3300 Hz, 7700 Hz*. Does Equation [32] correctly predict the aliased frequency you observe for each input?

In this exercise, we have the benefit of knowing the input signal's true frequency and so can use Equation [32] to figure out why the digitizing process is producing an incorrect lower-frequency output. In a real experiment, however, where we'd have no prior knowledge of the input signal's true frequency, we would have no other recourse than to assume that the digitized waveform has the same frequency as the incoming signal. That is, we'd assume the signal is not being aliased in the digitizing process. To ensure that this "no-aliasing" assumption is valid, it is imperative that an experimentalist first determines the digitizing system's Nyquist frequency (based on the known sampling rate) and then assures (e.g., though the use of a low-pass filter) that no signals with frequencies greater than the Nyquist frequency are input to the digitizer.

## DO IT YOURSELF

Build a computer-based spectrum analyzer that digitizes an incoming voltage waveform, takes the Fourier transform of this acquired data, and displays the resulting spectrum.

Complete the following steps.

1.  Write the **Spectrum Analyzer** VI:

    With **FFT (Magnitude Only)** open, use **File>>Save As...** to create a new VI called **Spectrum Analyzer** and save it in the **YourName\Chapter 10** folder.

    Switch to the block diagram of **Spectrum Analyzer**. Delete the **Waveform Simulator** and **Waveform Parameters** icons. Place **DAQ Assistant** on your diagram and configure it to perform an analog input voltage operation on a particular differential AI channel of your DAQ device (e.g., **ai0**) with **Acquisition mode>>N Samples** selected. Include the option of digital triggering, if you wish. On your block diagram, expand **DAQ Assistant**'s terminals to include **data, number of samples**, and **rate**. Complete the necessary modifications so that **DAQ Assistant** receives the values of **Number of Samples** and **Sampling Rate** from the **Digitizing Parameters** cluster, and then after obtaining *N* data samples at the instructed sampling rate, passes the acquired data array to **Scaled Time Domain Window.vi**. Finally, enclose all of the block diagram code within a **While Loop**, so that spectra are produced and plotted repeatedly, until a front-panel **Stop Button** is clicked. Save your VI.

2.  Operate the computer-based instrument:

    Using a function generator, input a sine-wave voltage with a frequency of about 100 Hertz input into the analog input channel of your DAQ device that was programmed into DAQ Assistant (e.g., **ai0**). If you configured DAQ Assistant for digital triggering, also input your function generator's sync output to the appropriate pins of your DAQ device. With **Number of Samples** and **Sampling Frequency** set to *1024* and *2000*, respectively, and your choice for **window**, run **Spectrum Analyzer** to see the spectral make-up of your voltage signal. How do the values of **Frequency** and **Amplitude** in **Estimated Values** compare with the known value (from the function generator's control panel) of your input signal? Try changing the input sine-wave frequency $f$ to values within the range $0 \le f \le 1000$ Hz. Does the instrument perform well within this frequency range?

3. Explore interesting inputs:

Increase the input sine-wave's frequency beyond the Nyquist frequency to see the aliasing effect. Explain your observations using Equation [32].

Try inputting a square wave with a frequency of about 180 Hz and amplitude of 1 Volt (i.e., oscillates between $-1$ V and $+1$ V). Fourier analysis of a square wave of frequency $f$ and unity amplitude predicts that this waveform is equivalent to the following sum of sine waves:

$$\frac{4}{\pi} \sum_{n \text{ odd}} \frac{1}{n} \sin[2\pi n f t] = \frac{4}{\pi} \left\{ \sin[2\pi f t] + \frac{1}{3} \sin[2\pi(3f)t] + \frac{1}{5} \sin[2\pi(5f)t] + \ldots \right\}$$

Note that this sum of sine waves only includes the odd harmonic frequencies $f$, $3f$, $5f$, and so on. With a square-wave input, is the output of **Spectrum Analyzer** consistent with this Fourier sum? Do you note any aliasing of the square wave's higher harmonics?

## Problems

1. Use **FFT (Magnitude Only)** to investigate aliasing. Program the **User-Defined** option of **Waveform Simulator** to produce the digitized signal $x$ consisting of the sum of four incoming sine waves of frequency $f_1 = 25$Hz, $f_2 = 70$Hz, $f_3 = 160$Hz, and $f_4 = 510$ Hz, each with amplitudes of 1, that is, $x = \sin[2\pi(25)t] + \sin[2\pi(70)t] + [2\pi(160)t] + \sin[2\pi(510)t]$. Then, on the front panel of **FFT (Magnitude Only)**, set **Number of Samples, Sampling Frequency, Shape,** and **window** equal to *4096, 100, User-Defined,* and *Rectangle,* respectively. Run **FFT (Magnitude Only)** and record the frequencies of the four peaks that you observe in the resulting spectrum. Identify each peak, that is, tell whether the peak is associated with a true frequency within the signal $x$ or with an aliased frequency produced by the digitizing process. For each aliased frequency, use Equation [32] to identify the true frequency ($f_1$, $f_2$, $f_3$, or $f_4$) that was aliased.

2. Use **FFT (Magnitude Only)** to explore the improved frequency resolution that results from long-time sampling of an input signal. Program the **User-Defined** option of **Waveform Simulator** under the assumption that the input signal $x$ from an experiment consists of two closely spaced sine waves of frequency 399 Hz and 401 Hz, each with amplitudes of 4, that is, $x = 4\sin[2\pi(399)t] + 4\sin[2\pi(401)t]$. Then, on the front panel of **FFT (Magnitude Only)**, set **Sampling Frequency, Shape,** and **window** equal to *2000, User-Defined,* and *Rectangle,* respectively.

   (a) Simulate sampling the input signal for longer and longer times by running **FFT (Magnitude Only)** with the following succession of values for **Number of Samples**—128, 256, 512, 1024, 2048, 4096—each time noting whether or not the 399 Hz and 401 Hz peaks are resolved in the resulting spectrum.

   (b) At what threshold value for **Number of Samples** do the two peaks begin to be resolved? From your knowledge of FFT theory, explain how this threshold value is determined by the various parameters values used in this investigation.

3.  Observe the Fourier Transform of a square wave. Program **FFT (Magnitude Only)** with the following choices: **Number of Samples** and **Sampling Frequency** equal to *4096* and *5000*; **Frequency, Amplitude,** and **Shape** equal to *180, 1,* and *Square,* respectively, to simulate a square wave with a frequency $f = 180$ Hz, which oscillates between 0 V and +1 V. Fourier analysis predicts that this waveform is equivalent to a DC component (of magnitude 0.5) plus a sum of sine waves as follows:

$$0.5 + \frac{2}{\pi}\sum_{n \text{ odd}}\frac{1}{n}\sin[2\pi nft] = 0.5 + \frac{2}{\pi}\left\{\sin[2\pi ft] + \frac{1}{3}\sin[2\pi(3f)t] + \frac{1}{5}\sin[2\pi(5f)t] + \cdots\right\}$$

Note that the sum of sine waves only includes the odd harmonic frequencies $f, 3f, 5f,...,$ with amplitudes of $2/\pi, 2/3\pi, 2/5\pi,....$ Finally, set **window** to *Flat Top* so that the Fourier peaks will have accurate amplitudes.

    Run **FFT (Magnitude Only)** and record the frequency and amplitude of every peak that you observe in the resulting Fourier spectrum. Then, identify the harmonic to which each peak corresponds, including those peaks that are present due to aliasing. Finally, comment on whether each observed peak has the amplitude expected from Fourier analysis.

4.  Use **FFT (Magnitude Only)** to explore the *frequency doubling* (and *tripling*) phenomenon that occurs when a pure sine-wave signal is input to a nonlinear detector/amplifier (e.g., radio wave on diode detector, monochromatic light on nonlinear crystal). To investigate in this effect, use the **User-Defined** option of **Waveform Simulator** to simulate the following process: a pure sinusoid $x(t) = A\sin(2\pi ft)$ is passed through a slightly nonlinear amplifier producing an output signal $a(x)$ given by

$$a(x) = \alpha x + \beta x^2$$

where $\alpha$ and $\beta$ are constants with $\alpha$ significantly larger than $\beta$. Remember .^ is Mathscript's element-wise power operator.

    Run **FFT (Magnitude Only)** with some choice of $f$, $\alpha$, and $\beta$ ($f = 250$ Hz, $\alpha = 10$, and $\beta = 1$, might be a good place to start) to find the spectrum of $a(x)$.

    (a)  In terms of $f$, what frequencies do you observe in the spectrum?
    (b)  The $x^2$ term in $a(x)$ produces a product of sine waves that can be rewritten using the following trigonometric identity:

$$\sin\theta\sin\phi = \frac{1}{2}\left[\cos(\theta - \phi) - \cos(\theta + \phi)\right]$$

        With the help of this relation, predict the frequencies that should appear in the spectrum of $a(x)$. Are these the frequencies that you observed?
    (c)  Explore what happens when the amplifier is even more nonlinear so that $a(x) = \alpha x + \beta x^2 + \gamma x^3$. In terms of $f$, what frequencies do you observe now in the spectrum?

    (In experimental optics, laser light of frequency $f$ is commonly input on a nonlinear crystal in order to produce output light of frequency $2f$ and $3f$.)

5.  Observe the *sidebands* that exist in the frequency spectrum of an *amplitude modulated* (AM) wave as follows.

(a) On the block diagram of **FFT (Magnitude Only)**, replace the subVI **Waveform Simulator** (and its input control clusters) with **AM Wave** (written as Chapter 3's **Do It Yourself** project). Then, with *1024, 2000, 250* and *50* for **Number of Samples, Sampling Frequency, Signal Frequency,** and **Modulation Frequency**, respectively, run **FFT (Magnitude Only)**. The spectrum that you observe consists of a central peak and "sidebands." In terms of $f_{sig}$ and $f_{mod}$, what is the central peak's frequency and the spacing of the sidebands from the central peak?

(b) In an AM radio receiver, the AM wave received from an antenna is "demodulated" by passing it through a nonlinear diode detector. By making an appropriate modification of the block diagram of **AM Wave**, simulate this demodulation method by applying a nonlinear amplification $a(x) = x^2$ to your AM wave $x(t)$. Run **FFT (Magnitude Only)**, and identify all of the frequencies in the resultant spectrum in terms of $f_{sig}$ and $f_{mod}$. (Note that one of these frequencies is the modulation frequency $f_{mod}$ and that by proper [low-pass] filtering it could be selected out for distribution to the radio's speaker. In optics, this technique is implemented to produce two optical laser frequencies separated by a small "radio-frequency" difference.)

6. Given an input consisting of the sum of many AC voltages with various frequencies, a *lock-in amplifier* can selectively measure the oscillatory amplitude of a single one of those frequencies. The theory of operation of this instrument can be simulated as follows. First, assume that the "input signal" $y_{sig}$ to the lock-in consists of the sum of three sine waves with frequencies 100 Hz, 200 Hz, and 300 Hz and amplitudes 4, 6, and 8, respectively, that is, $y_{sig} = 4 \sin[2\pi(100)t] + 6 \sin[2\pi(200)t] + 8 \sin[2\pi(300)t]$. Then, if a user programs the lock-in to measure the amplitude of the 200 Hz sine wave, the instrument internally produces a "reference signal" $y_{ref} = 2\sin[2\pi(200)t]$ and then the product $x = y_{sig}\, y_{ref}$. Finally, the instrument filters $x$ to determine the magnitude of the DC component present in $x$ and outputs this DC value.

(a) Use **FFT (Magnitude Only)** to gain insight into the above-described theory. First, program the **User-Defined** option of **Waveform Simulator** to produce the given $y_{sig}$ and $y_{ref}$ and then the product $x = y_{sig}\, y_{ref}$. Remember .* is Mathscript's element-wise multiplication operator. Next, run **FFT (Magnitude Only)** with **Number of Samples** and **Sampling Frequency** equal to *1024* and *2000*, respectively, and **Shape** set to *User-Defined*. You will find several frequencies, including $f = 0$, are present in $x$. How does the DC component compare to the amplitude of the 200 Hz sine wave present in $y_{sig}$?

(b) Change the reference signal to $y_{ref} = 2 \sin[2\pi(300)t]$ and rerun **FFT (Magnitude Only)**. What is the resulting DC component equal to now?

(c) From the trigonometric identity $\sin\theta \sin\phi = \dfrac{1}{2}\big[\cos(\theta - \phi) - \cos(\theta + \phi)\big]$ can you explain the frequencies that you observe in $x$?

7. Write a program called **FFT (Express)**, which uses the **Spectral Measurements** Express VI (found in **Functions>>Express>>Signal Analysis**) to perform a Fast Fourier Transform on a given input from your **Waveform Simulator** VI. Place two Waveform Graphs on the front panel. Label one **FFT (Peak Magnitude)** with $x$- and $y$-axes as **Frequency** and

**Amplitude**; label the other **FFT (Phase)** with x- and y-axes as **Frequency** and **Phase (Degrees)**. Complete the block diagram as shown below. Here the output of Waveform Simulator is packaged together as a *waveform* data type, a data type that can be directly input to an Express VI. The **Build Waveform** icon, found in **Functions>>Programm ing>>Waveform**, combines an initial time **t0** (called a *time stamp*), the sampling time increment **dt**, and the waveform array. The time stamp is created using **Get Date/Time In Seconds** (found in **Functions>>Programming>>Timing**). When the dialog box for **Spectral Measurements** opens, choose **Selected Measurement⊳> Magnitude (Peak)**, **Window>>Hanning**, and **Phase>>Convert to degree**.

On the front panel of **FFT (Express)**, choose *1024* and *2000* for **Number of Samples** and **Sampling Frequency**, respectively.

(a) Program your VI to produce a *Cosine* function with amplitude of *4.0* and frequency of *250* Hz, and then run it. Use the two plots to determine the resulting values of amplitude and phase at $f = 250$ Hz (it may be helpful to turn off autoscaling on the x-axis and manually scale it to display a restricted range of frequencies). Then program the VI to produce a 250 Hz sine function of amplitude 4 and run the program. Use the two plots to determine the values of amplitude and phase at $f = 250$ Hz that result from this input.

(b) In this chapter, you found that the complex-amplitude of a 250 Hz cosine and sine function at $f = 250$ Hz is $A = +2$ and $A = -2i$, respectively. Rewrite these two complex numbers in the form $A = |A|e^{i\theta}$, and thereby predict the phase $\theta$ expected for the FFT of a cosine and a sine function. How do these predicted values compare to the values you observed in part (a)?

(c) Program the **User-Defined** option of **Waveform Simulator** to be $x = 4\cos[2\pi(250) t + \pi/6]$, then run **FFT (Express)**. Are the values for amplitude and phase at $f = 250$ Hz on the two front-panel plots in agreement with the values you expect? Repeat this procedure for the input $x = 4 \sin[2\pi (250)t + \pi/6]$.

8.  Using a function generator, input a square-wave voltage with a frequency of about 180 Hertz input into the low-pass op-amp filter shown in Figure 10.4 with $R = 15\text{k}\Omega$ and $C=0.015$ $\mu$F, whose 3dB-frequency is given by $f_{3dB} = 1/(2\pi RC)$. Attach the output of this filter to the analog input channel of your DAQ device that was programmed into DAQ Assistant (e.g., **ai0**) on the **Spectrum Analyzer** block diagram. With **Number of Samples** and

**Sampling Frequency** set to *1024* and *2000*, respectively, run **Spectrum Analyzer** to see the spectral make-up of your filtered voltage signal.

(a) Calculate $f_{3dB}$. Given this frequency, which harmonics of the square wave do you expect to observe and what should be the ratio of their amplitudes compared to the first harmonic? Which harmonics do you actually observe and what are their amplitudes (in comparison with the amplitude of the first harmonic)? Comment on any deviation between your expectations and your observations.

(b) Are any of the observed peaks due to aliasing of harmonics with frequencies higher than $f_{nyquist}$? If so, should you increase or decrease $C$ in your filter in order to eliminate these peaks?

**Fig. 10.4  Low-pass filter for Problem 8.**

# Chapter 11 ▪ ▪ ▪ ▪ ▪ ▪ ▪ ▪ ▪ ▪ ▪ ▪ ▪ ▪ ▪ ▪ ▪ ▪ ▪ ▪

# Data Acquisition and Generation using DAQmx VIs

---

**BEFORE BEGINNING THIS CHAPTER**

Fundamental concepts of data acquisition are presented in Chapter 4. After reviewing that material, use the **Measurement & Automation Explorer (MAX)** to identify the DAQ device (e.g., model PCI-6251 with name *dev1*) connected to your computer. From the pinout diagram for your device, determine the pins associated with the differential analog input channel ai0, digital triggering channel PFI0, analog output channel ao0, and GROUND. Have appropriate cabling on hand to connect to these pins as well as a function generator, DC voltage source, and oscilloscope.

---

Since your work in Chapter 4, you have been able to control the operation of a multifunction data acquisition (DAQ) device using programs based on LabVIEW's **DAQ Assistant**. As you know, DAQ Assistant is a sophisticated high-level Express VI that, through the use of the dialog windows, can be configured to perform all manner of data acquisition and generation tasks (e.g., analog input, digital output). In this chapter we will show that after DAQ Assistant has been configured for a particular task, below its glossy, easy-to-operate user interface, LabVIEW automatically generates customized code that accomplishes the requested task. This customized code is based on a collection of low-level icons called the *DAQmx VIs*. Here, you will learn how write your own programs using the DAQmx VIs and explore the added flexibility and performance this low-level approach to programming offers in comparison to the use of DAQ Assistant.

## 11.1 DAQmx VIs

A large assortment of data acquisition (also called *read* or *input*) operations as well as data generation (aka *write* or *output*) operations is within the capability of any multifunction DAQ device manufactured by National Instruments. When a programmer chooses to perform one such operation, that choice (e.g., analog input voltage on channel ai0) along with its particular timing (e.g., sampling rate) and triggering (e.g., analog trigger) selections, is termed a *task*. To program a DAQmx task, one follows a template of five sequential steps: *create and configure the task, start the task, read or write data, stop the task*, and *clear the task*. Each of the DAQmx icons found in **Functions>>Measure I/O>>NI-DAQmx** is designed to execute one of these required steps and so, by properly configuring a sequence of these icons, a complete DAQmx task can be coded. Thus, most DAQmx-based data acquisition and generation programs are organized on the block diagram as shown in the next illustration.

In this diagram, the task is explicitly started by **DAQmx Start Task.vi** and stopped by **DAQmx Stop Task.vi**. However, if **DAQmx Read.vi** (or **DAQmx Write.vi**) is not preceded by **DAQmx Start Task.vi** and followed by **DAQmx Stop Task.vi**, **DAQmx Read.vi** (or **DAQmx Write.vi**) will automatically start and, after it has completed its work, stop the task. Hence, a DAQmx-based diagram can often be written in the following simplified form.

Many of the DAQmx icons are *polymorphic VIs*. When placed on a block diagram, a polymorphic VI has an associated *polymorphic VI selector* which consists of a drop-down menu of possible operational modes for the VI. For example, **DAQmx Read.vi** can be programmed to execute an analog input voltage or a digital input operation by simply making the appropriate selection on its polymorphic VI selector. To make the selection, either click on the selector with the 🖑, or pop up on the selector and use the **Select Type** option. Through this process, **DAQmx Read.vi** can be configured to operate in over 40 different modes. Below, we show how the 🖑 is used to select

**Analog>>Single Channel>>Multiple Samples>>1D DBL**, which instructs **DAQmx Read.vi** to perform an *N*-sample Analog Input operation on one channel and output the data in the form of a 1D double-precision floating-point array.

## 11.2 SIMPLE ANALOG INPUT OPERATION ON A DC VOLTAGE

As a step toward understanding how to program a DAQmx-based diagram, let's build a VI that turns your computer into a voltmeter. For this VI, we will use three DAQmx icons, so we begin by briefly describing the function of each of these block-diagram objects.

First, consider the polymorphic **DAQmx Create Virtual Channel.vi**. When placed on the block diagram, this VI can be used to define ("create") close to 50 different types of tasks (e.g., determine the frequency of the waveform at the counter input). The next illustration shows the selection on the polymorphic VI selector required to create a "voltmeter" task, that is, a task that digitizes voltage at a particular AI channel.

The Help Window of this polymorphic VI adapts to the selection on its polymorphic VI selector. After being configured as **Analog Input>>Voltage**, **DAQmx Create Virtual Channel.vi**'s Help Window appears as shown below. Note that the only required input (shown in bold typeface) is **physical channels**, all other inputs are optional. Given the analog input channel (or channels) specified at **physical channels**, the job of this icon is create a *virtual channel*, which consists of configurational information including the AI channel name(s) as well as the allowed input voltage range and input terminal configuration (e.g., differential mode). This virtual channel becomes part of the information that is packaged together to define a task. If a task already exists and is provided at the **task in** input, the newly created virtual channel is added to that task. However, if **task in** is left unwired, a new task is created based on the newly created virtual channel. The (modified or newly created) task is then passed out of the **task out** output for use by other DAQmx icons.

The polymorphic **DAQmx Read.vi** performs the actual data acquisition. To configure this icon to perform a single-sample analog input voltage operation on one channel and output the data in the form of a double-precision floating-point number, select **Analog>>Single Channel>>Single Sample>>DBL** on its polymorphic VI selector. After this selection has been made, the Help Window for this icon appears as shown next. Once presented with the task configurational information at its **task/channels in** input, this icon outputs the digitized voltage value as a double-precision numeric at its **data** output. Additionally, the task definition is available at the **task out** output terminal. As mentioned previously, this icon will start and stop the task, if **DAQmx Start Task.vi** and **DAQmx Stop Task.vi** do not appear explicitly on the block diagram.

Finally, the Help Window for **DAQmx Clear Task.vi** is given below. This icon discharges the task specified at its **task in** input, releasing any computing resources such as an allotment of RAM that the task has reserved. Once a task is cleared, it cannot be run again; a new, similarly defined task must be created and then run.

Note that all three of the above icons include error reporting via the error cluster, which appears at the **error in** and **error out** terminals.

A single voltmeter reading then can be accomplished by wiring these three icons together as follows.

This wiring scheme takes advantage of the principle of LabVIEW programming called *data dependency*. Simply stated, data dependency means that an icon cannot execute until data is available at *all* of its inputs. In this diagram, all of the inputs to **DAQmx Create Virtual Channel.vi** are wired (**physical channel**, the only required input, is explicitly wired; all of the unwired optional inputs are considered by LabVIEW to be implicitly wired to their default values). So, when this diagram is run, **DAQmx Create Virtual Channel.vi** executes immediately. Upon completion, **DAQmx Create Virtual Channel.vi** outputs a task at its **task out** terminal, which is passed through the wiring to DAQmx Read.vi's **task in** input. Because of data dependency, **DAQmx Read.vi** cannot execute until it receives the task from **DAQmx Create Virtual Channel.vi.** Then, after **DAQmx Read.vi** completes its execution, it passes the task from its **task out** terminal to DAQmx Clear Task.vi's **task in** input. Only then can **DAQmx Clear Task.vi** execute. Thus, through this programming trick, we are assured that the icons will execute in the desired sequence: **DAQmx Create Virtual Channel.vi** followed by **DAQmx Read. vi** followed by **DAQmx Clear Task.vi**.

Also, the correct manner of chaining together DAQmx icons for error reporting is shown above. If an error does occur at one point in the chain, subsequent icons will not execute and the error message will be passed to the **General Error Handler.vi**, which will display the message in a dialog box. **General Error Handler.vi** is found in **Functions>>Programming>>Dialog & User Interface**. Rather than using this dialog box approach to error reporting, one could alternately pass error information to a front-panel error cluster for viewing.

OK, we are ready to write our simple voltmeter program. Open a new VI, and using **File>>Save**, first create a folder called **Chapter 11** in the **YourName** folder, and then save this VI under the name **DC Voltmeter (DAQmx)** in **YourName\Chapter 11**.

Switch to the block diagram and write the following code, which continually reads and displays the voltage difference at the differential AI channel ai0 every 0.5 seconds until the front-panel **Stop Button** is clicked or an error occurs. Use **Create>>Constant** in the pop-up menu to create the wired inputs of **DAQmx Create Virtual Channel.vi**.

The **DAQmx Physical Channel Constant** created at the **physical channels** input can be programmed by clicking on its *menu button* with the . You will be presented with a list of available analog input channels on your particular DAQ device (as determined by MAX when it was last run). Otherwise, you can highlight the interior of the **DAQmx Physical Channel Constant** and manually enter the name of the desired AI channel (e.g., *Dev1/ai0*). Also note that this diagram configures the analog input channel to accept a minimum and maximum voltage difference of $V_{min} = -10$ and $V_{max} = +10$ Volts, respectively.

Switch to the front panel. Arrange the two objects as desired and rename the indicator **Voltage**, if you like. The resolution $\Delta V$ of the *n*-bit analog-to-digital conversion process performed by your DAQ device is given by $\Delta V = (V_{max} - V_{min})/2^n$. So for a device such as the PCI-6251 with $n = 16$, $\Delta V = (20 \text{ Volts})/2^{16} = 0.3$ mV. For $n = 14$ (e.g., USB-6009), $\Delta V = 1$ mV. Use this information to select the proper number of **Digits of precision** on the **Voltage** indicator. For example, if $\Delta V = 0.3$ mV, **Digits of precision** should be *4*. Save your work.

Connect a voltage difference of (approximately) +5 Volts to the differential analog input channel ai0 of your DAQ device. For example, on a PCI-6251 device in the differential mode, connect +5 Volts to pin 68 and GND to pin 34, while on a USB-6009, attach +5 V to pin 2 and GND to pin 3. On an NI ELVIS breadboard, the pins are labeled AI 0+ and AI 0– (or ACH0+ and ACH0–) in the Analog Input Signals section.

Run **DC Voltmeter (DAQmx)**. Does it read the input voltage correctly?

## 11.3 DIGITAL OSCILLOSCOPE

With a quick change on a polymorphic VI selector and an additional DAQmx icon or two, you can morph your DC voltmeter into digital oscilloscope program. This new VI—named **Digital Oscilloscope (DAQmx)**—will acquire $N$ equally spaced voltage samples of a time-varying analog input, and then quickly plot the array of data values. By repeating this process over and over, we'll achieve a real-time display of the waveform input.

With **DC Voltmeter (DAQmx)** open, use **File>>Save As...** to create the new program called **Digital Oscilloscope (DAQmx)** and store it in **YourName\Chapter 11**. On the front panel, delete the **Numeric Indicator** and replace it with a **Waveform Graph**, whose x- and y-axes are labeled **Time** and **Voltage**, respectively. Also, using **Select a Control...**, locate a **Digitizing Parameter** cluster in **YourName\Controls** and place it on the front panel (if you haven't stored a **Digitizing Parameter** cluster in **YourName\Controls** previously, construct the cluster now as described in section 3.11 "**Control and Indicator Clusters**"). Save your work.

Switch to the block diagram. Remove all broken wires. Then, using the 🖑, click on the **DAQmx Read.vi**'s polymorphic VI selector and select **Analog>>Single Channel>>Multiple Samples>>Waveform**. The next illustration shows how the 🖑 is used to make this selection, which instructs **DAQmx Read.vi** to perform an *N*-sample analog input operation on one channel and output the data in the waveform data type.

What is the *waveform* data type? Remember that previously (see Section 2.8) when using a Waveform Graph to plot an *N*-element 1D numeric array of data samples with a calibrated *x*-axis, we had to form the cluster shown below consisting of the initial *x*-axis value $x_0$, the *x*-axis increment $\Delta x$ between neighboring data samples, and the 1D data array itself.

Similar to the dynamic data type used by Express VIs, the waveform data type automatically packages these items ($x_0$, $\Delta x$, and the 1D data array) together in a single wire. Thus, by programming DAQmx Read.vi's output to be in the waveform (rather than 1D DBL) format, its **data** output terminal can be wired directly to the Waveform Graph terminal. The resulting waveform wire will produce the same calibrated plot as would result from the "bundling" code given above.

Place the **Waveform Graph** icon terminal within the While Loop and wire it to DAQmx Read.vi's **data** output. Note the banded brown and orange wire, which denotes the waveform data type. Also, delete the **Wait (ms)** icon and its associated **Numeric Constant** within the While Loop.

Our diagram now reads as follows: The one-time act of creating the task is done by **DAQmx Create Virtual Channel.vi**. This task is then passed into the While Loop where, during each iteration, **DAQmx Read.vi** acquires $N$ data samples at a given sampling frequency $f_s$, and then sends these data to the **Waveform Graph** for plotting. This data acquisition and presentation process repeats continuously until the While Loop ceases its execution in response to the **Stop Button** being clicked or an error having occurred. After exiting the While Loop, the one-time act of disposing of the task is done by **DAQmx Clear Task.vi**.

The unwired **Digitizing Parameters** cluster reminds us that we have a bit more work to do, namely, program the DAQ device with the desired values for $N$ and $f_s$. To accomplish this feat, we use the polymorphic **DAQmx Timing.vi** icon.

**DAQmx Timing.vi** offers a wide array of methods to control when an incoming analog signal is digitized. From among these possibilities, we will use **Sample Clock**, a "hardware timing" method in which a digital square wave with a very precise frequency (termed the *sample clock*) controls the rate at which samples of the incoming analog signal are acquired. Each tick, which corresponds to a rising (or else, falling) digital edge of the clock, initiates the acquisition of one sample. After selecting **Sample Clock** on its polymorphic selector, the Help Window of **DAQmx Timing.vi** appears as shown next. If the **source** input is left unwired, the built-in ("on board") sample clock on your DAQ device will be used to produce the required digital edges.

Place **DAQmx Timing.vi** on your block diagram as shown. Since the timing information being configured by this icon becomes part of the task definition, place it outside the While Loop so that it executes one time only. Here, **Number of Samples** and **Sampling Frequency** are wired to the **samples per channel** and **rate** inputs of DAQmx Timing.vi, respectively. Using the **Create>>Constant** pop-up selection, create the ring constant wired to the **sample mode** input of DAQmx Timing.vi, and then select its **Finite Samples** mode. In this mode, a finite number of samples (given by the value of **samples per channel**) is acquired each time DAQmx Read.vi executes. Save your work.

Return to the front panel and set **Number of Samples** and **Sampling Frequency** equal to *100* and *1000*, respectively. Connect the positive and negative (ground) outputs from a function generator to pins AI0+ and AI0– on your DAQ device, and then adjust the settings on the function generator so that it is outputting an analog sine wave with amplitude less than 10 V and a frequency of about 100 Hz. Run **Digital Oscilloscope (DAQmx)**. If all goes well, you will see a plot of about ten cycles of the 100 Hz sine wave.

I'm sure you're impressed by your computer-based instrument but, at the same time, concerned about an obvious flaw. Most likely, the sine-wave trace you are observing does not appear stationary but instead is either moving to the right or to the left. And, by slightly changing the function generator's sine-wave frequency, the plotted sine wave can be made to move first one direction, then the other. As explained in Chapter 4, the problem encountered here can be cured by properly *triggering* our digital oscilloscope, that is, by beginning all *N*-sample acquisitions on the same well-defined point of the incoming periodic signal's cycle. For **Digital Oscilloscope (DAQmx)** to be a useful program, we must build-in this triggering capability.

In a commercially available oscilloscope, triggering is accomplished in the following manner. The input signal is monitored by an analog *"level-crossing"* circuit. The purpose of this circuit is to determine each time the incoming signal passes through a specified voltage level and then immediately trigger the scope's data acquisition process. By using knobs on the scope's front panel, the oscilloscope user sets the circuit's threshold level and specifies whether acquisition should be initiated when the level is passed through starting from above (*negative*, or *falling*, *slope*) or starting from below (*positive*, or *rising*, *slope*) .

Many National Instruments DAQ devices are capable of performing the analog level-crossing triggering procedure described above. For example, the PCI-6251 has this capability; however, the lower-cost USB-6009 does not. In order that all readers can include triggering in their **Digital Oscilloscope (DAQmx)** VI, we'll implement an alternate mode—*digital edge triggering*—in our program because this mode is available on all NI DAQ devices.

A digital signal can be in one of two possible states, termed HIGH and LOW. For the digital ports on NI DAQ devices, these states conform to the *transistor-transistor logic (TTL) standard* with HIGH (close to) 5 Volts and LOW (close to) 0 Volts. As a digital signal alternates between its two states, the transition from LOW to HIGH is called the *rising edge*, while the HIGH to LOW transition is the *falling edge*. To execute a data acquisition operation under digital edge triggering, one connects a digital signal to the appropriate pins on a DAQ device. Then, by choosing the proper software settings, the desired data acquisition operation can be begun ("triggered") whenever the digital signal has a rising or, alternately, falling edge.

Stop **Digital Oscilloscope (DAQmx)** if it's running, then switch to its block diagram. We will add triggering capability to this VI using the polymorphic **DAQ Trigger.vi** icon, which can configure a task to include digital, analog, or more novel forms of triggering. To configure this icon in its digital triggering mode, select **Start>>Digital Edge** in its polymorphic selector. After being configured thusly, the Help Window for **DAQ Trigger.vi** appears as shown in the next figure. The required **source** input specifies the DAQ-device pin (e.g., PFI0) to which the digital triggering signal is connected.

Add **DAQmx Start Trigger.vi** to your block diagram as shown below. Use the pop-up selection **Create>>Control** to create the **edge** front-panel control. Also use

**Create>>Constant** to create a **DAQmx Terminal Constant** wired to **source**. Clicking on its *menu button* with the will present you with a list of available triggering pins from which to choose. Otherwise, you can highlight the interior of the **DAQmx Terminal Constant** and manually enter the name of the desired triggering pin on your DAQ device (e.g., */Dev1/PFI0*).

Return to the front panel. Arrange the objects to accommodate the new control and change its label from **edge** to **Trigger Edge**, if you wish. **Digital Oscilloscope (DAQmx)** is now configured for digital-edge triggering. Save your work.

Besides producing a sine wave with frequency *f* at its *main* output, your function generator creates a TTL digital signal of the same frequency *f*, which is available at its *sync* (sometimes called *TTL* or *trigger out*) output. The digital edges of this TTL signal coincide in time with particular points on the sine-wave's cycle. For example, on many generator models, the rising edge of the TTL signal occurs when the sine wave is at its peak. Thus, the sync output provides a convenient digital signal with which one can control the acquisition of sine-wave data (i.e., always begin at the same equivalent point on the cycle) via digital edge triggering.

Connect the sync output from your function generator to the digital triggering pins on your DAQ device. For example, sync's positive and negative (ground) terminals would connect to PFI0 (pin 29) and GND (pin 32), respectively, on a USB-6009 device. On a PCI-6251, sync's positive and negative terminals connect to PFI0 (pin 11) and GND (pin 12), respectively. On an NI ELVIS breadboard, the relevant pins are labeled PFI 0 (or TRIGGER) and GROUND.

On the front panel, set **Number of Samples, Sampling Frequency**, and **Trigger Edge** equal to *100*, *1000*, and *Rising,* respectively, and set the sine-wave frequency on your function generator to approximately 100 Hz. Run **Digital Oscilloscope (DAQmx)**. You should see a "stationary" plot of about ten cycles of the 100 Hz sine wave.

Stop **Digital Oscilloscope (DAQmx)**, and then change **Trigger Edge** to *Falling*. Run your VI. You should again see a "stationary" plot of about ten cycles of the 100 Hz sine wave, but this time beginning on a different point on the sine-wave cycle.

## 11.4 EXPRESS VI AUTOMATIC CODE GENERATION

Now that we've gained some experience with the DAQmx VIs, we're in a position to appreciate the automatic code generation feature of Express VIs by programming **DAQ Assistant** to do the same task we have just coded on the block diagram of **Digital Oscilloscope (DAQmx)**.

Place the **DAQ Assistant** Express VI on a blank block diagram. When the **Create New Express Task...** dialog window opens, make the selections **Acquire Signals>>Analog Input>>Voltage>>ai0**, then press the **Finish** button. You will recognize that these selections are the choices we programmed on the polymorphic VI selector and source input of **DAQmx Create Virtual Channel.vi** on the block diagram of **Digital Oscilloscope (DAQmx)**.

Next, when the **DAQ Assistant** dialog window opens, under the **Configuration** tab, in the **Signal Input Range** box, select **Max>>10** and **Min>>–10**. Also select **Terminal Configuration>>Differential** and, in the **Timing Settings** box, **Acquisition Mode>>N Samples**. Under the **Triggering** tab, in the **Start Trigger** box, select **Trigger Type>>Digital Edge, Trigger Source>>PFI0**, and **Edge>>Rising**. Again, you will recognize that these selections are the associated choices we programmed for the **DAQmx Create Virtual Channel.vi**, **DAQmx Timing.vi**, and **DAQmx Triggering.vi** icons on the block diagram of **Digital Oscilloscope (DAQmx)**.

Exit the **DAQ Assistant** dialog window by pressing the **OK** button. In the following moments, you will see a dialog window indicating the LabVIEW is building a DAQmx-based VI capable of carrying out the task that you have selected. After LabVIEW has created this (hidden) lower-level code, you will be returned to the Express VIs block-diagram icon.

To view the DAQmx-based code that LabVIEW created for the task that you configured in the dialog windows, pop up on the Express VI's icon and select **Open Front Panel**, and then **Convert** in the dialog window that follows. When the front panel opens, switch to the block diagram. There, you will find DAQmx-based code that is very similar to the block diagram you wrote for **Digital Oscilloscope (DAQmx)**. If desired, you can modify, save, and run this VI just like you would with any other LabVIEW program.

## 11.5 LIMITATION OF EXPRESS VIs

If the DAQ Assistant Express VI can automatically write the required low-level code for any data acquisition and generation operation requested, why would one ever wish to write his or her own DAQmx-based code manually? The answer to this question is that the Express VI developers designed these high-level functions to carry out well-defined, commonly needed tasks. If your need fits these particular tasks exactly, then Express VIs are the (painless) way to go. However, if your needs deviate somewhat from the Express VIs' defined functionality or if your project demands the utmost in software efficiency, the Express VI approach may not be able to offer the flexibility and performance level you require.

As a simple demonstration of Express VI limitations, open **Digital Oscilloscope (Express)**, the DAQ Assistant–based digital oscilloscope program you wrote in Chapter 4 and stored in **YourName\Chapter 4**. As you may remember, this program is digitally triggered and expects an analog signal input at AI Channel ai0, so configure a function generator to produce a (close to) 100 Hz sine wave and input the sine wave and the generator's sync output to ai0 and PFI0 pins of your DAQ device, respectively. On the VI's front panel, select **Number of Samples** and **Sampling Frequency** equal to *100* and *1000*, respectively, and then run **Digital Oscilloscope (Express)**. You will see (about) ten cycles of the digitized sine wave.

Now, let's say that you wanted to "zoom in" on one cycle of the sine wave, which can be accomplished by simply increasing the sampling frequency by a factor of ten, while keeping the number of samples the same. With **Digital Oscilloscope (Express)** still running, change **Sampling Frequency** to *10000* (to secure this choice, you will have click on the ✓ or press <*Enter*>). You will find that the VI does not appear to respond to this new choice for the sampling frequency.

The lack of response to an attempt at changing its sampling frequency during runtime is due to an intrinsic feature of DAQ Assistant. DAQ Assistant is hardwired to set its various digitizing parameters (including its sampling frequency) only during its first execution after the **Run** button has been pressed. To demonstrate this feature of DAQ Assistant, halt **Digital Oscilloscope (Express)** by clicking its **Stop Button**. Then, with **Sampling Frequency** set to *10000*, press the **Run** button. You will find that DAQ Assistant has accepted the new sampling frequency, and a single cycle of the digitized sine wave will be displayed.

Unfortunately, due to DAQ Assistant's fundamental design, there is no way to modify **Digital Oscilloscope (Express)** to accommodate runtime changes to the **Sampling Frequency** control. In fact, in the interest of optimizing the performance of our VI, we programmed **Digital Oscilloscope (DAQmx)** to have the same behavior as **Digital Oscilloscope (Express)**. That is, so that the digitizing parameters were not needlessly updated with each While Loop iteration, we placed all configurational DAQmx icons outside the loop.

The advantage of writing DAQmx-based code, however, is flexibility. If you now decide to make **Digital Oscilloscope (DAQmx)** capable of responding to runtime changes of its front-panel inputs, you can do so as follows. First, pop up on the While Loop and select **Remove While Loop**.

Then modify the block diagram so that **DAQmx Timing.vi** and **DAQmx Trigger.vi** are included within a newly placed While Loop, as shown next. Save your work.

Run this modified version of **Digital Oscilloscope (DAQmx)**. Try changing the front-panel values of **Number of Samples, Sampling Frequency**, and/or **Trigger Edge** during runtime, and enjoy the instant response to your command.

## 11.6 IMPROVING DIGITAL OSCILLOSCOPE USING STATE MACHINE ARCHITECTURE

In modifying **Digital Oscilloscope (DAQmx)**, we sacrificed performance efficiency in order to gain runtime response. Namely, most of the DAQmx icons contained on the block diagram simply set up (**DAQmx Create Virtual Channel.vi, DAQmx Timing.vi, DAQmx Trigger.vi**) or shut down (**DAQmx Clear Task.vi**) data acquisition operations and, thus, need only be called during one, or a small subset, of the multitude of **Digital Oscilloscope (DAQmx)**'s While Loop iterations. These set-up and shut-down actions are collectively termed *overhead operations,* and, as the program is now written, two of these overhead

397

operations (**DAQmx Timing.vi**, **DAQmx Trigger.vi**) are executed each time **the While Loop iterates**. Since your data-taking system is not able to digitize the incoming stream of real-time data during overhead operations, needlessly including them increases the program's *dead time*, the percentage of each iteration period during which the system is not sensitive to the incoming data. With increasing dead time, more and more cycles of the incoming repetitive data will stream past your system's input undetected. In certain situations, such as when collecting a large number of data cycles with the intent of adding them together so as to average out random noise, a programming inefficiency that causes you to miss a significant percentage of the incoming data cycles can easily extend the completion time of your experiment by minutes (or sometimes even hours).

If you care to correct the above-described problem, write the following program, which only executes set-up and shut-down operations when necessary (e.g., **DAQmx Trigger.vi** executes only when a change has been made to the front-panel **Trigger Edge** control). With **Digital Oscilloscope (DAQmx)** open, use **File>>Save As...** to create a new VI named **Digital Oscilloscope (DAQmx State Machine)** in **YourName\Chapter 11**. Consolidate all three controls in one control cluster by placing **Trigger Edge** in the **Digitizing Parameters** control cluster.

Switch to the block diagram and find a fairly large blank work area where you can build your new code. You can either delete the icons presently on the block diagram (without deleting the terminals for the front-panel objects) or stockpile these icons in some out-of-the-way place and then reuse them as you construct your new code.

The block diagram we will write is based on the *state machine* architecture. A state machine consists of a nested While Loop and Case Structure. The While Loop executes continuously until its  is set to TRUE, and with each loop iteration, one of the Case Structure cases (termed a *state*) executes. The state that executes during a particular iteration performs some operation (e.g., reads data) and, in addition, selects which state will be executed during the next iteration. The block-diagram template for a state machine is shown in the next illustration. The state selection is made using an Enum Constant, which is stored in a shift register. Information that needs to be passed from one state to another is similarly stored in other shift registers.

For our digital oscilloscope state machine, we require five states: *Create Task*, *Check Settings*, *Change Settings*, *Read Data*, and *Clear Task*. Thus, we need to create a block-diagram Enum Constant programmed with these five items.

To create the required **Enum Constant**, you can simply obtain this icon in **Functions>>Programming>>Numeric**, place it on the block diagram, select **Edit Items...** in its pop-up menu, and then program the five state names in the dialog window that appears. After closing the dialog window, the completed Enum Constant will appear on the block diagram. You can obtain multiple copies of it by cloning (pressing <*Ctrl*> while clicking on object with the ).

I'll outline an alternate method for creating the required Enum Constant that involves the **Control Editor**, and includes options that can be very handy. Open the **Control Editor** as follows: First, select **File>>New...** at the top of an open VI window or the **Getting Started** window. The **New** dialog window will open. In the **Create New** box, find **Custom Control** in the **Other Files** folder and double-click on it. The **Control Editor** window will open, which appears similar to the front panel of a VI.

With the **Control Editor** window open, activate the **Controls Palette**, then place an **Enum** (found in **Controls>>Modern>Ring & Enum**) on it.

Pop up on the **Enum** and select **Edit Items...**, then program the **Items** box with the five state names.

After programming the five states, press the **OK** button. You will be returned to the **Control Editor** window, which now contained the completed Enum. You may wish to resize the Enum to make all of the text within it visible.

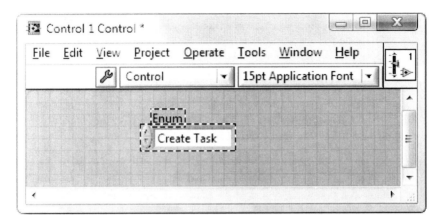

Finally, using **File>>Save**, save this customized control under the name **Oscilloscope States** in **YourName\Controls** (if this folder doesn't already exist, create it). The extension **.ctl** will be added automatically. Then close the **Control Editor** window.

For future reference, the above method describes saving our customized object in the **Control** mode. In this mode, the **Enum Constant** we place on the block diagram is an independent copy of the **Oscilloscope States** file in **YourName\Controls**. One can, instead, select to save the customized object in the **Type Definition** (or, the related **Strict Type Definition**) mode by selecting **Type Def.** as shown below before saving the customized control.

When the **Oscilloscope States** file in **YourName\Controls** is saved in the **Type Definition** mode, it becomes a master file that is linked to every related Enum Constant placed on the block diagram. Thus, any change made to the **Oscilloscope States** file will be transmitted to every related Enum Constant. This feature can be quite labor saving, say, when writing a state machine and finding well into the project that an additional state must be added.

Begin coding the block diagram of **Digital Oscilloscope (DAQmx State Machine)** by constructing the **Create Task** state as shown in the next diagram. To obtain a copy of your customized Enum, click on **Select a VI...** in the Functions Palette. In the dialog window that opens, navigate to the **YourName\Controls** folder and double-click on the file name **Oscilloscope States.ctl**. You will be returned to the block diagram, where you can place the customized **Enum Constant**.

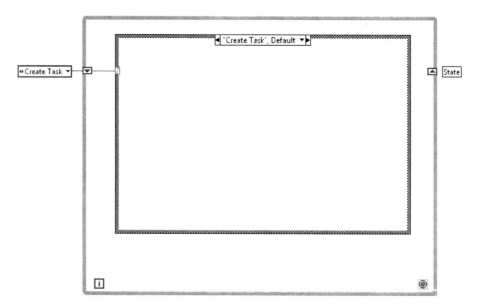

When the **Enum Constant** is connected to the Case Structure's selector terminal, two cases will be activated that are associated with the first two items programmed on the **Enum Constant (Create Task** and **Check Settings)**. Pop up on the case selector label and choose **Add Case for Every Value**. The Case Structure will then create a case for each of the five states.

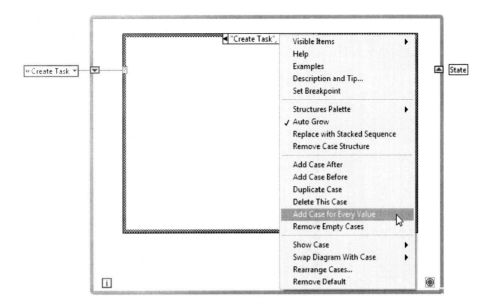

Place a **DAQ Create Virtual Channel.vi** icon in the **Create Task** state, configure it for **Analog Input>>Voltage** and complete the rest of the code shown in the next illustration. Once on the block diagram, the desired item of the **Enum Constant** can be selected by clicking on it with the 🖑.

Switch to the **Check Settings** state and place the **Digitizing Parameters** icon terminal here. Complete the code shown, which reads the current cluster values and stores them in a shift register for use in subsequent iterations.

Complete coding the **Check Settings** state, which compares the current **Digitizing Parameters** cluster to that from the previous iteration to determine if a change has occurred in the front-panel settings. Because we are checking if the whole current cluster equals the whole previous cluster, pop up on the **Equal?** icon and select **Comparison Mode>>Compare Aggregates**. In this mode, the **Equal?** output is a single Boolean value (in the **Compare Elements** mode, each pair of associated elements in the two clusters is compared and the output is an array of Boolean values). If no change has occurred (**Equal?** is TRUE), **Take Data** is chosen as the next state to execute. Alternately, if a change has occurred (**Equal?** is FALSE), **Change Settings** is chosen as the next state. The **Select** icon used here is found in **Functions>>Programming>> Comparison**. Also, to create the **Cluster Constant**, which initializes the shift register, pop up on the shift register and select **Create>>Constant**.

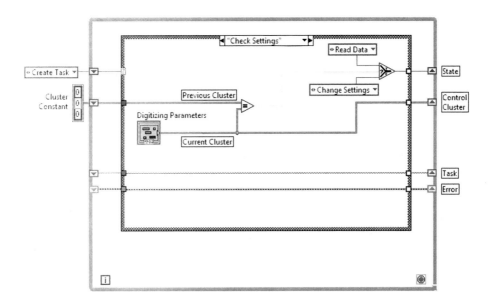

Switch to the **Change Settings** state and code it as shown. The order of **Unbundle By Name**'s output terminals is chosen (using the 🖑) to avoid crossed wires.

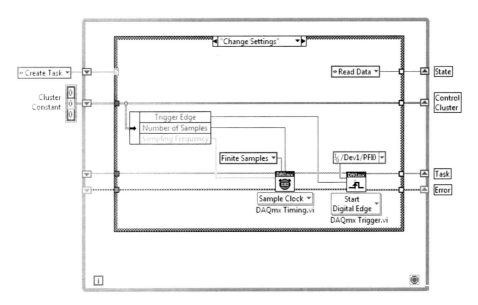

Next, code the **Read Data** state. Here, if the **Stop Button** has been clicked (so that its value is TRUE), **Clear Task** is chosen as the next state to execute. Alternately, if the **Stop Button** has not been clicked (so that its value is FALSE), **Check Setting** is chosen as the next state (which commences another cycle of data taking).

Finally, switch to the **Clear Task** state and complete the diagram as shown. With this code, after the **Clear Task** state executes, the While loop will cease execution on its next iteration because an **Enum Constant** with a value of **Clear Task** will be passed to the **Equal?**'s upper terminal, causing a TRUE to be output to the **Or** icon, hence passing a TRUE to the ⬛. If an error occurs at any time during the execution of the state machine, this too will cease execution of the VI.

Note that two of the Case Structure's output tunnels are not solidly colored, indicating that these terminals are not wired in all five cases. These unfilled tunnels each constitute a programming error that must be rectified before the VI can run.

In clicking through the five cases, we find that the *task*-related output tunnel is unwired in only the **Clear Task** state. Since there is no task output for **DAQmx Clear Task.vi**, we can simply wire this tunnel, as shown next.

For the *cluster*-related output tunnel, the **Create Task** state is the culprit. Resolve this problem by including the wire shown.

The program is finished now. Return to the front panel and save your work. Run the VI and see if it works.

## 11.7 ANALOG OUTPUT OPERATIONS

By simply reconfiguring the DAQmx icons with which we are now familiar, analog output operations can be carried out. Let's first write a program that simply outputs a requested voltage value. On a new VI, place a **Numeric Control** labeled **Voltage** and a **Stop Button**. If you'd like, hide the Stop Button's label (toggle it off via **Visible Items>>Label**) and make the Voltage control's **Digits of precision** appropriate for the resolution of your particular DAQ device. Save this VI in the **YourName\Chapter 11** folder under the name **DC Voltage Source (DAQmx)**.

Switch to the block diagram and write the following code, which will output the requested analog voltage at AO channel ao0 until the **Stop Button** is clicked or an error occurs. The values for **minimum value** and **maximum value** in the given diagram are appropriate for a USB-6009 DAQ device. If using an *M* series DAQ device such as the PCI-6251, you can choose **minimum value** and **maximum value** to be *–10* and *10*, respectively. Using the polymorphic VI selectors, configure **DAQmx Create Virtual Channel.vi** and **DAQmx Write.vi** to be in the **Analog Output>>Voltage** and **Analog>>Single Channel>>Single Sample>>DBL** mode, respectively. In the **RSE** output terminal configuration, the voltage output at the ao0 pin is referenced to the DAQ device GROUND.

Return to the front panel and save your work. Connect the positive and negative terminals of a voltmeter or oscilloscope to AO 0 (pin 14 on a USB-6009; pin 22 on a PCI-6251) and AO GND (pin 13 on a USB-6009; pin 54 on a PCI-6251), respectively, of your DAQ device. On an NI ELVIS breadboard, AO 0 is labeled AO 0 (or DAC0) in the Analog Outputs section and AO GND is labeled GROUND. Run the VI with various values for **Voltage**. You will find that when **Voltage** is set to outside of the range allowed by **minimum value** and **maximum value**, the program ceases operation.

When you are done, run **DC Voltage Source (DAQmx)** one more time with **Voltage** set to *0*, so the channel ao0 is left in the zero-voltage state.

## 11.8 WAVEFORM GENERATOR

If your DAQ device supports hardware-timed analog output operations (*M* Series devices such as the PCI-6251 do, the USB-6009 doesn't), try building the following hardware-timed waveform generator called **Waveform Generator (DAQmx)**. This VI is based on the **DAQmx Write.vi** icon configured in its **Analog>>Single Channel>>Multiple Samples>>1D DBL** mode. When in this mode, the Help Window for **DAQmx Write.vi** is as shown in the following illustration. Here, a 1D array of *N* double-precision floating-point numbers describing a few cycles of a periodic waveform is input at **data**. The icon then writes this *N*-element array to a memory buffer. If **autostart** is set to

TRUE, the icon then instructs the DAQ device to begin outputting the sequence of analog voltages values as dictated by the 1D array within the memory buffer.

In **Waveform Generator (DAQmx)**, the "waveform-describing" array will be created using **Waveform Simulator**, whose operation is controlled by the **Digitizing Parameters** and **Waveform Parameters** control clusters. Starting with a blank VI, build the following front panel. **Digitizing Parameters** and **Waveform Parameters** are found in **YourName\Controls**. Save this VI under the name **Waveform Generator (DAQmx)** in **YourName\Chapter 11**.

Switch to the block diagram and code it as shown next. Wire Waveform Simulator's **Displacement** array to the **data** input of DAQmx Write.vi. Also, wire **Sampling Frequency** to the **rate** input of DAQmx Timing.vi and configure this icon's **sample mode** to be **Continuous Samples**.

When this diagram is run, **DAQmx Write.vi** will write the *N*-element **Displacement** array into the memory buffer associated with the defined AO task, and the array will also be plotted on the front-panel **XY Graph** for the user to view. The continuous AO operation will then be started. In this operation, the DAQ device will output the pre-scribed sequence of voltages at selected analog output channel in a circular fashion (i.e., after the last array value is output, the next output is the first array value). The rate at which the output voltage is changed ("updated") from one value to the next is controlled by **Sampling Frequency**. While this continuous analog output operation executes, the value of the front-panel **Stop Button** is checked every 100 ms. When the button is clicked, the AO task is stopped and cleared.

Return to the front panel and save your work. Then, connect the positive and nega-tive (ground) input of an oscilloscope to the AO 0 and AO GND pins of your DAQ device, respectively.

Program **Waveform Generator (DAQmx)** to create two cycles of a 5-V 100 Hz sine wave, where each cycle is defined by 100 samples. That is, in the **Waveform Parameters** control cluster, set **Frequency, Amplitude**, and **Type** equal to *100*, *5*, and *Sine*, respectively; in the **Digitizing Parameters** control cluster, set **Number of Samples** and **Sampling Frequency** to be *200* and *10000*, respectively. Run the VI. If your oscilloscope is triggered properly, you will observe a 100 Hz sine wave produced continuously, until you click the VI's **Stop Button**.

What happens if you don't write a whole number of sine-wave cycles to the memory buffer? Stop the VI. Then program it to write 1½ (rather than two) cycles of the 100 Hz sine wave to the memory buffer. That is, keep all of the front-panel settings as before, except change **Number of Samples** to *150*. Run the VI. Can you explain the waveform observed on the oscilloscope based on your knowledge of the circular fashion in which the memory buffer values are sequenced?

## DO IT YOURSELF

Build a (crude) DAQmx-based frequency meter VI called **Frequency Meter (DAQmx Count Edges)**, which will determine the frequency of a digital signal input to your DAQ device.

1.   Configure a counter input task that uses *counter 0* on your DAQ device (e.g., *Dev1/ ctr0*) to count falling digital edges at its input (e.g., *PFI0*).
2.   Within a While Loop, write code that executes the following procedure each iteration: start the counting task, wait for 100 ms, read the number of counts *N* accumulated during the 100 ms, calculate the frequency $f=N/(0.1\,\text{second})$ and send it to the front panel for display, then stop the task.
3.   Configure the While Loop so that its operation ceases when a front-panel Stop Button is clicked or an error occurs.

     After you have completed **Frequency Meter (DAQmx Count Edges)**, connect the sync output of a function generator to the input of counter 0 (and GROUND) on your DAQ device, then use it to measure the frequency of this digital signal.

     (OPTIONAL) All DAQ devices, including the USB-6009, are equipped to count falling edges of digital signals; thus, everyone can carry out the above project. If you have a more advanced DAQ device (such as the PCI-6251), construct another frequency meter VI in which the DAQmx task is configured as **Counter Input>>Frequency** (as opposed to **Counter Input>>Count Edges**). Then, rather than measuring time with a Wait (ms) icon, the time measure can be carried out by a much more accurate clock on-board your DAQ device.

### Problems

1.   Observe the resolution $\Delta V$ of an analog input signal digitized by your DAQ device and verify that your observation is consistent with the prediction of Equation [1] in Chapter 4, that is, $\Delta V = V_{span}/2^n$. Attach a slowly varying analog signal (e.g., 1 Hz triangle wave) and the sync output from a function generator to the analog input channel and digital triggering pins programmed into **Digital Oscilloscope (DAQmx)**. Run **Digital Oscilloscope (DAQmx)** with a fast sampling rate and fairly small number of samples so that the input signal is effectively constant over the trace that you obtain. Stop the VI, and then by changing the *y*-axis scaling, zoom in on the acquired data (alternately, you can use a **Probe** to view the numerical values of the data samples). Under "high magnification" you should find that the data samples of this "effectively constant" input signal are distributed in discrete

levels (due to electronic noise). Measure the spacing $\Delta V$ between two of these adjacent levels. Does your observed value for $\Delta V$ agree with the predicted value?

2. By passing an analog signal through a low-pass filter prior to inputting it to a digitizer, the aliasing effect can be suppressed. To demonstrate this procedure, attach an analog sine-wave signal and the sync output from a function generator to the analog input channel and digital triggering pins programmed into **Digital Oscilloscope (DAQmx)**.

   On the front panel of **Digital Oscilloscope (DAQmx)**, set **Number of Samples** and **Sampling Frequency** equal to *100* and *1000*, respectively, and set the sine-wave frequency *f* on your function generator to approximately 100 Hz. Run **Digital Oscilloscope (DAQmx)**. If all goes well, you will see a "stationary" plot of about ten cycles of the 100 Hz sine wave. Turn off autoscaling on the *y*-axis.

   Next, set *f* on your function generator to approximately 1900 Hz, and then fine-tune *f* until **Digital Oscilloscope (DAQmx)** displays an aliased waveform of about 100 Hz. Use Equation [2] in Chapter 4 to explain why the input with *f* = 1900 Hz appears as a 100 Hz sine wave when digitized due to aliasing.

   Finally, without changing the frequency setting on your function generator, pass the *f* = 1900 Hz sine-wave voltage through the low-pass op-amp filter shown in Figure 11.1 with $R$ = 15 k$\Omega$ and $C$ = 0.1 $\mu$F, prior to inputting it to the DAQ device. When **Digital Oscilloscope (DAQmx)** is now run, by what factor is the amplitude of the aliased signal attenuated in comparison to the unfiltered situation? What is the 3dB-frequency $f_{3dB} = 1/(2\pi RC)$ of the low-pass filter? If you wanted the aliased signal to be attenuated even further, what component in the filter would you change? (This process can be optimized by using a low-pass filter with a steeper roll-off than the simple [first-order] circuit used here.)

**Fig. 11.1   Low-pass filter for Problem 2.**

3. Build a two-channel digital oscilloscope. With **Digital Oscilloscope (DAQmx)** open, use **File>>Save As…** to create a new VI named **Dual Digital Oscilloscope (DAQmx)**. On the block diagram, you must change the single physical channel selected (e.g., *Dev1/ai0*) to a list of two channels. Use the Help Window of **DAQmx Create Virtual Channel** to learn the proper syntax for making such a list. Besides programming this list of physical channels there is only one other change you must make on the block diagram to complete your program.

   Once the VI is finished, attach a voltage waveform and the sync out digital signal from a function generator to the two AI channels of your DAQ device that are programmed on the block diagram. Additionally, attach the sync out digital signal to the PFI0 channel of your DAQ device Run **Dual Digital Oscilloscope (DAQmx)** and verify that it simultaneously displays traces of the two inputs.

4.  Thermocouples are widely used as temperature sensors. A thermocouple is constructed by joining the ends of two dissimilar metals, for example, a copper and a constantan wire for a Type T thermocouple. This junction produces a millivolt-level voltage, which has a well-documented temperature dependence, where the temperature is measured relative to a "cold junction" reference temperature. Conveniently, this cold junction can be provided by a compact electronic device called a Cold Junction Compensator (CJC), which effectively makes the reference temperature equal to 0 °C.

    Connect a thermocouple to a CJC and then connect the plus and minus output of the CJC to the AI+ and AI− pins of your DAQ device. Using **DC Voltmeter (DAQ)** as a guide, write a program called **Thermocouple Thermometer (DAQmx)** that, every 250 ms until a Stop Button is clicked, reads the thermocouple voltage, converts this value to the corresponding temperature in Celsius, and then displays this temperature in a front-panel indicator. When using a Cold Junction Compensator, **CJC Source** and **CJC Value** are *Constant Value* and *0*, respectively. Program **Thermocouple Type** for your particular type of thermocouple (e.g., T).

    Run **Thermocouple Thermometer (DAQ)** and use it to measure room temperature as well as the temperature of your skin.

5.  Use **Digital Oscilloscope (DAQmx)** to measure the time constant of a *RC* circuit. Attach a TTL digital square wave from the sync output of a function generator to the PFI0 channel of your DAQ device as well as across the *RC* series circuit as shown in Figure 11.2 with $R = 4.7\,k\Omega$ and $C = 0.1\,\mu F$. When the square wave transitions from its LOW-to-HIGH (HIGH-to-LOW) state, the capacitor will be charged (discharged) through the resistor. Set the square wave frequency to a low frequency (say, 10 Hz) to allow the capacitor to fully charge and discharge after each transition. To observe the capacitor discharge, attach the analog input channel for which **Digital Oscilloscope (DAQmx)** is configured across the capacitor. Then, choosing appropriate values for **Number of Samples, Sampling Frequency**, and **Trigger Edge**, observe the discharge of the capacitor and accurately measure its time constant $\tau$. By definition, $\tau$ is the time taken for the capacitor's voltage to discharge to $e^{-1} = 0.37$ of its initial value. Compare the value you measure to the theoretically expected value of $\tau = RC$.

    To assist in measuring voltages along the discharge curve, you may wish to activate a cursor by popping up on the Waveform Graph and selecting **Visible Items>>Cursor Legend**. When the cursor legend appears, pop up on it and select **Create Cursor>>Single Plot**. Once activated, by moving the cursor to particular data points on the plot, the *x*- and *y*-coordinates of this point of interest are displayed in the cursor legend.

**Fig. 11.2** *RC* **series circuit for Problem 5.**

6. Write a program called **Waveform Generator (DAQmx State Machine)** that allows the parameters of the analog waveform being output by your DAQ device to be changed while the program is running. Use the state machine architecture for your block diagram. When a change in front-panel setting is detected, the task must be first stopped before changes are made to the sampling frequency.

7. Using the information given in Chapter 9's **Do It Yourself** project, build a digital thermometer with a DAQmx-based (rather than DAQ Assistant Express VI-based) program called **Digital Thermometer (DAQmx)**.

8. Using the information given in Chapter 10's **Do It Yourself** project, build a spectrum analyzer with a DAQmx-based (rather than DAQ Assistant Express VI-based) program called **Spectrum Analyzer (DAQmx).**

9. In this problem, you will create a task using the Measurement & Automation Explorer (MAX). This approach provides an alternate method to that used within the chapter text, where tasks were created on the block diagram. Use MAX to create a digital oscilloscope task similar to your **Digital Oscilloscope (DAQmx)** VI through the following steps: With MAX open, right-click on **Data Neighborhood** and select **Create New...>> NI-DAQmx Task**. In the dialog windows that follow, choose selections **Acquire Signals>>Analog Input>>Voltage>>ai0**, name your task *Oscilloscope*, and then press the **Finish** button. Within the MAX window, you will then find the **Configuration** tab displayed for your task. There, select **Acquisition Mode>>N Samples, Samples to Read>>100**, and **Rate (Hz)>>1k**. Next, click on the **Triggering** tab, and select **Trigger Type>>Digital Edge**. Finally, click on the **Save** button near the top of the window. The creation of your task is now complete.

   (a) To implement your task, open a blank VI and save it under the name **Digital Oscilloscope (MAX Task)**. On the block diagram place a **DAQmx Task Name**, found in **Functions>>Measurement I/O>>DAQmx**. Using the 🖑, click on its menu button and select *Oscilloscope*. Then, complete the following diagram.

   Using a function generator, attach a 100 Hz sine-wave voltage waveform and sync out digital signal to the AI 0 and PFI0 channels of your DAQ device. Run **Digital Oscilloscope (MAX Task)** and verify that it obtains a single $N = 100$ sample trace.

   (b) Alternately, the DAQmx Task Name can generate code for you automatically. To implement this feature, open a blank VI and save it under the name **Digital Oscilloscope (MAX Task Autocode)**. On the block diagram place a **DAQmx Task Name**, using the 🖑, click on its menu button and select *Oscilloscope*. Then, pop up on the DAQmx Task Name and select **Generate Code>>Configuration and Example**. A **DAQmx Read.vi** icon wired to a configurational subVI will be automatically created. Open the subVI and you will find code that is familiar. With the 100 Hz sine-wave and digital sync signal inputs from the function generator, run **Digital Oscilloscope (MAX Task Autocode)** and verify that it obtains a single *N* sample trace. With some simple modifications, you can transform this VI into a program that is equivalent to **Digital Oscilloscope (DAQmx)**, if you wish.

10. In this problem, you will create a "virtual channel" using the Measurement & Automation Explorer (MAX). This approach provides an alternate method to that used within the chapter text, where virtual channels were created on the block diagram using the **DAQmx Create Virtual Channel.vi** icon. Use MAX to create an analog input voltage virtual channel such as used in your **Digital Oscilloscope (DAQmx)** VI through the following steps:

    With MAX open, right-click on **Data Neighborhood** and select **Create New…>> NI-DAQmx Global Virtual Channel**. In the dialog windows that follow, choose selections **Acquire Signals>>Analog Input>>Voltage>>ai0**, name your virtual channel *AI Voltage*, and then press the **Finish** button. Within the MAX window, you will then find the **Configuration** tab displayed for your virtual channel where you can choose parameters related to the virtual channel, such as the Signal Input Range and Terminal Configuration. If any changes are made from the default values, then click on the **Save** button near the top of the window. The creation of your virtual channel is now complete.

    To implement your virtual channel, open **Digital Oscilloscope (DAQmx)** and, using **File>>Save As…**, create a new VI named **Digital Oscilloscope (MAX Virtual Channel)**. On the block diagram, delete the **DAQmx Create Virtual Channel.vi** icon (and the various items wired to it) and replace it with a **DAQmx Global Channel**, found in **Functions>>Measurement I/O>>DAQmx**. Using the 🖑, click on its menu button and select *AI Voltage*. Then, simply modify the diagram as shown below.

    Using a function generator, attach a 100 Hz sine-wave voltage waveform and sync out digital signal to the AI 0 and PFI0 channels of your DAQ device. Run **Digital Oscilloscope (MAX Virtual Channel)** and verify that it performs exactly like **Digital Oscilloscope (DAQmx)**.

# Chapter 12 ▪ ▪ ▪ ▪ ▪ ▪ ▪ ▪ ▪ ▪ ▪ ▪ ▪ ▪ ▪ ▪ ▪

# PID Temperature Control Project

In this chapter, you will apply your LabVIEW programming skill in the construction of a feedback-based system that controls the temperature of an aluminum block with great precision. This feat will be accomplished by controlling the current through a *thermoelectric (TE) device*, which performs as a "heat pump" between the aluminum block and a large temperature reservoir. When current flows in one direction through the TE device, the device pumps heat from the block to the reservoir, hence cooling the block. When current flows in the other direction, the TE device pumps heat from the reservoir to the block, heating the block. Your job will be to write a top-level program called **PID Temperature Controller** that uses the *PID (Proportion-Integral-Derivative) control algorithm* to produce the current flow required to maintain the aluminum block at a desired *"set point" temperature* $T_{\text{set-point}}$. If your program is written correctly, **PID Temperature Controller** should be able to maintain the block's temperature at $T_{\text{set-point}}$ with fluctuations less than about 0.05°C.

## 12.1 VOLTAGE-CONTROLLED BIDIRECTIONAL CURRENT DRIVER FOR THERMOELECTRIC DEVICE

Build the circuit in Figure 12.1. It is designed to provide the rather large bidirectional current flow necessary to power both the heating and cooling capabilities of the thermoelectric (TE) device. The voltage level $V_{in}$ serves as a "selector code word," which dictates the magnitude and direction of the current flow through the TE module. If the power source that supplies the ±8 Volts in your circuit has built-in ammeters, use them to monitor the current flow through the TE device, eliminating the need to insert the ammeter shown in the diagram. See Appendix I for a discussion of important practical issues regarding this circuit, especially the need (and method) for dissipating the prodigious amount of heat it produces.

Use **DC Voltage Source (Express)** or **DC Voltage Source (DAQmx)** found in **YourName\Chapter 4** and **YourName\Chapter 11**, respectively, to apply various voltages at $V_{in}$. First, test that when $V_{in}$ is positive (negative), an electric current flows in the proper direction through the TE device to heat (cool) the aluminum block in your experimental set-up. If a positive $V_{in}$ cools the block, reverse the connections to the TE device within your circuit. Second, find the positive and negative voltages $V_{sat}$, where $V_{sat}$ is the minimum voltage at $V_{in}$ for which the current to the TE device becomes saturated. The value of this saturation current is determined by the capability of the power supplies attached to the TIP transistors. We then define the range of acceptable values for $V_{in}$ to be between $\pm V_{sat}$.

**Fig. 12.1  Voltage-controlled bidirectional current driver.**

## 12.2 PID TEMPERATURE CONTROL ALGORITHM

Build a digital PID temperature controller that can control the temperature of an aluminum block to within less than 0.05°C of a given set-point temperature. The set-point temperature can be chosen anywhere within the range from 0 to 40°C.

An algorithm for such a controller is as follows:

1. Read the block's temperature $T_{block}$ using a thermistor as the temperature sensor.
2. Compare $T_{block}$ with the desired set-point temperature $T_{set-point}$.
3. Based on this comparison, decide what value of $V_{in}$ will most appropriately command the TE device to provide the heating or cooling needed to bring $T_{block}$ closer to $T_{set-point}$.
4. Apply this voltage at the $V_{in}$ input of the voltage-controlled bidirectional current driver circuit.
5. Repeat this process continuously to obtain the desired temperature control of the aluminum block.

The *Proportional Control* method offers a simple procedure for deciding what voltage should be applied at $V_{in}$ in your temperature-control algorithm. In this method, the difference between the desired set-point temperature $T_{set-point}$ and the actual sample temperature $T_{block}$ is defined to be the error $E \equiv T_{set-point} - T_{block}$. Then the control voltage $V_{in}$ is simply taken to be directly proportional to $E$:

$$V_{in} = AE \qquad \text{(Proportional Control)} \qquad [1]$$

where $A$ is a constant called the *gain*. The value for the gain is chosen empirically such that, when $T_{set-point}$ is changed, the optimal value for $A$ will cause the system to ramp to the new set-point and then stabilize near there (see the following discussion) quickly.

Here is a practical consideration you'll confront when trying to implement Equation [1]: When the sample temperature is far from the set-point, the error $E$ may become large enough to generate a calculated value for $V_{in}$ that falls outside its acceptable range of $\pm V_{sat}$. The solution to this problem is to truncate the expression given in Equation [1] so that the magnitude of $V_{in}$ is never allowed to be greater than $V_{sat}$. The graphical representation for the Proportional Control algorithm then is as shown in Figure 12.2.

Although pleasingly simple, the Proportional Control method is intrinsically flawed. Here's why. Assume the sample is initially at room temperature and that you select $T_{set-point}$ to be above room temperature. Turning on the Proportional Control algorithm, the initial error $E$ will be positive, resulting in a command to the TE device to heat the sample. So far so good. As time goes on, the algorithm will issue the proper heating instructions to bring the sample closer and closer to the desired set-point temperature. When $T_{block}$ nears $T_{set-point}$, the positive error $E$ will become small, causing the Proportional Control to ease off on the applied heating power so that $T_{set-point}$ is gently approached. Then, at the decisive moment when $T_{block}$ equals $T_{set-point}$, the error $E$ becomes zero and the Proportional Control turns off power to the TE device. There is the flaw.

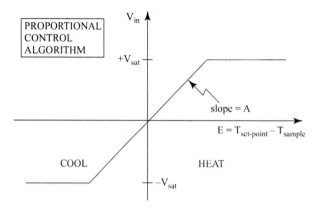

PROPORTIONAL
CONTROL
ALGORITHM

$V_{in}$

$+V_{sat}$

slope = A

$E = T_{set\text{-}point} - T_{sample}$

COOL

HEAT

$-V_{sat}$

**Fig. 12.2**

Unfortunately, when at an elevated temperature, the sample will constantly be losing heat to its surrounding (room-temperature) environment through the heat-transferring processes of conduction, convection, and radiation. Thus, to maintain a sample at a set-point above room temperature, heat must constantly flow into the sample to counteract the heat it is losing to its environment. Since the Proportional Control turns off the heating when $T_{block}$ equals $T_{set\text{-}point}$, the sample will never be able to stabilize at the desired temperature. Rather, the sample will stabilize at some temperature $T_0 < T_{set\text{-}point}$. The positive error $E$ produced at the magic temperature $T_0$ commands just the right amount of heating by the TE device to cancel out the heat being lost by the sample to the environment at that temperature. By similar reasoning, a set-point below room temperature will result in the sample stabilizing at a temperature $T_0 > T_{set\text{-}point}$.

Luckily, there is an easy remedy to the above-described defect in the Proportional Control algorithm—simply include a constant term $V_0$ on the right-hand-side of Equation [1]:

$$V_{in} = AE + V_0 \qquad\qquad [2]$$

When this expression is implemented, the proportional term will ramp the sample to the set-point temperature as before. Then, once at the set-point (where $E = 0$), $V_0$ will instruct the TE module to provide the constant heating (or cooling) necessary to counteract heat losses (or gains) to the environment, thereby stabilizing the sample at $T_{set\text{-}point}$. Of course, the value of $V_0$ must be precisely chosen so that the environmental influences on the sample, when at $T_{set\text{-}point}$, are perfectly neutralized.

Now the best news: There is an elegant way to build intelligence into the control algorithm so that the correct value for $V_0$, appropriate to the chosen set-point, will be found automatically. In the *Proportional-Integral (PI) Control* method, rather than defining $V_0$ as a fixed constant, an integral term is used to construct the correct constant during runtime according to the following expression:

$$V_{in} = A E + B \int E \, dt \qquad \text{(PI Control)} \qquad [3]$$

Here, the integral keeps a running sum of all the error values that occur over the entire execution time of the algorithm. During the times when $T_{block}$ is below $T_{set-point}$, a positive contribution will be made to the sum. When $T_{block}$ is above $T_{set-point}$, a negative contribution will be made. Due to the self-correcting manner in which these contributions are made, the second term in Equation [3] will eventually converge to the constant $V_0$, which allows the sample to stabilize at the set-point $T_{set-point}$. From that point on, the error $E$ will equal zero and the value of the integral (and thus constant $V_0$) will no longer change.

Inclusion of a derivative term to damp out oscillations provides one further refinement to the control algorithm. The expression for this so-called *Proportional-Integral-Derivative (PID) Control* algorithm then is given by

$$V_{in} = A E + B \int E \, dt + C \frac{dE}{dt} \qquad \text{(PID Control)} \qquad [4]$$

where $A$, $B$, and $C$ are constants. For the experimental situation of discrete data sampling, where the error $E$ is determined every $\Delta t$ seconds, the value of the control voltage $V_{in}$ after the $n^{th}$ sampling can be approximated by

$$V_{in} = A E_n + B \, \Delta t \sum_{m=0}^{n} E_m + \frac{C}{\Delta t} \{ E_n - E_{n-1} \} \qquad \text{(Discrete-Sampling PID)} \qquad [5]$$

where the summation $\sum_{m=0}^{n} E_m$ is over *all* of the error values determined since the control algorithm was turned on.

## 12.3 PID TEMPERATURE CONTROL SYSTEM

Start construction of your digital temperature control system by rebuilding the circuit of the thermistor-based digital thermometer described in Chapter 9's **Do It Yourself** project. The thermistor, which is embedded in the aluminum block, will be used to monitor $T_{block}$. If needed, add the 1 μF capacitor as shown in Figure 12.3 to shunt high-frequency noise signals to ground and the unity-gain buffer to prevent the voltage-sensing circuitry from loading the thermistor circuit at low temperatures (when the thermistor resistance is large). Then connect $V_{out}$ and GND to the inputs of an analog input channel of your LabVIEW system.

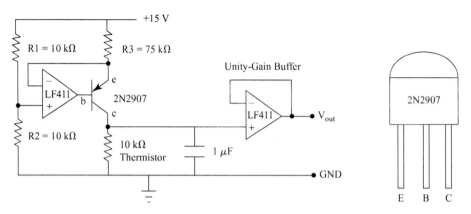

**Fig. 12.3   Hardware circuit for digital thermometer. Constant current of 0.1 mA flows through thermistor.**

Next, create a folder called **Chapter 12** within the **YourName** folder and save a VI called **PID Temperature Controller** in **YourName\Chapter 12**. Code **Temperature Controller** according to the following guidelines:

**A. Block Temperature Determination:** Noise-free temperature measurements of the block's temperature (assumed equal to the embedded thermistor's temperature) will be a "must" in order for your temperature-control algorithm to function properly. To average-out random fluctuations when reading the thermistor temperature, you might try the following: At each determination of the temperature, quickly sample the thermistor temperature $N$ times, then analyze this data to determine the mean. Given the Steinhart-Hart coefficients for your thermistor, the **DAQ Assistant** or **DAQmx** VIs make acquiring the array of $N$ temperature samples simple. To do the job of finding the mean from this data set, look at the VIs available in **Functions>>Mathematics>>Probability & Statistics**.

**B. Control Algorithm:** To decide the proper heating or cooling needed by the TE device, implement the Discrete-Sampling PID Control algorithm in software. Remember that $\Delta t$ in Equation [5] is the time between successive determinations of the error $E$; think carefully about what this time is in your particular VI. Also make sure to include a truncation feature, that is, if Equation [5] yields a value of $V_{in}$ with magnitude greater than $V_{sat}$, then truncate the magnitude of $V_{in}$ to $V_{sat}$. You will have to determine the optimum value for the constants $A$, $B$, and $C$ empirically. If your set-up is similar to the one described in Appendix I, try $A = 7$, $B = 1$, and $C = 0.5$.

**C. Output to the TE Device:** Use **DC Voltage Source (Express)** or **DC Voltage Source (DAQmx)** as a subVI on your block diagram to output the PID-determined value for $V_{in}$ over one of LabVIEW's analog output channels to the input of the TE device's voltage-controlled bidirectional current driver circuit.

**D. Front Panel:** Use your creativity here. Remember you have a wide array of controls, indicators, and graphs available. A front-panel control that resets the summation

$\sum_{m=0}^{n} E_m$ to zero when a button is clicked is an optional feature that you might find handy.

When you get the temperature controller to work, show it off to your friends and instructor!

# Chapter 13■  ■  ■  ■  ■  ■  ■  ■  ■  ■  ■  ■  ■  ■  ■  ■  ■  ■

# Control of Stand-Alone Instruments

---

**HARDWARE REQUIREMENTS FOR THIS CHAPTER**

To perform the exercises in this chapter, you must have a National Instruments General Purpose Interface Bus (GPIB) device connected to your computer. This device might be a PCI-GPIB board plugged into a PCI expansion slot or a GPIB-USB device attached to a USB connector. A stand-alone instrument equipped with an IEEE 488.2 compliant interface (ideally the Agilent 34401A Digital Multimeter) is also needed. This instrument is connected to a PCI-GPIB board using a GPIB cable; a GPIB-USB device connects directly to the instrument.

If, while performing the chapter exercises, the communication between the GPIB device and instrument breaks down (e.g., caused by an accidental error in your programming), communication can often be restored either by turning the instrument off and then on again, or by restarting your computer.

---

In previous chapters, you have used LabVIEW software to transform a personal computer (connected to an appropriate National Instruments DAQ device) into several handy laboratory instruments. In particular, you programmed this system to become a DC voltmeter, digital oscilloscope, spectrum analyzer, waveform generator, and digital thermometer. Pause to consider the following tantalizing prospect: Perhaps the only instrument required in a modern-day laboratory is a DAQ device-equipped computer controlled by LabVIEW software. That is, by simply writing a collection of appropriate VIs, it might be possible for you—the contemporary scientific researcher—to satisfy all of your laboratory instrumentation needs with this single LabVIEW-based data acquisition and generation system. This system's tremendous flexibility would then obviate the need to purchase an expensive collection of stand-alone electronic equipment such as power supplies, function generators, picoammeters, spectrum analyzers, and oscilloscopes.

The functioning VIs that you have written in previous chapters demonstrate that the above "tantalizing prospect" can, at least in certain situations, be realized. But don't discard your stand-alone instruments just yet. The timing, speed, sensitivity, and simultaneous data-taking requirements of many contemporary research experiments are beyond the capabilities of your DAQ device. For instance, while the LabVIEW-based digital oscilloscope we constructed worked well for observing audio-range frequencies (less than 20 kHz), it would prove miserably inadequate at displaying the several nanosecond-wide voltage pulses emanating from a photomultiplier tube. In this latter situation, a stand-alone digital oscilloscope with a very fast analog-to-digital converter (on the time scale of several gigasamples per second) would do the job nicely. Thus stand-alone instruments play a central role in state-of-the-art research, and so it might not surprise you to find that they, too, fall under the scope of LabVIEW.

Over the past few decades, a message-based communications standard has evolved by which stand-alone instruments can be software-controlled using a personal computer. In this communications scheme, a particular instrument obeys an array of manufacturer-defined ASCII character commands that represent all the possible ways of manually pressing buttons, turning dials, and viewing output data on its front panel. While the hardware conduit (called an *interface bus*) through which these ASCII messages are passed between the PC and laboratory instrument can take on various guises (including RS-232, GPIB, Ethernet, and USB), there is a single set of LabVIEW icons available to control this communication process. This icon set is named *VISA* (short for Virtual Instrument Software Architecture) and is found in **Functions>>Instrument I/O >>VISA**.

In this chapter, you will learn how to use VISA icons to control the message-based communication between a stand-alone instrument and your computer. You will explore generic features of this communication process such as the Standard Commands for Programmable Instruments (SCPI) language and various synchronization methods while writing code that controls a particular stand-alone instrument—the *Agilent 34401A Digital Multimeter*—using a particular interface bus—the *General Purpose Interface Bus*.

## 13.1 THE VISA SESSION

When using VISA icons to facilitate message-based communication between a computer and a particular stand-alone instrument, the instrument is termed a *VISA resource* and the communication activity is called a *VISA session*. To *"query"* a VISA resource (i.e., send it a command, then receive back its response), the required VISA session consists of the following four steps: *open the session, write the command message to the resource, read the response from the resource, close the session*. In **Functions>> Instrument I/O>>VISA** (and its subpalette **VISA Advanced**), the following four icons are available to perform the four given steps—**VISA Open**, **VISA Write**, **VISA Read**, and **VISA Close**. To understand how to wire these four icons together to query an instrument, we will first briefly describe the function of each individual icon.

The Help Window for **VISA Open** is shown in the following illustration. The job of this icon is to begin a VISA session between your computer and the resource defined at its **VISA resource name** input. The **VISA resource name** consists of text that specifies the interface type being used (e.g., GPIB or USB), the address of the resource (a number we'll discuss in a few minutes), and the resource type. For our work, the resource type will be a stand-alone instrument denoted by *INSTR*. To pass the **VISA resource name** to other VISA icons, this quantity is available at the **VISA resource name out** output terminal.

**VISA Write**, whose Help Window follows below, performs the actual ASCII message transfer from your computer to the stand-alone instrument. Once presented with the open session's **VISA resource name**, this icon writes the ASCII string at its **write buffer** input to the instrument. This string is one of the commands recognized by the instrument and, when received by the instrument, configures it properly for a desired data-taking measurement. Additionally, the VISA resource name is available at the **VISA resource name out** output terminal.

Context Help

**VISA Write**

VISA resource name ~~~~~~~ VISA resource name out
write buffer ~~~~~~~ return count
error in (no error) ~~~~ error out

Writes the data from **write buffer** to the device or interface specified by **VISA resource name.**

Next, the Help Window for **VISA Read** is shown. This icon transfers the response message from the stand-alone instrument into your computer's memory. When given the open session's **VISA resource name**, this icon receives the ASCII response string consisting of (a maximum of) **byte count** number of bytes and outputs this string at its **read buffer** terminal. This string typically contains the results of a data-taking measurement performed by the instrument. Additionally, the VISA resource name is available at the **VISA resource name out** output terminal.

Finally, **VISA Close**'s Help Window is given below. This icon closes the VISA session specified at its **VISA resource name** input.

Note that all four of these VISA icons include error reporting via an error cluster, which appears at the **error in** and **error out** terminals.

The four-step VISA session to query an instrument is accomplished by wiring these four VISA icons together as follows.

In this example, the message *IDN?* is sent over a GPIB interface to an instrument at address 22. *IDN?*, a command recognized by most instruments, instructs the instrument to identify itself. After receiving this command, the instrument's response (which is an ASCII string consisting of identification information) is received by the computer over the GPIB and displayed in the **read buffer** front-panel string indicator.

Similar to the File I/O and DAQmx icons that you have studied previously, the wiring scheme of VISA icons takes advantage of the principle of LabVIEW programming called *data dependency*. Simply stated, data dependency means that an icon cannot execute until data is available at *all* of its inputs. In the previous diagram, all of **VISA Open**'s required inputs (given in bold on the Help Window) are wired. So, when this diagram is run, **VISA Open** executes immediately. Upon completion, **VISA Open** outputs a VISA resource name at its **VISA resource name out** terminal, which is passed through the wiring to VISA Write's **VISA resource name** input. Because of data dependency, **VISA Write** cannot execute until it receives the VISA resource name from **VISA Open.** In a similar way, **VISA Read** cannot execute until **VISA Write** completes, and so on. Thus, through this programming scheme, we are assured that the icons will execute in the desired sequence: **VISA Open** followed by **VISA Write** followed by **VISA Read** followed by **VISA Close**.

Also the correct manner of chaining together VISA icons for error reporting is shown above. If an error does occur at one point in the chain, subsequent icons will not execute and the error message will be passed to the **error out** indicator cluster.

Because VISA-based programming is so robust, you can write highly dependable data-taking programs with just the information already presented. However, with a bit more grounding in the message-based communication scheme, you'll be able to create programs in which you can have near-total confidence. The following paragraphs will take you to the next level of sophistication in stand-alone instrument control.

## 13.2 THE IEEE 488.2 STANDARD

When remote control of laboratory instruments first became possible, there was a chaotic period during which, more or less, each instrument manufacturer defined its own communications protocol through a unique blend of parallel and serial modes, positive and negative polarities, and assorted handshaking signals. In 1965, Hewlett-Packard (now named Agilent) ended this cacophony by designing a universal instrument interface called the Hewlett-Packard Interface Bus (HP-IB) and offered it as the only option on all of its new computer-programmable instruments. Because of its high transfer rates, HP-IB quickly gained popularity with other instrument manufacturers and, in 1975, was accepted as an industry-wide standard known as IEEE 488 or, more commonly, the General Purpose Interface Bus (GPIB). In 1987, an improved version of this standard called IEEE 488.2 was adopted, which enhanced and strengthened message-based communication by specifically defining an instrument's minimally required communication capabilities, a protocol for message exchange, a generic set of commonly needed commands, and a status reporting system. Today, most computer-controlled laboratory instruments are IEEE 488.2 compliant, even those that communicate over interface buses other than the GPIB (such as Ethernet and USB).

## 13.3 COMMON COMMANDS

One important innovation of the IEEE 488.2 standard was the introduction of a standardized set of "*common commands*" for the many generic operations that all instruments must perform. The mnemonics for these common commands begin with asterisks to delineate them from the other device-specific commands recognized by a particular instrument. All IEEE 488.2 compliant instruments, at the very least, are required to recognize the subset of 13 common commands given in Table 13.1. Many of these commands are related to the reporting of events using two status registers called the SBR and SESR, which will described in detail starting in the next paragraph.

**Table 13.1**   Common Commands for IEEE 488.2 Compliant Instruments

| MANDATORY COMMON COMMANDS | FUNCTION |
| :---: | --- |
| *IDN? | Reports instrument identification string. |
| *RST | Resets instrument to known state. |
| *TST? | Performs self-test and reports results. |
| *OPC | Sets operation complete (OPC) bit in SESR upon completion of command. |
| *OPC? | Returns "1" to the output buffer upon completion of command. |
| *WAI | Waits until all pending operations complete execution. |
| *CLS | Clears status registers. |
| *ESE | Enables event-recording bits in SESR. |
| *ESE? | Reports enabled event-recording bits in SERS. |
| *ESR? | Reports value of SESR. |
| *SRE | Enables a SBR bit to assert the SRQ line. |
| *SRE? | Reports SBR bits that are enabled to assert the SRQ line. |
| *STB? | Reports the contents of the SBR. |

## 13.4 STATUS REPORTING

Another IEEE 488.2 innovation is a standardized scheme for *status reporting*. This status reporting system is available to inform you of significant events that occur within each instrument connected to an interface bus. In this scheme, each instrument is equipped with two status registers, called the *Standard Event Status Register (SESR)* and the *Status Byte Register (SBR)*. Each bit in these registers records a particular type of event that may occur while the instrument is in use, such as an execution error or the completion of an operation. When the event of a given type occurs, the instrument sets the associated status register bit to a value of one, if that bit has previously been enabled (see the following). Thus by reading the status registers, you can tell what events have transpired.

The Standard Event Status Register, which is schematically shown here, records eight types of events that can occur within a data-taking instrument.

### Standard Event Status Register (SESR)

| 7 | 6 | 5 | 4 | 3 | 2 | 1 | 0 |
|---|---|---|---|---|---|---|---|
| PON | URQ | CME | EXE | DDE | QYE | RQC | OPC |

The eight events associated with the eight bits of the SESR are described in Table 13.2. In our work, the OPC bit will be most useful.

**Table 13.2** Eight SESR Events

| BIT | ASSOCIATED EVENTS OF SESR |
|---|---|
| 7 (MSB) | **PON** (Power On): Instrument was powered off and on since the last time the event register was read or cleared. |
| 6 | **URQ** (User Request): Front-panel button was pressed. |
| 5 | **CME** (Command Error): Instrument received a command with improper syntax. |
| 4 | **EXE** (Execution Error): Error occurred while instrument was executing a command. |
| 3 | **DDE** (Device Error): Instrument is malfunctioning. |
| 2 | **QYE** (Query Error): Attempt was made to read the instrument's output buffer when no data was present, or a new command was received before previously requested data had been read from the output buffer. |
| 1 | **RQC** (Request Control): Instrument requests to be controller. |
| 0 (LSB) | **OPC** (Operation Complete): All commands prior to and including an *OPC command have been executed. |

The SESR exists as an event-signaling tool for you to use in your programs. However, this status register completely lacks initiative and will not perform any work unless you request it to do so. Thus, when initiating communications with an instrument, one of the messages that you may wish to send is an instruction that activates the subset of event-reporting SESR bits that are of interest to you. For instruments that conform to the IEEE 488.2 standard, this activation process is accomplished via the *ESE* (Event Status Enable) command. For example, suppose you wished the QYE bit to be activated and thus record any execution errors in the SESR's bit 2. Since $00000100_2 = 4_{10}$, the QYE bit can be activated by performing a VISA Write of the ASCII command *ESE 4* to the instrument. In our work to come, we will activate the OPC bit with the command *ESE 1*.

The Status Byte Register, which is schematically shown next, records whether data is available in the instrument's output buffer, whether the instrument requests service, and whether the SESR has recorded any events.

**Status Byte Register (SBR)**

| 7 | 6 | 5 | 4 | 3 | 2 | 1 | 0 |
|---|---|---|---|---|---|---|---|
| — | RQS | ESB | MAV | — | — | — | — |

The functions of the eight bits of the SBR are described in Table 13.3. The SBR bits are studious, performing their status reporting duties without need of a request from you.

**Table 13.3**    Functions of the Eight SBR Bits

| Bit | Function of SBR Bit |
|---|---|
| 7 (MSB) | May be defined for use by instrument manufacturer. |
| 6 | **RQS** (Request Service): The instrument has asserted the SRQ line because it requires service from the GPIB controller. |
| 5 | **ESB** (Event Status Bit): An event associated with an enabled SESR bit has occurred. |
| 4 | **MAV** (Message Available): Data is available in the instrument's output buffer. |
| 3—0 | May be defined for use by instrument manufacturer. |

An instrument can be configured to assert a *Service Request (SRQ)*, which is a digital signal carried on a dedicated wire within the 24-wire GPIB cable, in response to either of two events—an event detected by the Standard Event Status Register or the presence of previously requested data in the instrument's output buffer (that is, the assertion of the ESB or MAV bit, respectively). This configuration process is accomplished on IEEE 488.2 instruments by using the *SRE* (Service Request Enable) command. For example, if you wish an event detection by the SESR to trigger a request for service by the instrument, initialize the instrument by writing the ASCII command *SRE 32* to the instrument. Since $32_{10} = 00100000_2$, the setting of the SBR's fifth (ESB) bit will then be the criterion for the instrument asserting a SRQ. If, instead, you wish the presence of data in the output buffer (signaled by the MAV bit being set) to trigger a SRQ, then write *SRE 16*. Finally, *SRE 0* will disable the instrument ability to assert a SRQ.

The relationship between the Standard Event Status Register, the instrument's output buffer, and the Status Byte Register (along with the common commands that configure and query each) is illustrated in the following diagram.

As an alternative to the use of the SRQ, *serial polling* is a common method for determining the status of an instrument. In a serial poll process, the interface bus queries an instrument and the instrument responds by returning the value of the bits in its Status Byte Register. A serial poll is easily accomplished in LabVIEW using **VISA Read STB**, found in **Functions>>Instrument I/O>>VISA**. This icon's Help Window is shown next.

## 13.5 DEVICE-SPECIFIC COMMANDS

Finally, each stand-alone instrument is designed for a specialized purpose and has its own idiosyncratic methods for accomplishing its objectives. Thus, every programmable instrument comes with a set of *device-specific commands* that allow the user to control its functions remotely and to transfer the information it produces into a computer's memory. The array of device-specific commands for an instrument is listed in its user manual. This set of commands is defined by the instrument's maker... and therein lies a problem. When surveying the user manuals for programmable instruments of varying models and manufacturers, you will find a great diversity in the style of the various command sets. Some (especially those associated with older model instruments) are an alphabetized collection of cryptic one- or two-character strings (the designers' thinking was obviously "shorter commands yield quicker and, therefore, better computer-instrument communication"). At the other extreme are the user-friendly sets, with similar commands logically grouped, each represented by an easy-to-read-and-remember mnemonic.

As programmable instruments have come into wider use, it has become apparent that development costs and unscheduled delays can be diminished markedly by simplifying the instrument programmers' task whenever possible. Thus, user-friendly device-specific command sets are the rule, rather than the exception, for instruments being currently manufactured. Commonly, these command sets are organized in a hierarchical *tree structure*, similar to the file system used in computers. Each of an instrument's major functions, such as **TRIG**ger, **SENS**e (alternately, **MEAS**ure), **CALC**ulate, and **DISP**lay, define a *root* and all commands associated with that root form its *subsystem*. So, for example, to configure the Agilent 34401A Digital Multimeter to measure a DC voltage whose value is expected to fall within the range of ±10 Volts (an action within its **SENS**e subsystem), the appropriate command is as follows:

SENSe:VOLTage:DC:RANGe 10

Here, *SENSe* is the root keyword and colons (:) represent the descent to the lower-level *VOLTage*, then *DC*, then lowest-level *RANGe* keywords. Finally, *10* is a parameter associated with *RANGe*. While the full command mnemonic given here can be sent to the instrument, it is only absolutely necessary to send the capitalized characters.

In 1990, a consortium of equipment manufacturers defined the *Standard Commands for Programmable Instruments (SCPI)* in an effort to standardize the device-specific command sets of computer-controlled instrumentation. While this standard has not been universally adopted, it is not uncommon to discover that your post-1990 instrument is SCPI-compliant. As a means of categorizing generally applicable command groups, the SCPI standard posits the following model for a generic programmable instrument. An instrument that performs measurements on an input signal is assumed to have the root functions shown in the next diagram. Here, for example, **SENSe** includes any action involved in the actual conversion of an incoming signal to internal data, such as setting the range, resolution, and integration time, while **INPut** consists of actions that condition the signal prior to its conversion, such as filtering, biasing, and attenuation.

**SIGNAL MEASUREMENT INSTRUMENT**

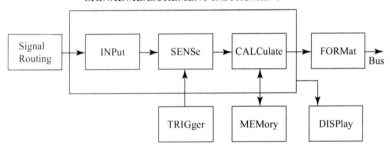

Alternately, an instrument that generates signals is modeled by the following diagram.

**SIGNAL GENERATION INSTRUMENT**

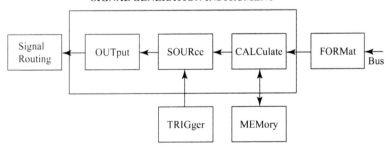

The SCPI command set is organized in a hierarchical tree structure using the syntax illustrated above by the Agilent 34401A Multimeter command. You'll learn more about the SCPI command syntax as you work your way through this chapter. But maybe now is a good time to dive in and actually control a stand-alone instrument.

## 13.6 SPECIFIC HARDWARE USED IN THIS CHAPTER

In designing this chapter, I faced the following problem. There are thousands of computer-controllable stand-alone instruments available for purchase from the myriad of worldwide scientific instrument makers. Each of these instruments communicates using one (or many times, a few) of the handful of available interface buses. I have a small subset of these instruments in my laboratory and I can use them there to practice the art of message-based communication. You also, hopefully, have a small subset of such instruments available to practice with in your own laboratory. What's the problem? Well, because of the high cost and specialized nature of such equipment, the probability that my subset and your subset have some common instrument is most likely very small. The unfortunate thing about this situation is that each stand-alone instrument is designed to take specialized measurements and understands its own unique set of ASCII commands (which are defined by its maker and are listed in its user manual). Thus, before attempting to control a particular instrument using an interface bus, the programmer must have a detailed understanding of the measurement that that instrument is designed to take, the procedure that it implements in doing its work, and the command list that it recognizes. All of these considerations greatly constrain the writing of a set of generic laboratory exercises that everyone can perform.

That said, I still was faced with the fact that I had to choose a particular instrument and interface bus to work with in this chapter's exercises. For the interface bus, I chose the GPIB because it is the interface you will almost certainly encounter in your computer-based laboratory work. Although USB and Ethernet are gaining in popularity with scientific instrument makers (due to the fact that most PCs come equipped with these interfaces), GPIB is currently (by far) the most widely used interface bus for laboratory equipment. With an estimated 10 million GPIB-equipped instruments in use in research and industry worldwide, the GPIB will most likely retain its popularity for many years to come. For the instrument used in this chapter's exercises, I chose the Agilent 34401A Digital Multimeter for the following reasons. First of all, this instrument measures voltage, current, and resistance—vanilla-flavored quantities that require no specialized knowledge to understand (unlike, for instance, the control of grating angle and slit size in a spectrometer). Second, for such a high-quality and useful GPIB-equipped instrument, its price tag of approximately $1000 makes it extremely affordable. Every lab should have one and many do! Third, this instrument is both IEEE 488.2 and SCPI compliant. Thus, the following exercises, rather than being narrowly tied to one specific device, can be much more universally applicable by demonstrating the generic features of these widely used standards, such as command syntax and status reporting.

In the best circumstance, an Agilent 34401A (earlier purchased units of this same instrument are named the Hewlett-Packard 34401A) is already available for your use or, with a modest investment, you can purchase this worthwhile instrument. Then, without need for modification, you can straightforwardly work your way through the given exercises to learn the basics of message-based communication. If, instead, you have some other interface-equipped instrument available, try reading the following pages to understand the generic issues being investigated. Then, by consulting the user manual, it may be fairly easy, for instance, to use the interface-appropriate VISA resource name and substitute an ASCII command string here and there in order to adapt the exercises to your particular instrument and interface bus. If neither of the above describes your situation, simply read through the following pages. I believe you will learn some valuable features of instrument control that will serve you well in future work.

## 13.7 MEASUREMENT & AUTOMATION EXPLORER (MAX)

To carry out the exercises in this chapter, you must have a National Instruments GPIB device connected properly to your computer, which in turn is connected (e.g., by a GPIB cable) to a GPIB-equipped stand-alone instrument. Also the GPIB device's driver software (called *NI-488.2*) must be correctly installed. To verify that these conditions are met, we will use the handy utility *Measurement & Automation Explorer*, which is nicknamed *MAX*.

To open MAX, either select **Tools>>Measurement & Automation Explorer…** (if you have an open VI or **Getting Started** window), or else double-click on MAX's desktop icon (if available). After MAX opens, double-click on **Devices and Interfaces** in the **Configuration** box. This action will command MAX to determine all of the National Instrument devices present within your computing system.

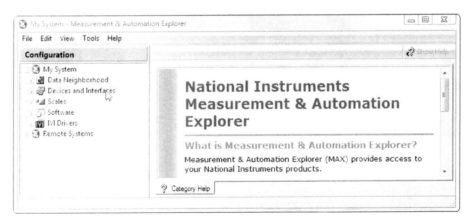

MAX will list the findings of its device survey in hierarchical tree fashion as shown next. If a GPIB device is connected correctly to your computer, a folder labeled **GPIB0** will appear in the resulting list (your system may have a different number than *0* in the folder's label). To find all of the stand-alone instruments properly connected to this device, right-click on the **GPIB0** folder and select the **Scan for Instruments** option. Alternately, you can click on the **Scan for Instruments** button near the top of the window.

In a few moments, MAX will complete the scan. To view its results, double-click on the **GPIB0** folder. For the case shown below, one stand-alone instrument was found and information about it is stored in the folder labeled **Instrument 0**.

Because up to 15 instruments can be connected to a single GPIB device, each instrument has an identification number called its *GPIB address*. A GPIB address can be any integer from 0 to 30 and is typically defined via a hardware DIP switch setting within the instrument or a sequence of button-pressing and/or knob turning on its front panel. The instrument's user manual will describe the method for setting its address.

The GPIB address of the instrument found in the **Scan for Instruments** operation is found by double-clicking on the **Instrument 0** folder. After the double-click, in the box associated with the **Attributes** tab, we find that the instrument's GPIB address is

22 (the address of your instrument may be different). A text description identifying the instrument also appears.

Clicking on the **VISA Properties** tab, we find that the correct **VISA resource name** for this instrument is *GPIB0::22:INSTR* and also are told (under **Device Status**) that the instrument is working (i.e., communicating) properly.

To verify that the instrument is indeed properly communicating over the GPIB, right-click on the **Instrument 0** folder and select **Communicate with Instrument**. Alternately, you can click on the **Communicate with Instrument** button near the top of the window.

An interactive dialog box will appear. Here, after typing an ASCII message in the **Send String:** box, a mouse-click on the **Query** button will carry out a write-then-read action. That is, the message in **Send String:** will be written over the GPIB to the selected instrument, and then the instrument's ASCII response will be read back over the GPIB to the computer and displayed in the **String Received:** box. When the dialog window opens, its **Send String:** box is preloaded with *IDN?*, the IEEE 488.2 common command for an instrument to identify itself.

Click on the **Query** button. If your GPIB communication is configured properly, the identification string received from the instrument will appear in the **String Received:** box. For the Agilent 34401A Digital Multimeter used here, this identification string identifies the instrument's manufacturer and model number followed by some integers that denote the version numbers of installed firmware that controls the multimeter's three internal microprocessors.

For future reference, this interactive dialog window is a handy tool for use in determining correct command syntax when developing message-based communication programs for a new instrument.

Exit this dialog window and then close MAX.

## 13.8 SIMPLE VISA-BASED QUERY OPERATION

Let's begin by writing a VISA-based program that carries out the query (i.e., write-then-read) action that you just completed using MAX.

On the front panel of a blank VI, place a **String Control** and a **String Indicator**, then label them **Command** and **Response**, respectively. As shown in the next diagram, you'll want to resize these objects so that they can display strings much larger than their default sizes allow. Using **File>>Save**, first create a new folder named **Chapter 13** within the **YourName** folder, then save this VI under the name **Simple VISA Query** in **YourName\Chapter 13**.

Switch to the block diagram and place a **VISA Open** icon (found in **Functions>> Instrument I/O>>VISA>>VISA Advanced**) there. Pop up on its **VISA resource name** input and create a **VISA Resource Name Constant** using **Create>>Constant**.

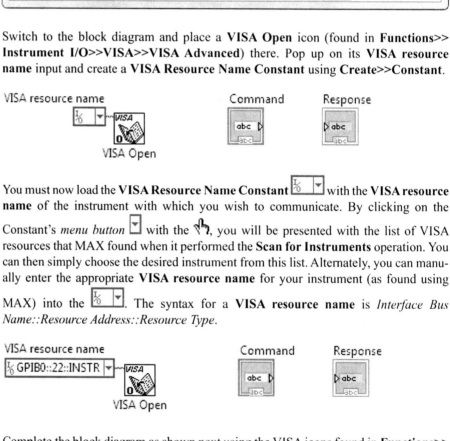

You must now load the **VISA Resource Name Constant** with the **VISA resource name** of the instrument with which you wish to communicate. By clicking on the Constant's *menu button* with the , you will be presented with the list of VISA resources that MAX found when it performed the **Scan for Instruments** operation. You can then simply choose the desired instrument from this list. Alternately, you can manually enter the appropriate **VISA resource name** for your instrument (as found using MAX) into the . The syntax for a **VISA resource name** is *Interface Bus Name::Resource Address::Resource Type*.

Complete the block diagram as shown next using the VISA icons found in **Functions>> Instrument I/O>>VISA** (and its subpalette **VISA Advanced**). Wire the **Command** control terminal to VISA Write's **write buffer** input and the **Response** indicator terminal to VISA Read's **read buffer** terminal. When executed, **VISA Read** will read up to *N* bytes from the selected resource, where *N* is equal to the integer wired to its **byte count** input. In a moment, we will read the Agilent 34401A identification string. According to this instrument's user manual, its identification string can be up to 35 characters long. Thus, wire the **byte count** input to an integer (**U32**) greater than or equal to 35 (I used *50*) as shown. Create the **error out** indicator cluster using the pop-up menu option **Create>>Indicator**.

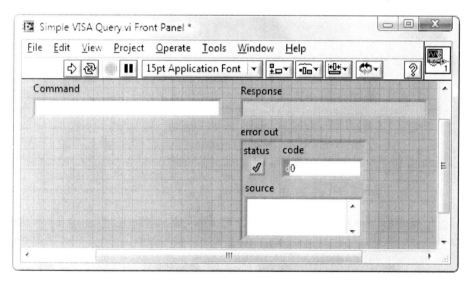

Return to the front panel, arrange the objects as you wish, and then save your work.

Enter *\*IDN?* into **Command**, then run your VI. If all goes well, **Response** will display the instrument's identification string upon completion of the VI execution, as shown.

**Simple VISA Query** will leave the multimeter in *remote mode* with its triggering circuitry "idled." You can return to *local* mode, which continuously "triggers" measurements, by pressing the instrument's front-panel SHIFT/LOCAL key.

## 13.9 MESSAGE TERMINATION

At the conclusion of a message-transfer process, some method must be used to signal that the complete message has been passed. The IEEE 488.2 standard appoints the ASCII LF (line feed, also called new line) as its special *end of string* (EOS) character. That is, when receiving a message string, the LF character is always interpreted by the receiver as the last byte of a message. Thus, appending LF to a command string is one method of signaling message termination in IEEE 488.2 communication. Alternately, the IEEE 488.2 standard allows the assertion of an *end or identify* (EOI) while the last character in the string is being passed as another acceptable termination method. The EOI is a digital signal on a dedicated wire within the GPIB cable. When using VISA icons to control an IEEE 488.2 compliant instrument, message termination is taken care of automatically, allowing you to remain blissfully ignorant of this lower-level activity. If you'd like to view an example of this (usually invisible) message termination activity, pop up on the **Response** indicator on the front panel of **Simple VISA Query**, then select '**\**' **Code Display**. The \n character you see at the end of the identification string is the backslash code for LF. The multimeter appended this termination character to its identification string to the alert the receiver (in this case, the GPIB device) that the message has ended.

To deactivate backslash coding, pop up on **Response** and select **Normal Display.**

## 13.10 GETTING AND SETTING COMMUNICATION PROPERTIES USING A PROPERTY NODE

In addition to message termination, there are other low-level functions connected with message-based communication. Many of these low-level functions have an associated parameter setting, which is termed a *VISA property*. VISA assigns default values for these properties and, as long as the VISA-based programs that you write fall within the scope of these default settings, the VISA icons will automatically take care of these low-level functions without any programming effort needed by you (as demonstrated by the message termination example shown above). At times, however, you will most likely write programs that fall outside the scope of the default VISA property settings and so you will need to assign nondefault values to these quantities. Reading ("getting") and writing ("setting") VISA property values can be done within your programs using a *Property Node* (and also can often be done in MAX).

As a concrete example of a VISA property, consider **VISA Read**'s *timeout*, which is a fail-safe feature of **VISA Read** that prevents a program from running endlessly if an error occurs. If, for example, an instrument which is being queried doesn't seem to be responsive (perhaps a nameless experimenter forgot to flip on the instrument's power switch), **VISA Read** will only wait for the instrument's response for a certain number of milliseconds (given by the value of the VISA property named **Timeout**) before aborting the read operation and issuing an error message.

The default timeout value for VISA Read on your system can be determined using a **Property Node**. The Help Window for a **Property Node**, which is found in **Functions>>Instrument I/O>>VISA>>VISA Advanced**, is shown next.

Write the following VI, which reads the current **Timeout** value on your system. Open a new VI, and save it under the name **Get Timeout Value** in **YourName\Chapter 13**. On a new block diagram, place a **Property Node** and then, using **Create>>Constant**, wire the **VISA resource name** for your instrument to the Property Node's **reference** input as shown.

Next, using the 🖑, click on the **Property** terminal and select **General Settings>>Timeout Value**. You might explore what other Properties appear in this menu, many of which are specific to a particular interface bus.

Note that, within the **Timeout** terminal, a small arrow at the right points outward from the terminal's interior. This outward-directed arrow indicates that the **Timeout** terminal is configured as an indicator, that is, it reads ("gets") the current **Timeout** value. Using **Create>>Indicator**, create a front-panel indicator to display the value of **Timeout**.

Switch to the front panel, change the indicator label to **Timeout Value (ms)**, then save your work. Run the VI. As shown below, the (default) **Timeout** value for my system is 3000 ms = 3 seconds.

You can also use a Property Node to set the value of a VISA property. To demonstrate this procedure, with **Get Timeout Value** open, use **Save As...** to create a new program called **Set Timeout Value**, and store it in **YourName\Chapter 13**. When run, **Set Timeout Value** will change VISA Read's **Timeout** to a value input from its front panel.

On the block diagram of **Set Timeout Value**, pop up on the **Timeout** terminal and select **Change To Write**.

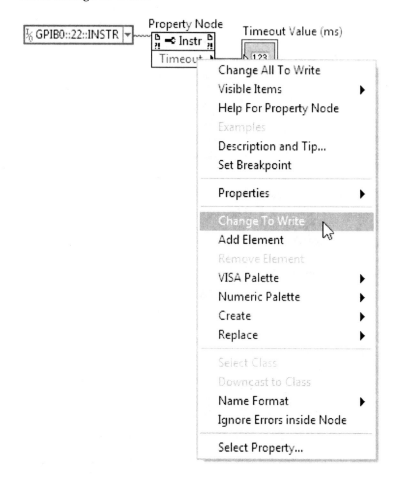

Note that, within the **Timeout** terminal, the small arrow is now at the left pointing inward toward the terminal's interior. This inward-directed arrow indicates that the **Timeout** terminal is configured as a control, that is, it writes ("sets") the **Timeout** value. Delete the **Timeout Value (ms)** indicator terminal, then using **Create>>Control**, create a front-panel control labeled **Timeout Value (ms)**.

Timeout Value (ms)

Return to the front panel and save your work. Set **Timeout Value (ms)** to be *10000*, and then run **Set Timeout Value**. Next, run **Get Timeout Value**. Is **Timeout** now equal to 10000 ms = 10 seconds? Try setting **Timeout** equal to *8000*. You will find that only certain values for **Timeout** are allowed. LabVIEW takes the value you input to **Set Timeout Value** as a suggestion (rather than an order) and sets **Timeout** to the nearest allowed value.

## 13.11 PERFORMING A MEASUREMENT OVER THE GPIB

Now that **Simple VISA Query** has given us a template for the VISA query process, let's try controlling a real measurement. Hook up some known DC voltage difference, say 5 or 6 Volts, between the HI and LO Voltage Inputs of the Agilent 34401A Multimeter. This instrument's user manual instructs us that delivering the following sequence of ASCII commands will result in one DC voltage sample being acquired, and then loaded into the instrument's output buffer (which is part of its interface circuitry):

CONF:VOLT:DC<Space>10,0.000001
INIT
FETC?

Here is the meaning of this secret code. First, the Agilent 34401A can be programmed to perform 11 different types of measurement functions, including DC voltage, AC voltage, DC current, AC current, resistance, and frequency. Given these options, the first command instructs the instrument that we desire to take a DC voltage measurement. The full command is *CONFigure:VOLTage:DC <Space><Range>,<Resolution>* (this command actually executes a collection of commands drawn from the Agilent 34401A's **INP**ut, **SENS**e, **TRIG**ger, and **CALC**ulate root subsystems). The command mnemonic *CONFigure:VOLTage:DC* is constructed in the hierarchical tree structure, typical of SCPI-compliant instruments. *CONFigure* is the root-level keyword and colons (:) represent the descent to the lower-level *VOLTage*, then lowest-level *DC* keywords. While the full command mnemonic can be sent to the instrument, it is only absolutely necessary to send the capitalized characters. Separated from the command mnemonic *CONF:VOLT:DC* by a *<Space>*, the numerical values for two measurement parameters— *<Range>* and *<Resolution>*—are specified. *<Range>* selects among the instrument's five available voltage measurement scales. Each scale offers a different sensitivity with *<Range>* giving the maximum measurable value on a particular scale. The five available ranges are 100 mV, 1V, 10 V, 100 V, and 1000 V. In our situation of measuring a signal of approximately 5 V, the 10 V scale is appropriate. <Resolution> specifies the precision of the measurement, with the options of three levels of accuracy—4½, 5½, and 6½ digits (the ½ digit means that the most significant decimal place can only take on a value of "1" or "0"). Thus on the 10 V scale, voltages can either be resolved at the level of 0.001, 0.0001, or 0.00001 Volts. The trade-off in requesting higher accuracy is that the measurement takes a longer time. In the command sequence above, the highest resolution of 6½ digits is selected by setting <Resolution> equal to *0.00001* when <Range> equals *10*. Note the syntax of the *CONF* command, which obeys the conventions of the SCPI language: A *comma* (,) separates the parameters from each other and a <Space> separates the mnemonic from the parameters.

Once the multimeter has been configured for the desired measurement function as described in the previous paragraph, the data-taking process is begun by sending the *INITiate* command (from the **TRIG**ger root subsystem). Upon receipt of *INIT*, the multimeter will acquire the requested voltage sample, and then store this value in its internal memory. Finally, the *FETCh?* command (from the **MEM**ory root subsystem) instructs the instrument to transfer the reading in its internal memory to its interface-related output buffer.

We'd like now to place this command sequence into **Simple VISA Query**. Since there are three commands to be sent, it appears that we must modify the VI to include a sequence of three successive implementations of **VISA Write**. While you are free to do so, a much easier solution is available. The SCPI language allows the programmer to concatenate several commands into one long multicommand string that can be sent in a single **VISA Write** statement. The syntax for this concatenation process is as follows:

- Use a semicolon (;) to separate two commands within the string.
- Begin a command with a colon (:), if it has a different root-level than the command preceding it. The first command in the concatenated string and IEEE 488.2 common commands (which begin with an asterisk) do not require a leading colon.

Since each of our three commands has a different root-level, applying the above rules results in the following concatenated string:

*CONF:VOLT:DC<Space>10,0.00001;:INIT;:FETC?*

Type this command into the **Command** control on the front panel of **Simple VISA Query** as shown next. Run the VI. Your computer will instruct the multimeter to acquire a 6½ digit voltage reading, retrieve this value, and then display it on the front panel in the **Response** indicator. Cool, eh?

Note that, while extra digits are displayed, the value within **Response** is only accurate to the fifth decimal place.

As shown above, the Agilent 34401A reports its data samples in the form of an ASCII character string using the exponential format SD.DDDDDDDDESDD, where S is a positive or negative sign, D is a numeric digit, and E is an exponent. For future reference, note that the string that represents a data sample is 15 bytes long. If you want to use this reading as input to a mathematical calculation (a common situation), you will need to convert the string representation into a numerical format. Such conversion operations can be easily accomplished in LabVIEW using the array of conversion icons found in **Functions>>String**. In the present case, use **Fract/Exp String To Number** in **Functions>>Programming>> String>>String/Number Conversion**. The Help Window for this icon is given next.

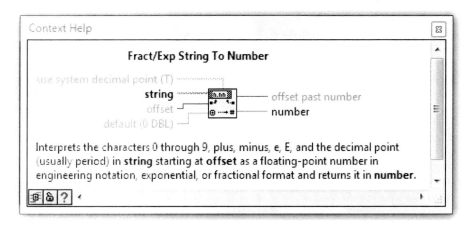

Place a **Numeric Indicator** on the front panel of **Simple VISA Query** and label it **Numeric Voltage**. Use **Display Format...** in this indicator's pop-up menu to make its **Digits of precision** equal to 5, and disable **Hide trailing zeros**. Then modify the block diagram as follows.

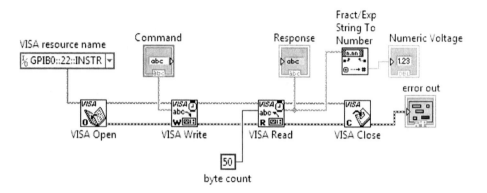

Run the VI to verify that the string-to-number conversion icon performs as expected.

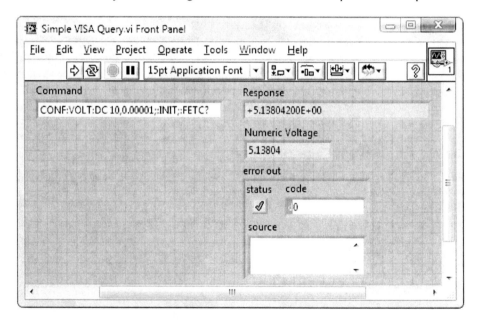

## 13.12 SYNCHRONIZATION METHODS

Although most ASCII commands are completed quickly after being received by a programmable instrument, some commands start a process that requires a significant amount

452

of time (such as acquiring a large amount of data or moving an object from Point A to Point B). The time required for such processes must be taken into account when writing a data acquisition program, else, upon execution, the program may request data before it is available, induce undesirable motion, or cause some other chaotic outcome.

As an example, in its default configuration, the Agilent 34401A multimeter acquires one data sample after receipt of the *INITiate* command, then stores this measured value in its internal memory. However, through use of the *SAMPle:COUNt <Space><Value>* command, the multimeter can be instructed to take and store multiple data samples upon receiving *INITiate*. The Agilent 34401A is configured to acquire 100 DC voltage samples with 6½-digit resolution via the following concatenated string of commands:

*CONF:VOLT:DC<Space> 10,0.00001;:SAMP:COUN<Space> 100;:INIT;:FETC?*

The *FETCh?* command will load the 100 acquired samples from the multimeter's internal memory (which, by the way, can hold up to a maximum number of 512 measured values) into the instrument's interface-related output buffer.

Let's write a VI that uses the given command string to gather a sequence of 100 voltage samples. Open **Simple VISA Query**, then use **Save As...** to create a new VI called **Simple VISA Query (Long Delay)**. Delete **Numeric Voltage** from the front panel and enlarge **Response** so that it can display a very long string (which is the concatenation of 100 voltage values). Type the command given above into the **Command** control. Once entered into **Command**, you can keep this command permanently loaded there by selecting **Edit>>Make Current Values Default**.

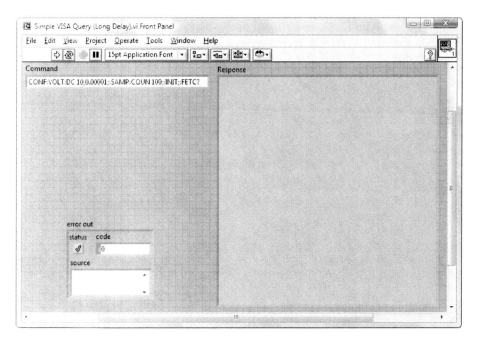

Switch to the block diagram. Delete **Fract/Exp String To Number**. After the given command string is written to the multimeter by **VISA Write**, **VISA Read** will receive a string containing the 100 voltage samples. Since each voltage sample is reported as a 15-byte string and a (single) delimiting ASCII character will be needed to separate each sample, this 100-sample string is expected to be about $(100 \times 15) + 100 = 1600$ bytes long. Input an integer larger than 1600 to **byte count** as shown.

Run **Simple VISA Query (Long Delay)**. Count down the seconds 3. ..2. ..1. .. Disappointed? You will find that your VI does not display even one voltage sample (let alone the expected 100 values), but rather produces an error.

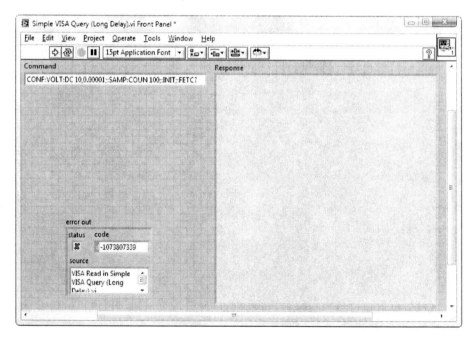

To find out what produced this error, pop up on the **code** indicator within the **error out** cluster and select **Explain Error**.

A dialog box appears, where we are told that a "timeout expired" at **VISA Read** before the requested operation (i.e., take 100 data samples) could be completed.

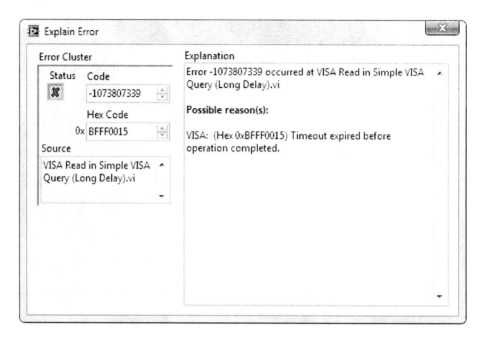

After some head scratching and checking of the Agilent 34401A user manual, the following explanation then emerges for the error we observed when running **Simple VISA Query (Long Delay)**. Simply stated, voltage sampling takes time. In its default configuration, it takes the Agilent 34401A multimeter 10 power line cycles (PLC) for each voltage sample. Additionally, the Agilent 34401A has an autozero feature, which is enabled by default. This feature operates as follows: After each voltage measurement, the multimeter internally disconnects the input signal and takes a zero reading. The instrument then subtracts the zero reading from the preceding measured value to prevent offset voltages in the multimeter's internal circuitry from affecting measurement accuracy. Since the zero reading also takes 10 PLC, each complete voltage sample by the multimeter takes 20 PLC. Assuming this instrument is plugged into a 60 Hz power source (that is, 60 PLC per second), 100 voltage samples will take about

$$100 \times \left( \frac{20 \text{ PLC}}{60 \text{ PLC/s}} \right) = 33.3 \text{ s}$$

There's the problem! A few moments ago, we found that the default timeout value for **VISA Read** is 3 seconds, but the measurement we have initiated takes over 30 seconds.

Thus, long before the requested data is available, **VISA Read** terminates the execution of **Simple VISA Query (Long Delay)**.

There are a couple of crude solutions to this dilemma. First, on the block diagram of **Simple VISA Query (Long Delay)**, you can insert a single-frame **Sequence Structure** into the VISA execution chain that simply contains a **Wait (ms)** icon, wired to produce a delay of about 33 seconds between the issuance of the data-taking command and the order to read the gathered data samples. The resulting diagram would appear as follows.

Second, for a slightly more elegant fix, you can use a **Property Node** to change the **Timeout** value for **VISA Read** from its default value (on my system) of 3000 ms to something larger than 33 seconds. Use this approach to modify the block diagram of **Simple VISA Query (Long Delay)** as shown below. Here, the **Timeout** value is chosen to be 60 seconds.

Return to the front panel of **Simple VISA Query (Long Delay)**. With the command to perform 100 samples programmed into **Command**, run the VI. About 33 seconds later you should see something like the following front panel. Note that the delimiter used by the Agilent 34401A to separate neighboring data values is a comma.

In the preceding example, we found that with a detailed knowledge of the measurement process being implemented, it was possible to troubleshoot a malfunctioning VISA-based VI. Please note that lack of communication (in particular, the GPIB device not correctly knowing when the instrument's data will be available) is the root problem that led to the malfunction.

Fortunately, powerful tools exist that allow one to monitor the status of tasks being performed by a programmable instrument. For IEEE 488.2 compliant instruments, these tools are the Standard Event Status Register (SESR) and Status Byte Register (SBR) that were discussed at the beginning of this chapter. With proper use of the SESR and SBR, many potential data-taking glitches, such as the one just experienced, can be avoided.

The status reporting capabilities of the SESR and SBR can be employed in several ways. We will explore two commonly used techniques—the Serial Poll and Service Request Methods. The core operation for both of these methods is the same—the completion of an assigned task triggers the Operation Complete (OPC) bit in the Standard Event Status Register to be set, which in turn sets the Event Status Bit (ESB) of the Status Byte Register.

In the Serial Poll Method, the setting of ESB is detected by directly checking the Status Byte Register, whose state is obtained by serial polling the instrument. The complete step-by-step process of this method is shown in the following diagram.

**Serial Poll Method**

In the Service Request Method, the Status Byte Register is configured such that, when its ESB is set, the Request Service bit is induced to be set also. This action then causes the instrument to assert a SRQ, which alerts the GPIB device that the assigned operation is complete. This method is pictured here.

**Service Request Method**

| | Standard Event | | | Status Byte | | |
|---|---|---|---|---|---|---|
| | Event Register | Enable | | | | |
| 7 | Power On | No | | | | |
| 6 | User Request | No | | Status Byte Register | Enable | |
| 5 | Command Error | No | | | | |
| 4 | Execution Error | No | (+) | Request Service | ⊠ | |
| 3 | Device Error | No | OR | Event Status Bit | Yes | |
| 2 | Query Error | No | | Message Available | No | (+) OR |
| 1 | Request Control | No | | | | |
| 0 | Operation Complete | Yes | | | | |
| | *ESE 1 | | | *SRE 32 | | |

We'll write VIs that implement both of these approaches to status reporting.

## 13.13 MEASUREMENT VI BASED ON THE SERIAL POLL METHOD

Let's try the Serial Poll Method first. To configure the Agilent 34401A for status reporting using the Serial Poll Method, write the following VI called **Status Config (Serial Poll)** and save it in **YourName\Chapter 13**. First, code the VI's block diagram as shown next. Use the autocreation feature in pop-up menus to create all of the constants, controls, and indicators.

Switch to the front panel and arrange the object logically. Design an icon and assign the connector terminals consistent with the Help Window shown.

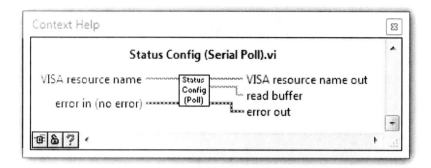

Here's how the VI works, assuming that the instrument referenced by **VISA resource name** is the Agilent 34401A multimeter. Within the chain of VISA icons, **VISA Clear** (found in **Functions>>Instrument I/O>>GPIB**) executes first. The Help Window for this icon is shown below.

Although not an absolute necessity for inclusion in **Status Config (Serial Poll)**, this VI performs the precautionary action of "clearing" the Agilent 34401A. **VISA Clear** instructs the multimeter to abort all measurements in progress, disable its triggering circuitry, clear its interface-related output buffer, and prepare to accept a new command string.

Next, **VISA Write** sends the concatenated command string *CLS;*ESE 1;*SRE 0;*OPC? to configure the Agilent 34401A for status reporting using the Serial Poll Method. Note that since the component strings are all IEEE 488.2 common commands, leading colons are not required in the concatenation. In this sequence of commons, *CLS clears the contents of the SESR and SBR. As described in the beginning of this chapter, *ESE 1 enables the SESR's OPC bit to set the ESB in the Status Byte Register and *SRE 0 disables the instrument from asserting a SRQ. Then, *OPC? requests the instrument to return a "1" to the instrument's output buffer after this command is completed. This last command is included simply as a method of checking that the entire sequence of commands has been executed.

Finally, **VISA Read** reads the contents of the instrument's output buffer. If all goes well, there should be a single ASCII character 1 read into the computer.

Test drive your VI as follows. Click on **VISA resource name** control's menu button with the 🖑.

Then, from the list presented, select the VISA resource name for your computer-controlled instrument.

Then, run **Status Config (Serial Poll)**. Upon completion, does the **Buffer Reading** string indicator display an ASCII character *1*?

Next, construct a VI called **Serial Poll** which continuously reads the Status Byte Register of an instrument until a given bit is set. A suggested coding of **Serial Poll** is shown in the following diagrams, and explanations of the unfamiliar icons are in the subsequent paragraphs. Save **Serial Poll** in **YourName\Chapter 13**.

**VISA Read STB**, found in **Functions>>Instrument I/O>>VISA**, is the workhorse of this VI. With each iteration of the While Loop, its **status byte** output returns the current values of the SBR's eight bits in the form of an integer. For example, if the SBR's fifth bit (ESB) is set, then **status byte** outputs the integer $32$, since $00100000_2 = 32_{10}$. The Help Window for **VISA Read STB** is shown next.

The individual bits of **status byte** can be checked through the use of **Number To Boolean Array** (found in **Functions>>Programming>>Boolean** with its Help Window shown next). This VI creates an array of TRUE and FALSE values that mirror the sequence of zeros and ones (starting from the least-significant bit) in the binary representation of the integer input **number**. For example, if **number** equals the decimal integer *48*, then the **Boolean array** output will be [F, F, F, F, T, T, F, F], since $48_{10} = 00110000_2$. **Index Array** can then be used to ascertain the value of a particular element in this array. **Serial Poll**'s While Loop will continue to iterate until **Which Bit?** becomes TRUE.

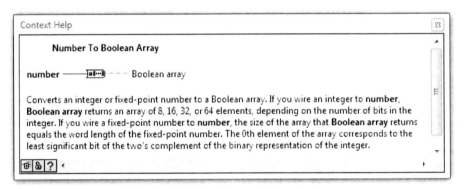

Run **Serial Poll** under **Highlight Execution** and, through your observations, gain a better understanding of its operation. Remember to input values for **VISA resource name** and **Which Bit?** on the front panel. When run in this isolated manner, the VI will most likely never be able to exit the While Loop, so you'll have to stop it using the **Abort Execution** button in the toolbar.

We're finally ready to write **VISA Query (Serial Poll)**. This top-level program implements serial polling to synchronize the GPIB activities necessary in acquiring 100 voltage samples using an Agilent 34401A multimeter.

Open **Simple VISA Query (Long Delay)**, then use **Save As...** to create **VISA Query (Serial Poll)**. The front panel can remain unchanged. If it's not already there by default, type the following command into the **Command** control:

*CONF:VOLT:DC<Space> 10,0.00001;:SAMP:COUN<Space> 100;:INIT;*OPC;:FETC?*

Switch to the block diagram and modify it as shown below with **Status Config (Serial Poll)** and **Serial Poll** used as subVIs.

Here is how this diagram works: The concatenated command string is sent to the instrument by **VISA Write**. After configuring the multimeter for the desired DC Voltage measurement function, the acquisition process is begun by the *INIT* command. The succession of 100 samples is acquired and temporarily stored in the multimeter's internal memory. After the hundredth sample is obtained, *OPC* instructs the instrument to set its SESR's OPC bit (which, in turn, sets the SBR's ESB), then *FETC?* loads the contents of the internal memory into the instrument's output buffer. At that point, **Serial Poll** detects the setting of ESB, which then triggers the instrument's output buffer to be read by **VISA Read**. One might be tempted to write the concatenated command with *OPC* after *FETC?*, rather than sandwiched between *INIT* and *FETC?*, as above. It is best, however, to avoid sending *OPC* after a query (a query is a command like *FETC?* that ends in a question mark) as such commands cause a message to be loaded into an instrument's output buffer. If the message exceeds the finite size of the output buffer, as happens in our present situation, the query must be immediately followed by **VISA Read** as the program executes in order to read the long message string over the bus successfully.

Return to the front panel, save your work, and then run **VISA Query (Serial Poll)**. Does the VI obtain the requested 100 DC voltage samples successfully? If so, try running it again with **Highlight Execution** activated for both **VISA Query (Serial Poll)** and its subVI **Serial Poll**. This exercise will illustrate the weakness of the Serial Poll Method, namely, the large volume of interface bus traffic required by this technique. During the 30-odd seconds while the 100 data samples are being gathered, the instrument is polled countless times by the GPIB device so that its status can be continuously monitored. While effective, the Serial Poll Method is rather inefficient because of its excessive use of the interface bus and processor time.

## 13.14 MEASUREMENT VI BASED ON THE SERVICE REQUEST METHOD

The Service Request Method provides status reporting with a minimum of interface bus activity. To configure the 34401A for status reporting using the Service Request Method, open **Status Config (Serial Poll)**, then create **Status Config (SRQ)** using **Save As...** and save it in **YourName\Chapter 13**. The front panel and terminal assignments can remain as is, but the icon should be redesigned as shown here.

Only two modifications of the block diagram are needed. First, by changing *SRE 0* to *SRE 32* in the command string sent to the instrument, the Agilent 34401A will assert the

SRQ line when the SBR's fifth (ESB) bit is set. The already present *ESE 1* command configures the instrument to set the ESB in response to the SESR's OPC (Operation Complete) bit being set. Second, in order for VISA icons to detect service request (SRQ) events during this VISA session, **VISA Enable Event**, with **Service Request** wired to its **event type** input, must be included in the diagram as shown. **VISA Enable Event** is found in **Functions>>Instrument I/O>>VISA>>VISA Advanced>>Event Handling.**

Save your work as you close this VI.

Open **VISA Query (Serial Poll)**, then use **Save As...** to create a new VI named **VISA Query (SRQ)**, and store it in **YourName\Chapter 13**. The front panel is fine as is.

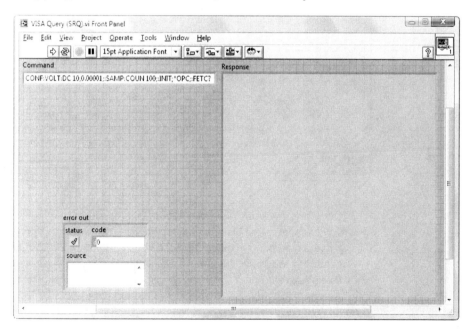

Switch to the block diagram and modify it as shown next.

Here, **VISA Disable Event**, found in **Functions>>Instrument I/O>>VISA>>VISA Advanced>>Event Handling**, must be included in order disable VISA servicing of SRQ events before the VISA session is closed.

**Wait for RQS.vi**, also found in **Functions>>Instrument I/O>>VISA>>VISA Advanced>>Event Handling** (Help Window shown below), sits idly until the instrument denoted by **VISA resource name** asserts a SRQ. However, there is a limit to the patience of this icon. It will only wait up to a total time of **timeout**, with a default value of 25000 ms = 25 seconds. Because our measurement requires over 33 seconds, a constant larger than 33,000 must be wired to the **timeout** input of **Wait for RQS.vi**, as shown in the above diagram.

Save your work, then run **VISA Query (SRQ)**. Does it successfully acquire the requested 100 DC voltage samples? Do you understand the operation of this program and how the Service Request Method manages to work with a minimum of interface bus activity?

To simplify the block diagram of **VISA Query (SRQ)**, you might consider packaging **VISA Disable Event** and **VISA Close** together in a subVI called **Close (SRQ)**, since both of these icons are involved in closing down the service request–based VISA session. To accomplish this feat easily, simply create a highlighting box around the two icons using the ℟.

Then select **Edit>>Create subVI**. A new subVI icon will appear wired on your diagram.

Double-click on this new icon to open it. Then, relabel the front-panel objects appropriately, design an icon, and assign the connector consistent with the Help Window shown (using the $4 \times 2 \times 2 \times 4$ pattern). Save this VI under the name **Close (SRQ)** in **YourName\ Chapter 13**.

Switch to the block diagram of **Close (SRQ)**. It should appear as follows.

Close **Close (SRQ)**, and return to the block diagram of **VISA Query (SRQ)**. You may have to delete the originally created subVI and load a new copy of **Close (SRQ)** there using **Functions>>Select a VI...** After that, the finished block of **VISA Query (SRQ)** will appear as shown next. Try running this VI to verify that it functions correctly.

## 13.15 CREATING AN INSTRUMENT DRIVER

An instrument driver is a collection of modular software routines that perform the operations required in the computer control of a programmable instrument. These operations include configuring, triggering, status checking, and sending data to and receiving data from the instrument. Above, **Status Config (Serial Poll)** and **Status Config (SRQ)** are examples of configuration VIs that would be useful to include as part of the Agilent 34401A software driver. You will now write another configuration VI, this time one that prepares the multimeter for taking a desired measurement.

The Agilent 34401A is capable of implementing 11 types of measurement functions: DC and AC voltage, DC voltage ratio (ratio of voltage at two different inputs), DC and AC current, 2- and 4-wire resistance (2-wire is the "normal" method for measuring resistance; the more involved 4-wire technique is necessary only when measuring very small resistance samples), frequency and period of an AC signal, continuity, and diode check. To gain experience with some of the LabVIEW tools available for developing instrument drivers, let's write a driver that offers the choice of configuring the Agilent 34401A for either a DC voltage, AC voltage, or 2-wire resistance measurement. You, of course, can be more ambitious and write your VI to control up to all 11 possible measurement functions.

Referring to the Agilent 34401A user manual, we find that our driver must allow a user to select one of the following three possible commands in order to configure the instrument for the desired measurement function:

CONFigure:VOLTage:DC <Space> <Range>, <Resolution>
CONFigure:VOLTage:AC <Space> <Range>, <Resolution>
CONFigure:RESistance <Space> <Range>, <Resolution>

Here, the possible values of *<Range>* for both the DC and AC voltage measurements are 0.1, 1, 10, 100, and 1000 Volts. For the resistance measurement, the allowed *<Range>* values are 100, 1k, 10k, 100k, 1M, 10M, and 100M ohms. In all cases, the measurement precision may be $4\frac{1}{2}$, $5\frac{1}{2}$, or $6\frac{1}{2}$ digits, which corresponds to *<Resolution>* being $10^{-4}$, $10^{-5}$ or $10^{-6}$ times the *<Range>* value, respectively.

We will write two programs called **Range and Resolution Decoder** and **Command String**, which will allow a user to construct the desired command string using front-panel controls. On **Range and Resolution Decoder**, given range and resolution choices from a user-friendly front-panel listing of the multimeter's available offerings, the program will convert these choices to the double-precision floating-point numeric format needed in **Command String**. **Command String** will construct the appropriate ASCII command string to be sent to the Agilent 34401A, based on selections made on its front-panel controls.

Create a new VI named **Range and Resolution Decoder** and save it in **YourName\Chapter 13**. Place four **Emun** controls (found in **Controls>>Modern>>Ring & Emun**) on the front panel and label them **Function, Voltage Range, Resistance Range**, and **Resolution**, respectively. Pop up each **Enum**, select **Edit Items...** and then program each with the items given in the following list.

**Function**: *DC Voltage, AC Voltage, Resistance*
**Voltage Range**: *100 mV, 1 V, 10 V, 100 V, 1000 V*
**Resistance Range**: *100 ohm, 1 kohm, 10 kohm, 100 kohm, 1 Mohm, 10 Mohm, 100 Mohm*
**Resolution**: *4 1/2 Digits, 5 1/2 Digits, 6 1/2 Digits*

Then place these four **Enum** controls in a **Cluster** shell (found in **Controls>>Modern>>Array, Matrix & Cluster**) labeled **Function Parameters** as shown next.

Switch to the block diagram, place a **Case Structure** there, and complete the code shown. Pop up on the Case Structure and select **Add Case for Every Value**, and then verify that it has three cases labeled **DC Voltage**, **AC Voltage**, and **Resistance**.

Select the **DC Voltage** case, and then place an **Index Array** icon within it. Pop up on Index Array's **n-dimension array** input and select **Create>>Constant** to create an **Array Constant** and label it **Voltage Ranges**. Next, program the index-0 through index-4 elements of this **Array Constant** as *0.1*, *1.0*, *10.0*, *100.0*, and *1000.0*, respectively. The **Array Constant** then will serve as a *look-up table* of the multimeter's allowed voltage ranges, given as double-precision floating-point numbers. Complete the code for the **DC Voltage** case shown below. Here, the integer associated with a selected **Voltage Range** on the front-panel Enum control provides the index of the desired look-up table element. This element is then output by **Index Array**.

Clone the **Voltage Ranges** Array Constant (mouse-click, while holding down *<Ctrl>*), and place the copy somewhere on the block diagram. Then switch to the **AC Voltage** case and (using your cloned **Voltage Ranges**), write the code shown.

Function Parameters

Finally, switch to the **Resistance** case and program it as shown. Here, the index-0 through index-6 elements of the **Resistance Ranges** Array Constant are *1.0E2*, *1.0E3*, *1.0E4*, *1.0E5*, *1.0E6*, *1.0E7*, *1.0E8*, respectively.

Function Parameters

Add a second **Case Structure** and complete the diagram as shown next. Remember to pop up on the Case Structure and select **Add Case for Every Value**. The **Range & Resolution** indicator cluster is created by popping up on **Bundle** and using **Create>>Indicator**.

Return to the front panel. Within the **Range & Resolution** indicator cluster, label the top and bottom **Numeric Indicator** as **Range** and **Resolution**, respectively. Design an icon and assign the connector terminals as shown. Save your work.

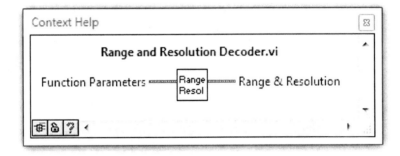

Run **Range and Resolution Decoder** and verify that it functions properly. For example, with **Function**, **Voltage Range**, and **Resolution** equal to *DC Voltage, 10 V*, and *6 1/2 Digits*, respectively, **Range** and **Resolution** should equal *10.0* and *0.00001*.

Next, open a blank VI, and save it under the name **Command String** in **YourName\ Chapter 13**. Switch to the block diagram and write the following code, which constructs the desired ASCII command string. The **Function Parameters** control cluster and **output string** string indicator are made using the autocreation feature in pop-up menus.

This diagram constructs the desired command string in a three-step process. First, all three possible commands begin with the keyword *CONF:*, so this sequence of ASCII characters is wired to the **string** input of **Pick Line** (found in **Functions>>Programming>> String>>Additional String Functions** with its Help Window given below). The value of the **line index** input (an integer given by the front-panel **Function** Enum control) then selects which of the three possible lines programmed into the **String Constant** wired to **multi-line string** is to be appended to *CONF:*. Create the three lines in this **String Constant** by the following sequence of keystrokes: *VOLT:DC<Space> <Enter>VOL T:AC<Space><Enter>RES<Space>*. Be sure to include the *<Space>* character at the end of each command string. You can make the invisible space and line feed characters visible by popping up on the **String Constant** and selecting **"\" Codes Display**. The correct entry will then appear as *VOLT:DC\s\nVOLT:AC\s\nRES\s*.

**Format Value**, from **Functions>>Programming>>String/Number Conversion** (Help Window shown next), then is used to attach two more string fragments, each with embedded ASCII-coded numbers that program the *<Range>* and *<Resolution>* settings of the multimeter. This icon takes the number at the **value** input, and converts it to an ASCII string representation with the format defined at the **format string** input. This ASCII string is appended to **string** and presented at **output string**. In the above diagram, the scientific notation format *%7.2e* (see section 5.5) is used for both *<Range>* and *<Resolution>* parameters. Note a comma (*,*) and semicolon (*;*) follow *<Range>* and *<Resolution>*, respectively.

Switch to the front panel and change the label of the String Indicator from **output string** to **Command**. Run the VI with a given choice of the controls within **Function Parameters**, and verify that the correct command string appears in the **Command** indicator. Save your work.

Add a **Push Button** (found in **Controls>>Modern>>Boolean**) and a **Numeric Control** to the front panel and label them **Autozero** and **Sample Count**, respectively. Change the representation of **Sample Count** to **U16**.

Switch to the block diagram, then include code to control the multimeter's autozeroing feature and to program the desired number of data samples to be taken. The *%5d* format in the *SAMPle:COUNt* command specifies a five-place decimal integer because the maximum allowed value for *SAMPLe:COUNt* (according to the Agilent 34401A user manual) is 50000. The format string entry for this command should be *:SAMP:COUN<Space>%5d.*

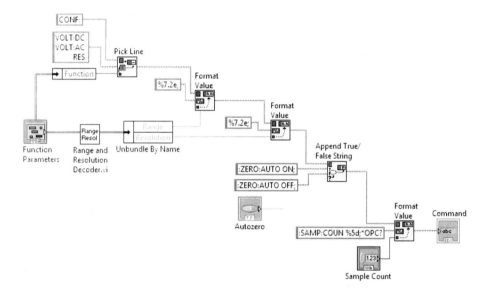

Autozero can be either turned on and off with the following commands:

ZERO:AUTO<Space>ON
ZERO:AUTO<Space>OFF

**Append True/False String** (found in **Functions>>Programming>>String>> Additional String Functions**), whose Help Window follows, provides an easy way to select which of these two choices is concatenated to the command string. Remember to include the leading colon and final semicolon in the **false string** and **true string** entries to assure proper command concatenation.

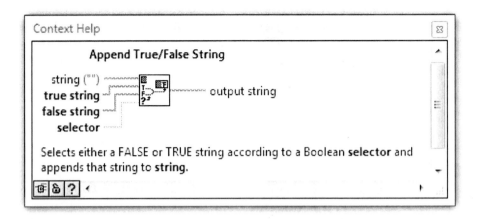

To guarantee that the instrument fully processes the sent command string before exiting the configuration VI (which you will write in a moment), conclude the string with *OPC?*.

Return to the front panel. Run the VI with a given choice of the front-panel controls to verify that the correct command string appears in the **Command** indicator. Then design an icon and assign the connectors consistent with the following Help Window. Save your work as you close the VI.

Finally, create a VI named **Measurement Config** and save it in **YourName\Chapter 13**. Switch to the block diagram and code it as shown. This VI will write the command string to the instrument. When this diagram runs, the **read buffer** indicator will display an ASCII "1" if the command string was successfully read by the instrument.

Switch to the front panel, arrange the objects there as you wish. Then design an icon and assign the connector terminals consistent with the Help Window shown below. Save your work.

Input the VISA resource name for your instrument into the **VISA resource name** control, and then run **Measurement Config** with front-panel control settings shown above so that the following command is sent to the instrument:

*CONF:VOLT:DC<Space>1.00E+1,1.00E-5;:ZERO:AUTO<Space>ON;:SAMP: COUN<Space> 5;\*OPC?*

If the command is successfully sent over the GPIB, an ASCII "1" will appear in **read buffer**. If the Agilent 34401A beeps, there is most likely an error in the sent command. Open the front panel of **Command String**, then run **Measurement Config** again. Check that the concatenated command in the **Command** indicator on **Command String**'s front panel has a form as given above; make sure all of the colons, semicolons, and spaces are included. If there is an error, correct it on the block diagram of **Command String**.

After running **Measurement Config,** the multimeter will be left in *remote mode*. You can switch to *local mode* by pressing the instrument's front-panel SHIFT/LOCAL key. The Agilent 34401A can then be triggered (equivalent to sending the INIT command over the GPIB) with the SINGLE/TRIG button. A star (*) annunciator will blink on the instrument's front-panel display as it acquires each voltage sample. Does this annunciator blink **Sample Count** times after the SINGLE/TRIG button is depressed?

Save **Measurement Config** as you close it.

Write a final modular VI for your Agilent 324401A instrument driver called **Take Data,** as shown below, and save it in **YourName\Chapter 13**. The leading *CLS* command assures that all bits in the SESR and SBR register are set to zero, prior to each data-taking process.

Note: **Take Data** cannot be run independently without generating an error. However, if you first run one of the other VIs that you have written (can you figure out which one?), then **Take Data** can be run successfully.

## 13.16 USING THE INSTRUMENT DRIVER TO WRITE AN APPLICATION PROGRAM

Ultimately, the merit of an instrument driver is measured by the ease with which you can use it to write an *application program* to fulfill some specialized need in your laboratory work. Let's quickly write an application program called **Data Sampler** that can be configured to take a multisample voltage or resistance measurement.

With **VISA Query (SRQ)** open, use **File>>Save As . . .** to create **Data Sampler** and save it in **Your Name\Chapter 13**. Rewrite the block diagram using your modular driver software as shown below.

Return to the front panel and arrange the object there as desired. Save your work.

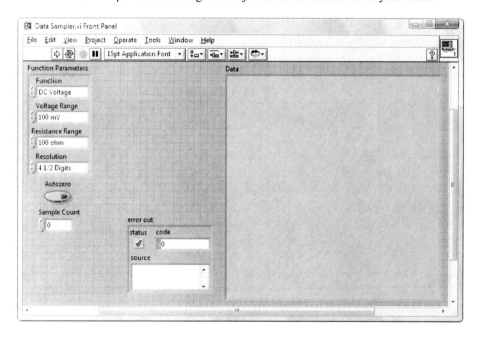

Run **Data Sampler** with various choices of front-panel settings, and then pat yourself on the back for a job well done.

Now that you know some of what goes into writing an instrument driver, here's some very good news. In many cases, the LabVIEW instrument driver you will need for a particular instrument in your laboratory has already been written and is available for your use free of charge. National Instruments provides an extensive library of downloadable instrument drivers at *http://www.natinst.com/idnet/*. You can also access this resource within LabVIEW by selecting **Tools>>Instrumentation>>Find Instrument**

**Drivers...** Most of these drivers are written using VISA icons and so, by using the interface-appropriate VISA resource name for your instrument, can be used to communicate over various interface buses—RS-232, GPIB, Ethernet, and USB. LabVIEW itself comes equipped with the VISA driver for the Agilent 34401A Digital Multimeter in **Functions>>Instrument I/O>>Instrument Drivers>>Agilent 34401**. Take a look at some of these subVIs and see if you can decipher them.

Finally, a useful Agilent 34401A instrument driver utility would perform the following task: Take the instrument's **Data** string (data samples delimited by commas and the string terminated by a LF character) and convert it to a numeric array and a spreadsheet format. One manifestation of that utility, called **Reformat Data String,** is shown next. On the front panel, the string control and string indicator have been resized and scrollbars have been activated by selecting **Visible Items>>Vertical Scrollbar** in the pop-up menus.

If interested, try writing **Reformat Data String**. It implements the **Search and Replace String** icon found in **Functions>>Programming>>String** (Help Window shown below) to coerce the original **Data** string into the spreadsheet format (by replacing comma delimiters and the LF terminator with tabs and an EOL, respectively). Do you understand how it works? Once written, include this program as a subVI in **Data Sampler** and watch it perform its magic.

## DO IT YOURSELF

Assume that you have a widget in your laboratory that is providing you with some interesting information about $X$, where $X$ might be the position of an object or the intensity of a light source. Additionally, say, the widget provides this information about $X$ in the form of a "voltage-code," that is, it produces an output voltage $V$ that is some known function of $X$. Then, with an Agilent 34401A multimeter and an appropriate application VI (called it **Time Evolution of X**), you can monitor $X$ (via measurement of $V$) as a function of time.

Using your Agilent 34401A instrument driver programs as subVIs, write **Time Evolution of X**. When run, this top-level VI continuously obtains a single DC voltage sample every **Wait Time** seconds (where **Wait Time** is given by the value on the similarly named front-panel control) until the front-panel **Stop Button** is clicked. While running, the VI provides real-time graphing of the *Voltage* versus *Time* data on a **Waveform Chart** with the Chart's *Time* axis properly calibrated. The front panel also provides the option of storing the all of the accumulated data in a spreadsheet file with *Time* and *Voltage* in the spreadsheet's first and second column, respectively. Define *Time* = 0 at the moment that the first voltage sample is acquired.

The front panel of **Time Evolution of X** should appear as shown below. All needed parameters without a front-panel control should be input on the block diagram. After building this VI, run it to observe a time-dependent voltage input (e.g., from a function generator) to the multimeter and save the resulting *Voltage* versus *Time* data in a spreadsheet.

A helpful tip: The *Time* axis of the Waveform Chart can be calibrated using a Property Node. Pop up on the Chart's icon terminal and select **Create>>Property Node>> X Scale>>Offset and Multiplier>>Multiplier**. Then set the **Multiplier** property appropriately.

## Problems

1.  Thermocouples are widely used as temperature sensors. A thermocouple is constructed by joining the ends of two dissimilar metals, for example, a copper and a constantan wire for a Type T thermocouple. This junction produces a millivolt-level voltage, which has a well-documented temperature dependence, where the temperature is measured relative to a "cold junction" reference temperature. Conveniently, this cold junction can be provided by a compact electronic device called a Cold Junction Compensator (CJC), which effectively makes the reference temperature equal to 0°C.

    Connect a thermocouple to a CJC and then connect the plus and minus output of the CJC to the HI and LO Voltage Inputs of the Agilent 34401A Multimeter. Then, write a program called **Thermocouple Thermometer (VISA)** that, every 250 ms until a Stop Button is clicked, reads the thermocouple voltage, converts this value to the corresponding temperature in Celsius, and then display this temperature in a front-panel indicator.

    To convert the thermocouple voltage to its corresponding temperature, use **Convert Thermocouple Reading.vi**, which is found in **Programming>>Numeric>>Scaling**, with its **CJC Voltage** input wired to *0* (the **CJC Sensor** and **Type of Excitation** inputs can be left unwired). Program **Thermocouple Type** for your particular type of thermocouple (e.g., T).

    Run **Thermocouple Thermometer (VISA)** and use it to measure room temperature as well as the temperature of your skin.

2.  As written in this chapter, **Serial Poll** is flawed in that, if the bit being monitored in the Status Byte Register is never set, this VI will loop endlessly. With **Serial Poll** open, use **Save As...** to create a new VI called **Serial Poll with Timeout**. Then modify the block diagram so that, if the bit being monitored is not set within 10 seconds, the While Loop is stopped.

3.  When LabVIEW is installed on your computer, a driver for the Agilent 34401A Multimeter is included. This driver is found in **Functions>>Instrument I/O>> Instrument Drivers>> Agilent 34401**. Use the icons from this "built-in" driver to write a program called it **Time Evolution of X (Built-In Driver)**, which carries out the task described in this chapter's **Do It Yourself** project. The icon VI Tree gives a helpful overview of the "built-in" driver.

4.  Regardless of the chosen resolution, the Agilent 34401A Multimeter always reports data sample values with eight digits to the right of the decimal point. Thus, some of these decimal-place values are not significant. With **Take Data** open, use **File>>Save As...** to create a new VI called **Take Data (Accurate Resolution)**. Add a **Function Parameters** front-panel control to this new VI (so that the selected resolution setting can be input) and then modify the block diagram appropriately so that the **data** output reports values with the actual resolution selected (e.g., 4½ digits, if that is the selected resolution).

5.  Use the **Instrument I/O Assistant** Express VI to query the Agilent 34401A Multimeter. Place an **Instrument I/O Assistant** (found in **Functions>>Express>>Input**) on the block diagram of a VI called **Simple VISA Query (Express)**. When this Express VI's dialog window opens, select the desired instrument, and then click on **Add Step**. In the **Add Step** dialog window that appears, double click on **Query and Parse**. In the Enter a command box, type

*CONF:VOLT:DC 10,0.00001;:INIT;:FETC?*

and then click **Run this step**. The command will be sent to the Agilent 34401A Multimeter and its string response will be displayed. Click the **Auto parse** button to convert the response string to numeric format and then close the dialog box by clicking the **OK** button. When returned to the block diagram, simply create an indicator for the icon's **token** output terminal.

Run **Simple VISA Query (Express)** and demonstrate that it successfully obtains a DC Voltage sample from the Agilent 34401A Multimeter.

# Appendix I ■ ■ ■ ■ ■ ■ ■ ■ ■ ■ ■ ■ ■ ■ ■ ■ ■ ■ ■ ■ ■ ■ ■

# Construction of Temperature Control System

To perform the project in Chapter 12, you will need access to an apparatus which controls the temperature of a small object through the use of a thermoelectric (TE) device. There is, of course, a multitude of ways to construct the required gadget. In this Appendix, I offer a design (total cost approximately $100) that has worked successfully for me.

First, I use a rectangular aluminum block of dimensions 2" × 1.5" × 5/16" (1" = 2.54 cm) as the object whose temperature is to be controlled. Hereafter, I'll simply call this object "the block." So that accurate temperature measurements can be taken on the block, a small hole of diameter 3/16" is drilled into one of its sides to a depth of approximately 1/2." A thermistor (preferably coated with thermal grease, see below) can then be inserted into this hole and held securely by the addition of a 1/4" long #6–32 nylon set screw. For the thermistor, I use an inexpensive 10 kΩ model from Epcos (Part Number B57863S0103F040, available from Digikey P/N 4952–149-ND, Telephone: (800) 344–4639, World Wide Web: http://www.digikey.com, $3) with soldered-on 12" lead wires for easy connection to a constant-current circuit. Finally, two #8–32 clearance holes are drilled (using #19 drill bit) into the block as shown in the diagram below. The 1.450" spacing is chosen so that a 30-cm wide TE module will fit between these holes.

Fig. A.1  Aluminum block design.

The required heating or cooling of the block is facilitated by placing it in contact with one side of a thermoelectric module. A TE module is a compact solid-state device that, via the Peltier Effect, acts as a heat pump. When current is caused to flow through the TE device in one direction, heat will be absorbed from the contacting block, cooling it, as shown below.

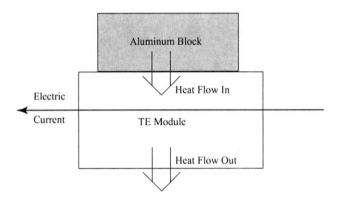

**Fig. A.2   TE module cools block.**

When current is made to flow through the TE module in the opposite direction, heat is pumped into the block, heating it.

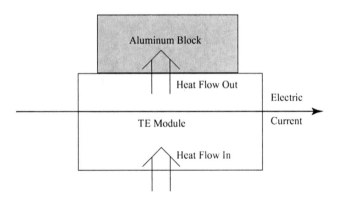

**Fig. A.3   TE module heats block.**

Note that, because the TE device is a heat pump, a not-pictured heat reservoir must be in contact with the bottom side of the TE module in both of the above pictures. With insignificant change to its temperature, this heat reservoir accepts (provides) the heat pumped through the TE module when the block is being cooled (heated).

Melcor (Telephone: 609–393–4178, World Wide Web: http://www.melcor. com) manufactures over 200 standard models of TE devices. The choice of a particular TE model is dictated by the heat-pumping capacity necessary and electric power supply limitations in your application. For our project, I use a Melcor Model CP 1.0–127–05L (approximately $15 each) TE device. Within the relatively small range of current flow between $\pm 2$ Amps, this TE module has enough heat-pumping ability to change the temperature of a $2" \times 1.5" \times 5/16"$ aluminum block significantly away from room temperature (including cooling it to below 0 °C and icing it up!).

How can we construct the required heat reservoir? Remember that a heat reservoir is simply an object that can accept or produce an amount of heat (up to a maximum value determined by your particular application) without significantly changing its temperature. A big slab of aluminum (for example, $5" \times 3" \times 3"$) possesses a large "thermal mass" and thus can perform as a fairly decent heat reservoir. By tapping two #8–32 screw holes 1.450" apart into such a slab, a TE module can be sandwiched between the block and the slab. Then using two 3/4" long #8–32 (insulating) Nylon screws, you can secure the three items together and, in the process, obtain good thermal contact between contacting surfaces. To guarantee efficient heat flow, coat the TE device's heat-pumping surfaces with thermal grease (available, for example, from Melcor, Jameco), before sandwiching the three items. This method for realizing a heat reservoir is inexpensive, but imperfect. If the TE module is run over moderately long time periods, the temperature of the aluminum slab will begin to significantly change from room temperature due to its rather poor efficiency at transferring heat with the surrounding room.

A much more stable heat reservoir can be constructed using a finned aluminum heat sink. Melcor and Aavid Thermalloy (Telephone: 603–224–9988 World Wide Web–http://www.aavidthermalloy.com/) offer a line of products called *extrusions*, which are plate-like pieces of aluminum with a many seamlessly attached fins. Due to the large surface area of its fins, an extrusion will very efficiently exchange heat with its surrounding room air, imbuing it with the desired properties of a heat reservoir. I have had good success using 6" lengths of Thermalloy Model 78780 Extrusion ($30 per 6") as the heat reservoir in the temperature control system of Chapter 12. Tapping two #8–32 screw threads 1.450" apart into the plate-like surface of the extrusion, the TE module with the block atop can be mounted, then secured with two #8–32 (insulating) Nylon screws. Again, use of thermal grease will enhance the conductance of heat between the block and heat reservoir. A final improvement results by mounting the extrusion on top of a small fan (for example, 120 VAC 4" Cooling Fan, Radio Shack Model 273-238–42, $20), which forces air flow through the extrusion's array of fins. Legs for the fan and small angle brackets to connect it securely to the extrusion can be fashioned from 1/16" aluminum strips. Attach these strips to the fan using the fan's mounting holes. A photograph of the fan with its homemade legs and brackets is shown in Figure A.4.

**Fig. A.4**  Cooling fan assembly.

Photos (from two different angles) of the entirely assembled temperature-control system follow in Figures A.5 and A.6. A small aluminum bar attached to the plate-region of the extrusion acts as a strain-relief for the TE module wires.

**Fig. A.5**  Fully assembled temperature control system hardware.

**Fig. A.6**   Fully assembled temperature control system hardware.

Alternately, a complete system similar to the one shown above can be purchased from Melcor (see, for example, Model DA-014–12-02) for about $150.

The bidirectional current driver circuit for the TE module is shown in Fig. A.7. For simplicity, the TIP transistors are represented in this diagram as simple bipolar transistors, but are, in actual fact, Darlington transistors. In regard to external connections, a Darlington transistor behaves exactly the same as to a single bipolar transistor with a very large current gain $\beta$ (hence, the acceptability of the simplified representation below). The Darlington's large $\beta$ (on the order of 1000) results from its two-transistor internal construction, in which the collector current of one transistor provides the base current to the other. In its ON state, the Darlington's base-emitter voltage is "two diode-drops" ($\approx 1.2V$). Because of the coupling between its two internal transistors, the Darlington is especially susceptible to thermal instabilities, necessitating proper heatsinking of these components for reliable service (see below).

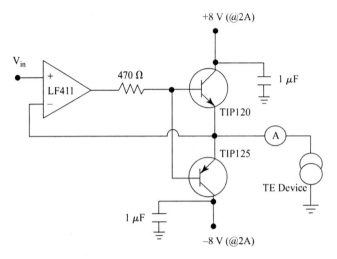

**Fig. A.7**   Voltage–controlled bidirectional current driver.

The ±8 Volts power sources attached to the TIP transistor's collectors must be capable of providing up to about 2 Amps (the ground of these power sources must be the same ground as the rest of the circuit, i.e., these grounds must be connected by, for example, cabling). The Laboratory DC Power Supplies sold by several electronic equipment manufacturers (for example, Tektronix PS280 or BK Precision 1760), which contain two identical DC supplies, are an ideal instrument for providing both the negative and positive voltage levels. These Laboratory DC Supplies typically have built-in ammeters which, in the above circuit, can be used to monitor the current flow through the TE device. The 1 μF capacitors in the diagram above remove any high-frequency noise present in the power supply voltage levels.

When this circuit is in operation, the TIP power transistors generate A LOT of heat. These hot transistors present the following problems: (1) if you touch them, they will hurt your fingers (an avoidable danger, once you're aware of it); (2) if the circuit is set up on a solderless breadboard, they will melt the breadboard's plastic (this is the voice of experience talking); and (3) if the transistors are allowed to get too hot, they will become thermally unstable and possibly destroy themselves (again, the voice of experience). Proper heatsinking of the transistors will solve these problems. Because this process is somewhat involved, I provide students with a widget that contains the two transistors mounted on a heat sink. The noise-suppressing 1 μF capacitors are included as well. A top and bottom view of the widget (total cost approximately $30) is shown in Figure A.8.

**Fig. A.8** Top and bottom view of widget.

The two TIP transistors have TO-220 packaging; however, I mount them on a TO-3 Heat Sink (Wakefield Engineering Part Number 401K, available from Newark Electronics (Telephone 800–463–9273, World Wide Web: http://www.newark.com, $13). The pattern of mounting holes (appropriate for a single TO-3 packaged power transistor) on this finned heat sink allows the two TIP transistors to be affixed side-by-side as well as providing convenient openings to pass wiring through. A TO-220 Heat Sink Mounting Kit (Jameco Part Number 34121, Telephone: (800) 831–4242, World Wide Web: http://www.jameco.com, $4 per 5-piece kit) must be used in mounting each TIP transistor to insulate its collector from the heat sink. The heat sink is then attached to a homemade chassis, fashioned from 1/16" aluminum sheet metal. Additionally, to facilitate necessary electrical connections, banana jacks and binding posts are mounted on the chassis.

The widget is then hardwired according to the following diagram (Figure A.9). All of the parts in this diagram may be purchased from Newark Electronics or from any most other electronics parts distributors.

**Fig. A.9**  Widget circuit.

# Index

9 780195 373950